The Chimpanzees of the Taï Forest

Behavioural Ecology and Evolution

Christophe Boesch and Hedwige Boesch-Achermann

OXFORD

UNIVERSITY PRESS

OXFORD

UNIVERSITY PRESS

Great Clarendon Street, Oxford OX2 6DP

Oxford University Press is a department of the University of Oxford.
It furthers the University's objective of excellence in research, scholarship,
and education by publishing worldwide in

Oxford New York

Athens Auckland Bangkok Bogotá Buenos Aires Calcutta
Cape Town Chennai Dar es Salaam Delhi Florence Hong Kong Istanbul
Karachi Kuala Lumpur Madrid Melbourne Mexico City Mumbai
Nairobi Paris São Paulo Singapore Taipei Tokyo Toronto Warsaw

with associated companies in Berlin Ibadan

Oxford is a registered trade mark of Oxford University Press
in the UK and in certain other countries

Published in the United States
by Oxford University Press Inc., New York

A Catalogue record for this book is available from the British Library

Library of Congress Cataloging in Publication Data
(Data applied for)
ISBN 0 19 850508 6 (Hbk)
ISBN 0 19 850507 8 (Pbk)

Typeset by EXPO Holdings, Malaysia

Printed in Great Britain
on acid-free paper by
Biddles Ltd, Guildford & King's Lynn

The Chimpanzees of the Taï Forest

Front cover picture: Loukoum cracks *Coula* nuts with a wooden hammer, while her 18-month old son, Lefkas, watches intently.

To all—people and chimpanzees—whose contributions made this book so rich, and to our parents and Lukas and Léonore for having accepted the adventure.

Preface

1 January 1980

We are invited to a cheerful New Year's Party at the *sous-préfet's* house at Taï village. It's strange to be out in the open after the first six months of life in the jungle, mixing with so many people who heartily enjoy consuming food and drink in large quantities and in audacious, hitherto unheard-of combinations. We are regarded as rather strange, though friendly, people and many seek to engage us in conversation. The chief of police eventually asks the fundamental question he has had in mind since we moved into the forest: 'Why are you doing this? It cannot be for money; who would pay anybody to live like a savage with the animals? What will you tell your father when he asks what you are doing with your life?' A good question — we asked him the same. He was not at all abashed. Whenever he arrests a thief, he is improving the quality of life for others. We then said that we might learn a lot about animals and man by observing the chimpanzees, and one day we would write a book about it and hopefully our parents would be happy with that. He agreed that it would be a good thing for a book to be written. That day has now arrived.

Why does a young couple choose exile in a tropical rainforest with no electricity, a difficult water supply and the nearest doctor some 250 km away? We were both fascinated by nature and wanted to gain some intimacy with a really pristine place, such as barely exists any more in Europe. However, a passion does not suffice to make a viable longterm scientific project. First of all, there was curiosity, and we wanted to understand what we saw. Christophe undertook his first research project of three months with the mountain gorillas under the supervision of Diane Fossey, and this experience merely served to confirm our interest in fieldwork. In addition before starting our work in the Taï forest, we visited Sumatra and observed some orang-utans, gibbons and siamang in their natural environment. Finally, we have fed ourselves with the literature of pioneer primatologists. Books like *In the Shadow of Man* by Jane Goodall and *The Year of the Gorillas* by George Schaller, as well as the books of Konrad Lorenz, had shaped our vision of nature. When we then started to work on the Taï chimpanzees, we were already aware of the special character of this environment and of the importance of making comparisons between populations and between species.

In this book, we report the observations on the Taï chimpanzees that we have accumulated during the last sixteen years. We consider our study a junior one in comparison with the classic studies on chimpanzees at Gombe Stream and in the Mahale Mountains in Tanzania, and the results of these earlier studies helped us to realize what in some of the behaviours we observed was unknown or different and thus should be studied with special care. With the accumulation of data over the years, it became possible for the first time to analyze also aspects of the biology of the chimpanzee that require long periods of observation, because this species lives so much longer than do commonly used animal models such as fruitflies, mice or small birds. Thus, two goals have informed the conception of this book. On the one hand we wanted to present new and original conclusions from long-term data collected over a period of sixteen years. And secondly, we wanted to collect in a comprehensible form the available information about the Taï chimpanzees and document in this way what is emerging as the main characteristic of the chimpanzee: its wide and generalized behavioural diversity. It becomes difficult to state that the chimpanzee behaves in such-and-such a way: rather we need to say that Gombe chimpanzees behave like this, whereas Mahale chimpanzees do that and Taï chimpanzees behave in yet another way. An ethnological approach to the chimpanzee seems required.

Fieldwork, beside the numerous physical challenges that have to be overcome, required observing the animals with sufficient rigour and advance planning to collect data systematically on the various topics we wanted to study. This was achieved by following the animals for 12 hours per day on an almost daily basis. Predetermined protocols have been used to guarantee that the required data are collected. However, science is not just about collecting data systematically; it is just as much about *not missing* the unexpected and the unknown. For this a certain intimacy with the animals you observe is the best tool available. During the weeks and the months we followed the chimpanzees in the forest, we learned to know them individually. Some of them have impressed themselves upon us more than others, and the trust and admiration that we developed for them has helped us to make some of our most important observations. Héra and Salomé, two experienced females of the study community, were particularly expert nut-crackers; and it was while concentrating on them that we discovered some of the most fascinating aspects of nut-cracking, such as the examples of teaching and the cooperation between mother and infant. Brutus, the long-term alpha male of the community, was an unusually skilled hunter, and we concentrated on him for years during the hunting season, which enabled us to gain a thorough insight into the preparation of the hunts. Our admiration for his skills grew the more we learned about them; and we took care not to let it influence our scientific objectivity. But at the same time it was this admiration, which made us observe him so closely, that led to our coming to understand the complicated hunting roles in use among the chimpanzees. Similarly, it was by staying with Brutus that we realized that his various drummings were specialized signals and that we were observing a case of symbolic communication, or that defending the territory was often an action that had been planned a long time ahead.

After the first introductory chapter relating the general background of our study, we present in the first chapters of this book biological aspects of the life of the chimpanzees, such as the demography (Chapter 2) and aspects of the life history of the two sexes (Chapters 3 and 4). Next, we present two chapters about the social life of the chimpanzees (Chapters 5 and 6). These are followed by three chapters on more specific behaviour patterns observed in the wild chimpanzees; inter-group aggression (Chapter 7), hunting behaviour (Chapter 8) and tool-use (Chapter 9). To conclude somewhat closer to home, we discuss the contribution of our observations to our knowledge of intelligence in the chimpanzee (Chapter 10), and we suggest a new evolutionary scenario for the chimpanzee and human that attempts to include the most recent findings acquired in relation to the first of these species (Chapter 11).

A long term study of the chimpanzee, like any long term enterprise, is possible only with the help of many people and organizations. First of all, we should like to thank the government of the Côte d'Ivoire, and all the public authorities that have helped us in our pursuit of this enterprise. The Taï National Park has been throughout the years subject to attacks from various directions, mainly in the shape of logging, farming and poaching, and this has threatened the survival of the forest and its fauna, including the chimpanzees. The Ivorian authorities have always undertaken the steps necessary to guarantee the survival of this precious park, and unfailingly supported the continuity of our project. The ministries of *Eaux et Forêts* and *la Recherche Scientifique* as well as the directors of the *Direction de la Protection de la Nature* that were in charge of the National Parks, have been our reliable key partners. The numerous park agents, especially those of the Taï *sous-préfecture*, have always done their best to help us and support the chimpanzee study. The *Centre Suisse de Recherches Scientifiques en Côte d'Ivoire* (CSRS) has been a key partner for the project from our very first visit in 1976 down to the present, and we thank all their successive directors and presidents for going out of their way to support us. The project was initially integrated in a UNESCO-MAB project supervised by the *Institut d'écologie tropicale* in Abidjan directed by Dr Jean-Louis Guillaumet and later by Dr Henri Dosso: we thank them for constant support. In our daily camp life we were helped in many ways by Theo Tiépkan Zoroa, in charge of the Station IET in Taï forest, and his staff, and we thank them all. We thank the Tanzanian National Parks and Jane Goodall for allowing us to observe the Gombe chimpanzees.

As novices in this field we profited from the support of many experts whose assistance made the start of this project possible; for this we want to thank especially the late Professor François Bourlière, Professor Hans Kummer, Professor André Aeschlimann, Professor Hans-Peter Huggel and Professor Rudolf Schenkel. No scientific project can survive without financial support; and here we have to stress the continuous, generous and unvarying support we received for nineteen years from the Swiss National Science Foundation. Very few public funding organizations dare to support a long term field project for so long, and we are immensely grateful for their understanding of the value of such a study and for their support in it. Additional funds helped us to work under the best conditions all these years, and we thank the Messerli Foundation, the Leakey Foundation, the Jane Goodall Institute, the Schultz Stiftung, the Freie Akademische Gesellschaft, the Roche Foundation, the Max Planck Society, and the Wenner Gren Foundation for their help.

Many people helped us to observe and study the chimpanzees of the Taï forest, and we are very grateful for their work. Firstly we would like to thank the first two Ivorian assistants who followed the chimpanzees with us, and who still do so today: Grégoire Nohon and Honora Kpazahi. Many students helped in collecting the long-term data used in this book; Martina Funk, Rainer Neumeier, Margareet Hoitink, Christian Falquet, Penny Simpson, Diane Doran, Pascal Gagneux, Andy Kurt, Claudia Steiner, Daniel Robert, Franca Donati, Thomas Pfluger, Miriam Behrens, David Jenny, and Frédéric Dind.

The content of this book has benefited decisively from the comments of colleagues: Steve Stearns, Hans Kummer, Toshisada Nishida, Alan Dixson and Duri Rungger invested much effort in commenting on the whole manuscript, for which we are especially grateful to them. We also thank Bill McGrew, Ernest Boesch, Diane Doran, Mike Tomasello, Kim Hill, Andy Whiten, and Richard Wrangham for commenting on individual chapters, as well as Cathy Crockford, Tobias Deschner, Ilka Herbinger, Roman Wittig and Martha Robbins.

Finally, we thank our families and friends for their constant and warm support of many kinds during all these years.

Contents

Brutus

1. *Chimpanzees, humans, and the forest*

Forest scene: July 1976: Valentin, a man of the local Guéré tribe, probably about 40 years old, had the reputation of being a healer, and one who not only used natural remedies but who also relied on magic forces for his art. He was recommended to us by the park authorities as someone who could help us in setting up our first camp and in finding our way through the forest. He had his own, in our view quite unorthodox, way of doing this. Also very intriguing were the stories he told when we said that we were here to live in the forest to learn all about the chimpanzees.

'That's easy', he said, 'the old people know a lot about the chimpanzee. They say he is our brother. In the old days he was of great help to us in warfare against our neighbours. Therefore they ask us to be grateful towards him and respect him. Nobody must kill him, even if he takes some of our fruit and vegetables. Anyway, he is very clever and you never see him because there is always a guardian. But when you hear a big 'palaver' in the forest, you know that the head of the family is distributing nuts. 'And do you know how they carry these nuts around?' Valentin asked us. Of course we did not. 'Well, they catch a monkey, take its skin and use it as a bag. Isn't that very smart?' Indeed, we were highly impressed and wondered whether we were dreaming. War, nuts, hunting — all the big words were there. Valentin had certainly set us thinking.

We shall see in this book what there is behind these legends — and what to think, for example, of an intact, neatly skinned red colobus fur we found on the ground behind a highly excited bunch of chimpanzees …

The chimpanzee is our closest living relative and has been recognized as such even in the earliest accounts of them in previous centuries. All taxonomic classifications, starting with Buffon and Linné, place the chimpanzees and humans very close together. A study of this species thus not only provides us with a fascinating account of the life of one of the most endangered great apes, but also opens a window on the differences and similarities existing between human beings and other animals. We shall keep in mind these two aspects throughout this book and come more specifically back to the second one in the last chapters.

The chimpanzee has been studied in the wild since the early 1960s, and we shall first review these studies of wild populations before going on to discuss our own project. Chimpanzees cannot be considered in isolation, and so we shall also describe some of the types of environment they inhabit, and the major threats that the chimpanzees face and which are contributing to their decline.

Studies of wild chimpanzees

Chimpanzees were studied in Africa for the first time as early as the beginning of the 1930s, but the first long-term field studies began in the 1960s. Shorter studies of chimpanzees that were not habituated to human observers and that were not individually identified have been abundant (see review in McGrew 1992), and the location of some of them are presented in Fig. 1.1. However, short-term studies are difficult to use for comparative purposes for two reasons. First, the absence of observation does not equal the absence of the behaviour in the population under investigation. For example, ant-dipping with sticks has been observed in East African chimpanzees. At Taï, despite the fact that we regularly checked the entrances of ant nests, it was only after eight years that we saw the chimpanzees dip for ants with sticks. Ant-dipping is a mainly female activity, and during the period that they were not tolerating our presence, they simply interrupted this behaviour before we could see it. Similarly, West African chimpanzees were supposed not to hunt, and this was proposed as being a major difference between them and the East African chimpanzee populations (Teleki 1975). When we started our project, we did not actually see them hunt for 24 months. We checked more than 380 fresh chimpanzee faeces during this period, but we found bones and animal remains in only one of them, supporting the idea that hunting must be rare. Towards the end of the second year, we suspected that they probably interrupted their hunting attempts as soon as we approached, and it is only their increasing tolerance that finally allowed us our first glimpses of them hunting. Once they were well habituated to us, after nine years, it appeared that Taï chimpanzees have one of the highest frequencies of hunting of all known chimpanzee populations (Boesch and Boesch 1989). There are many more examples which could be quoted to illustrate the danger of using negative results from studies of too short a duration to claim that a behaviour is really absent from the repertoire of a given population. In contrast, the positive results of short studies are of real value and we shall use them.

Second, the lack of individual identification, normal in short studies, makes the use of information on a social level very difficult. Chimpanzees live in a fission–fusion social structure (see Chapter 5), which means that at any one time in the field we follow only a small sub-group of the chimpanzee community. We may follow some individuals for a short period of time, then at irregular intervals of time encounter new sub-groups. Without individual identification, how can we decide whether we are following the same chimpanzees or even the members of the same community of chimpanzees? What criteria can be used if they are not recognized individually? It regularly happened at the beginning of the third year of our study, once habituation had started, that we encountered parties of chimpanzees which, we felt, were somehow more quick to run away than usual, and in what we felt at the time was outside their 'normal' territory. Based on this criterion we decided we had found members of another community that were in auditory contact with our study community. In the following months we could confirm that in all these cases we never saw any of the chimpanzees we knew individually in these parties of timid chimpanzees, and that all parties composed of individuals we knew were now obviously less afraid of us and remained within the boundary of what we considered their territory. With this in mind, we doubt the suggestion that originated from studies where chimpanzees were not identified (Kortlandt 1962; Reynolds 1965; Reynolds and Reynolds 1965), that

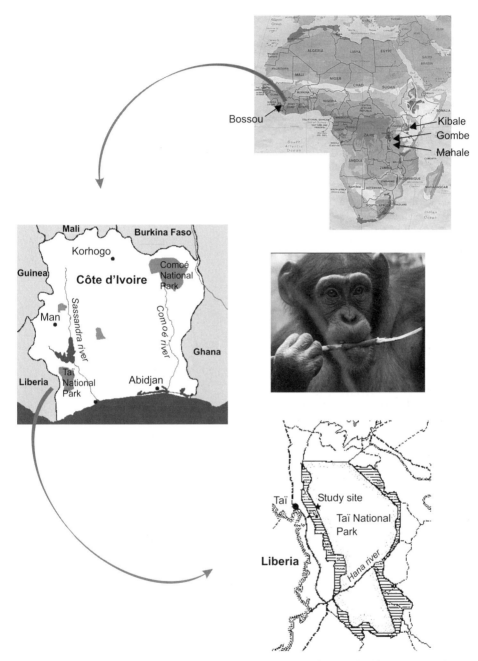

Fig. 1.1: Maps of Africa and Côte d'Ivore showing the location of the main chimpanzee study sites and the precise location of the study site within Taï National Park.

members of different communities interact peacefully. This inference was contrasted with the aggressive behaviour observed in studies which did have individually identified chimpanzees (Power 1991). Our experience of Taï suggests that this difference in behaviour is an artefact resulting from lack of individual identification. All studies with identified individuals show clear territoriality in chimpanzees (see Chapter 7).

The Gombe chimpanzees

In 1960, Jane Goodall started her pioneering long-term study of the chimpanzees in the Gombe Stream National Park on the shore of the Lake Tanganyika in Tanzania. A wealth of data has been published (Goodall 1968, 1973, 1983, 1986) and popularized in two books (Goodall 1970, 1995). All aspects of the behaviour and the long-term social relationships of the members of the community were followed in detail. At present the second and third generations of individuals are being followed. We shall regularly compare our observations with findings from this study.

Gombe chimpanzees live between the eastern shore of Lake Tanganyika and the escarpment crest of the Great Rift Valley. The terrain of this small national park is cut into many parallel 3 km long steep-sided ridges and valleys by a stream system that flows from a height of about 750 m into the lake (Goodall 1968, 1986). It consists of a mixture of five habitats, predominantly open woodland and thicket woodland, vine tangle, and closed forest (Collins and McGrew 1988). The closed forests are only found along the rivers, where tree height is about 30 m, whereas the woodlands, including the vine tangle, grow on the steep slopes of the valley and tree height is at most 10 to 15 m (see Goodall 1986 for more details on the habitats).

The Mahale chimpanzees

Five years after Jane Goodall had started her project, Toshisada Nishida began his long-term project on the chimpanzees of the Mahale Mountains National Park some 200 km south of the Gombe Stream National Park on the shore of Lake Tanganyika in Tanzania (Nishida 1968, 1983, 1987, 1990). This project is also still going on, and with Gombe it has provided the bulk of the data on wild chimpanzees in East Africa. The social interactions of the chimpanzees have been the central topic of the Mahale project and we shall refer to it frequently throughout this book. The habitat at Mahale is similar to Gombe but is more humid, with more woodlands and higher mountains. The very dense bush makes it difficult to move about.

In both studies chimpanzees were artificially provisioned, with bananas at Gombe and sugar cane at Mahale, as this speeded up their habituation to human observers and guaranteed regular and long observation periods. In the late 1960s, the quantity of bananas provided was diminished at Gombe and most observations were then made while following the chimpanzees in their daily forays through their territory (Goodall 1986). The chimpanzees of Mahale were artificially provisioned with sugarcane until 1987 (Nishida 1990). There has been an argument about how far artificial feeding might affect the behaviour of the chimpanzees. The Gombe study showed that it leads to a high concentration of chimpanzees near to the feeding place and also leads to aggression between the chimpanzees as they compete for access to the food provided (Wrangham 1974). It has been claimed that this had a major and lasting effect on most aspects of the behaviour of these chimpanzees, and that our view of chimpanzee behaviour is distorted if we rely only on results from Gombe and Mahale (Power 1991). However, from the Taï study chimpanzees, which were not artificially provisioned (except for a few weeks, in three distinct years, by enriching a cracking site with *Coula* nuts to allow good filming of the behaviour), it appears that most behaviours thought to be strongly affected by artificial provisioning, i.e. territoriality, dominance, community residence, and individual dispersion, are in fact all part of the chimpanzees' natural repertoire.

The Bossou chimpanzees

Yukimaru Sugiyama initiated a project, in November 1976, on the chimpanzees of Bossou in Guinea in the deciduous forest region some 300 km north of Taï National Park. Sugiyama concentrated on social interactions. The chimpanzees were not followed continuously, but only for three to six months a year. In this study, the centre of the chimpanzees' territory is a village surrounded by fields, and the small chimpanzee community is isolated from other chimpanzee communities (Sugiyama and Koman 1979*a, b*). This constitutes a special situation, making it difficult to judge to what extent the community size, ranging pattern, and social behaviour have been influenced by humans. This population is the only one beside Taï where the nut-cracking behaviour has been studied.

The Kibale and Budongo chimpanzees

In 1981, some short studies were begun on wild chimpanzees living in the Kibale forest in Uganda (Ghiglieri 1984; Isabirye-Basuta 1989). These became a long-term study in 1987, the chimpanzees being followed on a regular basis by many students under the direction of Richard Wrangham (Wrangham *et al.* 1992, 1996). The Kibale forest is deciduous and has been affected by intensive logging as well as by the introduction of some plantations within the territory of the community. Owing to the fact that the individuals are only partially habituated, many aspects of chimpanzee behaviour are not yet well described, whereas the ecological data are much more advanced. In the Budongo forest, chimpanzees have been studied on an irregular basis for some years starting in 1960 (Reynolds and Reynolds 1965; Sugiyama 1968; Suzuki 1971). This study site was revived in 1994 and has yielded some preliminary social information (Reynolds, in press). These two studies are of particular interest to Taï, because they are the only chimpanzee studies in forest habitats from East Africa and thus directly comparable.

To conclude, data of many studies on wild chimpanzees of varying duration are now available. The Gombe and Mahale studies are outstanding for being the most extensive and complete ones (more than thirty years each). In each chapter we shall attempt comparisons between the different chimpanzee populations to enhance what we believe to be the most fascinating challenge of chimpanzee biology, their behavioural diversity. One goal of this book is to try to sort out the factors in the environment and in the social life of chimpanzees that generate the variations observed across sites.

The Taï Chimpanzee project

When planning our project in 1976, the late French Professor François Bourlière draw our attention to two short reports. Both were on a tool-use, possibly performed by the chimpanzees of the Taï National Park in Côte d'Ivoire and the forests of Liberia, and so far unknown to the scientific world. Direct observations of the animals doing it were missing (Struhsaker and Hunkeler 1971; Rahm 1971). In addition to this obviously exciting prospect, we were looking for a chimpanzee population that could be of a special interest compared to the already existing chimpanzee projects. No study had been done so far on chimpanzees living in the dense tropical rainforest of Africa. As this was the habitat where the majority of wild chimpanzees live, this was a gap to be filled in the knowledge of chimpanzee biology.

After a preliminary study of eight months in 1976, we started for good a long-term study of the Taï chimpanzees in July 1979. Since the main topic at the beginning of the study was a feeding behaviour, nut-cracking (Chapter 9), we did not provision the chimpanzees with food, a controversial, although successful, method used to accelerate the slow habituation process of this timid species at other sites. The method we applied was to contact the chimpanzees preferably on our own or at most in twos, as often and as regularly as possible when they were in noisy groups, finding a silent individual being impossible anyway, except by chance. Searching and following them from dawn to dusk (chimpanzees are noisier early and late in the day) finally led to success. Gradually we noticed subtle signs of something other than pure terror towards us, that is they gave us a quick glance before running away, looked back at us before vanishing into the vegetation, and began to vocalize not too far away, whereas at first, if we disturbed them, we did not hear them again for the whole day. After two years, the chimpanzees showed the first signs of habituation by not running away from us systematically. We could now start to identify them, and by mid-1982, three years after having started the project, we were fairly sure that we had done so for all members of the community. There was a clear sex-difference in this habituation process, the females being generally less tolerant than the males. After five years, in 1984, we followed for the first time, in full sight, some of the males during their forays in the forest. Gradually and slowly, the females accepted being followed as well; but it was not until 1992 that the most mistrustful of them finally ignored the presence of human observers.

Right from the beginning we adopted rules to keep the stress imposed upon the chimpanzees by our presence as low as possible, and we tried hard not to interfere at all. Those rules are still in place today. Whenever we are with the chimpanzees we do nothing else apart from watching — not staring — and writing, displaying always what we consider to be a peaceful attitude. We remain silent, only whispering to one another when absolutely necessary. Talking aloud in the forest is poachers' behaviour, and poachers are a great danger to chimpanzees (see Chapter 2). We wear dull, inconspicuous clothes, avoid abrupt movements, never use, or even carry, a bush knife or any other weapon, and we never follow the chimpanzees when we are ill in order to exclude all possibility of contaminating them with a human pathogen. With rare exceptions, we follow them alone or in pairs, always keeping a distance of 5–8 meters, and do not react to any of their movements towards us, positive or negative. Lastly, we bury any garbage, or bring it back to camp.

Fig. 1.2:   In the tropical rainforest of the Taï National Park, a male chimpanzee sits on a buttress after having drummed on it.

The former of the two last rules became especially important the more the chimpanzees were habituated. Indeed, the males started to display closer to us and were quick to make use of any sign of fear we might show, ready to incorporate us in their displays. However, by strictly following this rule, the males' attempts to somehow integrate us in their social displays did not succeed and their behaviour has only rarely been directed toward us. We avoid particularly any physical contact, which is sometimes difficult with the curious and cheeky youngsters that now often come very close to the observer.

The very slow habituation progress, particularly of the females, made us hesitate to integrate other people into the project. We did so eventually in 1986 with Martina Funk from Zurich and Grégoire Nohon from Ponan in Côte d'Ivoire, both very keen and gifted observers. It was a very lucky choice and Grégoire Nohon is still with us. Since 1989 a team of two Ivorian observers has systematically collected data on the main study community and many students have followed different aspects of the chimpanzees' behaviour. We limited the number of students to two at a time plus two Ivorian field assistants to ensure steady observations while keeping the disturbance through human presence as low as possible. Since 1989, individuals have been followed from nest to nest on an almost daily basis. Meanwhile, with the help of two more teams of Ivorian assistants and several students, two neighbouring communities are now under study with the aim of understanding the interactions between the three chimpanzee communities.

The territory of the main study community consists of slightly undulating terrain with a maximum difference in altitude of about 120 m. The dense forest is uninterrupted over the whole area, except for the natural treefalls that let the sunlight reach the ground through the gaps in the canopy. A fine network of small rivulets covers the forest floor with fairly large areas of flat land along them that are regularly flooded during the rainy seasons. These areas are also distinctive for being covered by a less dense but still continuous forest. The territory of the chimpanzees is all encompassed within the boundary of the Taï National Park and consists mainly of undisturbed primary rainforest except for some parts at its western limit. Here, in 1975, a logging company illegally cut trees for some weeks. Fortunately, this was stopped before serious damage could be done.

The Taï chimpanzees

In all, 123 chimpanzees have been identified, 77 females and 46 males. Table A.1 in the appendix lists the names of all of them, together with their direct family relationship. This relationship was obvious in the cases where the infant was born while we were there, but for older youngsters we had to base our decision on the close association between the mother and the youngster as well as on the presence of behaviour suggesting a close relationship, such as support in conflicts, waiting for the youngsters, especially generous food sharing and/or sexual avoidance. We did not use physical resemblance to judge relatedness. We rapidly saw that this was not reliable as some adult females and males looked so similar that they would have to be considered brothers and sisters. This is difficult to explain as at Taï almost all females have transferred between communities before adulthood. The two latter criteria (food sharing, sexual avoidance) were used to tentatively determine the mother of the adult males Rousseau, Darwin, Kendo, and Snoopy.

Chimpanzees live in large social groups of thirty to over a hundred individuals belonging to all classes of age and sex. The main characteristic of the social life of the chimpanzees is a flexible social structure, with individuals leaving a group at any moment to associate with others. All the individuals who stay with one social group over a long period of time are said to belong to the same community. This community is regularly observed to split into separate parties. These sub-groups are of varying sizes, and include individuals who are temporarily within sight of one another. This feature, combined with the timid nature of wild chimpanzees, leads both to a particularly long habituation process and to uncertainties about community size. Indeed, some individuals can be absent from the parties of the more social members for months on end. This makes the dominance interactions between group members less rigid, as the position of any particular member of a party is dependent on the presence or absence of dominant individuals. Nevertheless, in chimpanzees a hierarchy is observed between the adult males, regulating access to food sources and sexually active females. The highest-ranking male, called the alpha male, is able to defeat all other group members, and group members regularly greet him by giving submissive pant-grunts. Female chimpanzees have a sexual skin around the genitals that swells for 10 days around the time of ovulation. This is a conspicuous signal of her willingness to mate and is restricted to this period of her cycle. Oestrous females are usually the centre of much attention by the males. Females have one infant about every five years, and normally a mother is accompanied by three of her infants: the oldest, about ten years old, will have only just started to move out of her sight for longer periods of time.

A general look at Table A.1 reveals that most of the individuals that we had identified at that time have now died; this low survival rate of the chimpanzees will be confirmed in Chapter 2. Is this situation unique to our study populations or does it reflect a general trend in forest chimpanzees throughout Africa? We are afraid the latter might be true and we shall now present some of the evidence concerning the chimpanzees' environment and the pressures from the human population that seems to account for the decrease in the chimpanzee population in forest areas.

History of the forests in West Africa

In West Africa, west of the Dahomey Gap, the tropical rainforest once covered some 40 million hectares, of which in the mid-1980s there were about 8 million hectares left (Martin 1989). The forest receives abundant rainfall twice per year (March to July and September to December). Inland, further north away from the sea, rainfall gradually decreases and the forest gives way progressively to the savannah and ultimately to the Sahara desert. This forest belt has varied extensively in its distribution during the last million years. During the last major drought in Africa, the forest was reduced to small pockets, and most of the continent was extremely dry (Hamilton, A. 1982). For example, the last drought reached its maximum 18 000 years ago in Côte d'Ivoire, and for 8000 years the rainfall was 75% lower than at present. The forest cover reached its present distribution only 10 000 years ago. Such prolonged periods of isolation between refuges are thought to favour speciation, and forest refuges became centres for the biodiversity of the present African forests. Two of them are located in West Africa: the first at the present border of Liberia and Côte d'Ivoire near the Taï National Park and the second near the border to Côte d'Ivoire in Ghana.

However, the forest block that used to cover most of Liberia, southern Côte d'Ivoire, and Ghana, has suffered in the last decades from a deforestation rate that has been among the fastest in the world. Humans have reduced the forest cover by some 80% and the trend is still in progress. In the late 1980s based on satellite pictures, we found that only 19.5% of the previous Guinean forest belt of the estimate of 8 million hectares mentioned above remained (Boesch *et al.* 1994; Marchesi *et al.* in press) of which 75% was degraded by timber exploitation and agriculture — a decrease of about a third within 5–8 years! Probably, in Côte d'Ivoire, the only remaining intact tropical rain forest is within Taï National Park.

This process of deforestation not only physically depleted the habitats available to wild chimpanzee populations, but also had a direct effect on the amount of rain falling on the forests. This is important since rainfall markedly affects food productivity of the forest. In the Taï forest, there are two rainy seasons between March and June and between September and November (Fig. 1.3), but no clearly defined dry season (less than 20 mm of rainfall per month). However, the whole region (Liberia to Ghana) has seen a steady decline in rainfall in the last four decades (Paturel *et al.* 1995). This was dramatically perceptible in 1983 when for the first time the dry wind from the Sahara, the 'Harmattan', reached the Taï region at the end of December and lasted for two weeks. Both the total amount of rainfall and the number of rainy days have decreased (Servat *et al.* 1997). By 1989, within the territory of the chimpanzee study community, there was at least one dry month per year. There was no rainfall at all at the beginning of the year and an unusual

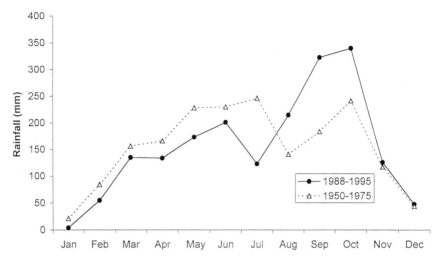

Fig. 1.3:   Rainfall in the Taï forest for two different periods.

decrease in temperature to a minimum of 13–15 °C, whereas the normal minimum is 18 °C, a fact that can only negatively affect the regeneration of moist tropical rainforest plant species. The trend is towards greater seasonality.

The Taï National Park

At present, the Taï National Park includes 435 000 hectares of forest and constitutes by far the largest remain of tropical rainforest in West Africa. Adjacent to it is the 'Réserve partielle de Faune du N'Zo' with an area of 75 000 hectares. The only other areas of rainforest within West Africa are the Sapo National Park in Liberia (131 000 hectares), and the Biosphere Reserve of the 'Massif de Ziama' in Guinea (116 000 hectares). The Taï National Park was registered as a Biosphere Reserve in 1978 and added to the World Heritage List of UNESCO in 1982. This status of the park acknowledges the fact that owing to its location and history it is a centre of biological diversity and thus also a specially rich habitat for the chimpanzees.

In the south-west of Côte d'Ivoire, including Taï National Park, Guillaumet (1967) identified 1300 species of vascular plants. Aké Assi and Pfeffer (1975) found 870 species of them within the park boundaries. This is a large number compared to the diversity found in the forests of temperate regions, but South-American and Asian forests are even more diverse. Of those species, 10% are strict endemics (86 species) found only in the region near the Liberio–Ivorian border. The present state of deforestation in the region, combined with the fact that Taï forest is the only large remnant in this region, means that this forest constitutes the last chance of survival for the endemic plants of the region. Our study site belongs to a part of the forest that receives an annual rainfall of about 1800 mm, which is an intermediate amount. Here, tree species of the evergreen and the semi-deciduous forest overlap and it is the richest in large tree species within the park. This point is critical for the chimpanzees since their diet consists mostly of the fruit of large trees.

The fauna of the Taï Forest

The richness of species within the park is equally impressive for animals as it is for plants. The eleven species of primates living within its boundary have made this park a conservation priority for primates in Africa (Oates 1985). The existing populations of pygmy hippos (*Cheoropsis liberensis*), as well as the zebra (*Cephalophus zebra*), and Jentink duikers (*Cephalophus jentinkii*) live exclusively in this forest region, and their survival is certainly tied to the survival of the park itself. Table A.2, in the appendix, presents the list of mammals seen within the park boundaries since 1979. The eleven primate species seen and heard many times a day clearly dominate the forest. It is one of the last strongholds of the red colobus population, badly endangered in most regions in tropical Africa. The red colobus seem, however, already restricted to the study site area, since reports from other areas of the Taï National Park have revealed some population decline. Duikers are the other group of mammals which are seen very frequently, but they are much more elusive. Many other species of animals are present within the park but are more rarely seen.

Human population growth in the region

The high rate of deforestation we described is a direct response to the constant increase in human population of the region around the Taï National Park. The south-west of the Côte d'Ivoire was presented as a 'green paradise' when the Ivorian authorities started the development of this region. It is difficult to find reliable figures on the human population growth during the early colonial period. However, the development of the populations living south of the Taï forest was followed from 1911 to 1971. The Krou, living south of the Hana river (see Fig. 1.1), increased by only 1.6% during these 60 years. The Bakwé population living around the town of Soubré, east of Taï forest, remained constant from 1922 to 1971. Overall, for an area of 17 312 km^2 south of the Taï forest, the population was about 20 000 or 1.16 inhabitants per km^2 (Schwartz 1993). The region was inhabited by small traditional socio-political units with a very low density of population which remained stable since the beginning of the century.

In the 1970s, the development of the south-west of the Côte d'Ivoire began, leading to a very strong surge of immigration during the last two decades (Schwarz 1993). The 'green paradise' of Côte d'Ivoire attracted a lot of people because of the prospect of easily grown cash crops like coffee and cocoa, and more recently oil-palm and hevea trees. In contrast, the more northerly part of Côte d'Ivoire and the northern neighbour countries suffered from a chronic deficit in rainfall. Since the start of our project, the Taï National Park has become an island within a rapidly growing agricultural landscape. The human population around the park increased from 23 000 in 1965 to 375 000 in 1988. Such a sixteen-fold population explosion is mainly the result of a high rate of immigration of people originating from other regions. As an example, the human density increased in the Taï sous-préfecture, close to our study site, from 8 inhabitants per km^2 in 1971 to 135 inhabitants per km^2 in 1991. We remember driving from the northern limit of the Park to the village of Taï in the early days of the project on a piste which was like a tunnel through the forest. Now you need to drive to the limits of the national park to see the forest for the first time.

In addition, the civil war that started in Liberia around 1990 produced a dramatic flow of refugees into the bordering area of Côte d'Ivoire. This resulted in a further population increase of 400% in the five villages near our field station in 1991 (19 347 refugees for 6337 residents). The situation in Liberia is improving, and in 1997 the refugees started slowly to move back to Liberia. This demographic pressure not only directly reduced the surface of the forest in this region, as the demand for land was so great, but poaching for meat, which always existed to some extent, became again a tremendous problem for the fauna around and within the park.

Human pressure on the forest

Poaching is the dominant problem for the fauna within the park, and despite our presence it has always been a threat for the chimpanzees. The first victims were, however, the elephants that were still numerous in 1979 when we started. The price of ivory was then so high that teams of poachers roamed the forest and the elephant herds were decimated. In Taï village, a kilogram of ivory was worth 50 000 CFA francs, which was far more than a month's salary for an official. With a small tusk weighing about 5 kg, the poacher easily acquired a year's salary by killing just one elephant. In 1981, a team of poachers killed six elephants within 24 hours at three different locations within our study site (five youngsters and a mother). Within a week, all the elephants in the territory of the chimpanzees left and have not since reappeared. Their striking tracks, the distinct roads across the forest to their favourite mud baths were quickly overgrown. In the 27 km^2 area that we regularly patrolled, we found twenty carcasses of elephants over two years. Of those only one had apparently died from natural causes, the skull still having two tusks, all the other skulls we found had been damaged by removing the tusks. In other words, 95% of the mortality of this elephant population was due to poaching. No animal population can survive such an increase in the mortality rate. Elephants were not the only ones to suffer so much from poaching. We regularly saw buffaloes during the early days of the project on the track to our camp, but after four years they were simply wiped out. We have not seen buffalo again.

Poaching does not always have such a devastating impact on wild populations as in the elephants' case, but is none the less dramatic in many cases. Local farmers poach for meat for their own consumption. Their usual technique is to place snare traps to capture forest antelopes, which are a favourite bush meat. Traps do not distinguish between species and the victims often die of starvation — a gratuitous and cruel death because the poachers check the traps too late or not at all. The chimpanzees quite regularly step into snares, and while adult chimpanzees can free themselves without too much damage, juveniles usually suffer badly and may die of necroses from the iron cable cutting into their flesh (see Chapter 2). Once, the whole social structure of the chimpanzee study community was seriously disturbed when the alpha male was handicapped by a snare cable around his wrist. On a larger scale, bushmeat is much sought after throughout the country. Today, animals eaten as bushmeat are hardly ever found outside the National Parks, and more and more commercial poachers come into the parks to provide towns with this meat. Poachers of this kind hunt for forest antelopes by night and for monkeys by day (mainly red and black-and-white colobus). To be most efficient, they make temporary camps within the forest, smoke the meat on the spot until they have plenty, and carry it back to fill their lorries. The impact on the fauna is dramatic. In many parts of the park, the density of monkeys has drastically decreased. Such poachers are ruthless; they do not spare the chimpanzees either and also endanger the observers following their study animals. We have both heard gunshots right next to us when following chimpanzees and have even been confronted by men with guns cautiously approaching the same nut-cracking chimpanzees. Poaching is very difficult to control in such a dense forest, and the situation in the Taï National Park is worrying. Unfortunately, there is nothing left to worry about in some other protected forests of Côte d'Ivoire; the forests themselves stand in good condition, but they are simply devoid of large fauna, wiped out by poachers from the neighbouring villages. Without adequate conservation measures, the situation in Taï forest may soon be identical.

The chimpanzee and its relatives

The chimpanzees' closeness to man was recognized long ago. Buffon (1792) classified them as *Pithecus* without differentiating between the chimpanzee and the gorilla. Soon all great apes were classified with humans as being part of the Hominoid family, with the gorilla and the chimpanzee grouped together in the Pongid sub-family. More recent analysis, however, has shown that the human and the chimpanzee are the most closely related of all the apes as they share 98.3% of their genetic material, while the similarity between chimpanzee and gorilla is lower at 96.5% (Sarich and Wilson 1967). The orang-utan (*Pongo pygmeus*) comes out as the third of the great apes, with the gibbon (*Hylobates lar*) and the siamang (*Symphalangus syndactylus*) forming the lesser great apes. Such analyses suggest that the chimpanzee and the human should be classified together separately from the gorilla and the orang-utan (see Chapter 11).

Molecular differences between two species can also be used to estimate the age of divergence between them, if it is possible to base the comparison on some known fixed date. Classically, primatologists and anthropologists thought that human ancestors diverged from the chimpanzee/gorilla ancestor some 12–14 million years ago (Leakey 1980; Pilbeam 1980). Geneticists found such a distant date to be incompatible with their data and proposed that the time of the chimpanzee/human divergence was around 5 million years ago, estimating 8 million years for the divergence between the gorilla and the chimpanzee/human ancestors (Sarich and Wilson 1967; Jones *et al.* 1992). This date has now been accepted by most scholars.

Two species of chimpanzees are currently recognized: the bonobo or the pygmy chimpanzee (*Pan paniscus*) and the chimpanzee (*Pan troglodytes*). The bonobo is limited to the forested regions south of the Congo river, in the democratic republic of Congo (former Zaïre). This species is the least well known so far, and what we know comes from two populations. The one in the Lomako region has been under study by various scientists (Badrian and Badrian 1984; Susman 1984; White 1988; Fruth and Hohmann 1994). Habituation was slow there, but recently more detailed data are being collected. More is known from the second population in Wamba, where detailed observation and filming have been possible at feeding sites where sugar cane was abundantly provided (Furuichi 1989; Kano 1992).

Classically three subspecies of the chimpanzee have been distinguished: the western subspecies from Senegal to Nigeria (*Pan troglodytes verus*), the central subspecies from Cameroon to Congo (*Pan troglodytes troglodytes*), and the eastern subspecies from

Uganda to Tanzania (*Pan troglodytes schweinfurthii*). However, the criteria for distinguishing them are vague, mainly based on morphological appearances, such as the presence or absence of a white beard, the bald head or the grey back in adults, and the white tuft in juveniles. All these criteria vary extensively between chimpanzees within a single community. We have for instance individuals in our study community in the Taï forest that conform to the criteria for all three subspecies. Recent molecular analyses (Gonder *et al.* 1997; Morin *et al.* 1994, Gagneux *et al.* 1999) have produced a more complicated picture with four distinct groups now being genetically recognized, a new one having been added with populations on both sides of the Niger river. The separation between the western and the central/eastern group appears to have occurred 0.6 to 1.6 million years ago, suggesting that something in the distinction of different subspecies might be sound (Morin *et al.* 1994a). It remains difficult to propose taxonomic distinction only on the basis of genetic differences. We shall see throughout this book that if behavioural differences exist between chimpanzee populations, they do not follow those proposed subspecies distinction. This suggests that these genetic differences might not be so important for understanding the diversity of the chimpanzee.

The Taï chimpanzee community has gone through many demographic changes (see Chapter 2), but has remained for the entire study period within the same area in the Taï forest (see Chapter 7).

2 Demography of the Taï chimpanzee community

Forest scene, Survival: 21 August 1996: Mognié, a six-year old female, was in the company of her mother Mystère with baby Mozart, Sirène and Brutus were also there. Suddenly, at 13.40, there was a violent outburst of screams, sign of a possible leopard attack. Grégoire Nohon heard Mystère bark very loudly and very aggressively. He hurried to the site to find all the chimpanzees in trees. Mognié was badly wounded. Claws had raked her chest and belly. Five holes were visible with a piece of intestine popping out of one. Her respiration could be heard through one of the holes in her chest. Forty minutes after the attack, all the chimpanzees moved westwards. Mystère licked some of the blood off Mognié's chest. Four minutes later, Mognié again climbed a tree to rest in a nest. At 16.45, they all moved further west. At 16.50 Mognié climbed into a Uapaca tree and Grégoire saw that the piece of intestine that had been popping out must be back in place, but the respiration was still strongly audible through this hole. Mystère and her baby Mozart remained with Mognié for two days. She entered Mognié's nest many times, but it was not possible to see what she did to her.

The morning of 23 August, Mystère left with baby Mozart to feed further away. For the next 17 days, Mognié remained on her own within the same two fruit-bearing trees lying down most of the time and eating some fruit. On the 11 September 1996 at 12.42, Mognié joined the group with some screams of excitement. Mystère ran towards her and they embraced screaming briefly. Mystère lifted Mognié's chin and looked at her wounds that looked fully healed except for a small pink spot in the largest wound, showing an incredible ability to heal from very dangerous wounds.

The observation with Mognié is a rarity, for often a wounded chimpanzee simply disappears for days or weeks. What happens to them? In this chapter, we investigate how chimpanzees' likelihood of survival varied for different age ranges and periods of time, and how the growth of the community was affected by such variations.

Demographic information on large mammals with long generation spans is difficult to obtain, and requires prolonged observation (Begon *et al.* 1990). Such data are needed to understand population dynamics, the probability of local extinction, the reproductive strategies of the sexes, and also the factors affecting social grouping and interaction. In other words, demography is the context within which we can study most aspects of the biology of a wild animal population, especially life histories and social behaviour. Social dynamics within a population depend upon its structure, for social interactions are affected by demographic differences between and within social units over time (e.g.

Altmann and Altmann 1970). The extent of competition for social and sexual partners is directly affected by the number of individuals of each age/sex class (Dunbar 1988). The sex ratio also influences mother–infant relations, since sex-specific survivorship may affect the maternal investment in male versus female offspring (Trivers 1974, 1985). This may affect the interbirth interval, especially since chimpanzee mothers maintain long-lasting relations with sons but only rarely with daughters (Goodall 1986; Nishida *et al.* 1990; see Chapter 4).

Conservation measures to protect any species also depend upon demographic information from many different populations, as population growth depends upon local conditions (Begon *et al.* 1990). In this chapter we present for the first time the long-term demographic information we obtained from a chimpanzee study community living in a moist evergreen tropical rainforest, the Taï National Park. Most remaining wild chimpanzee populations live in this kind of environment (Tutin and Fernandez 1984; Marchesi *et al.* 1995; Teleki 1989). To understand the effect of the habitat on the demography of the chimpanzees, we shall make some detailed comparisons with the demography of chimpanzee populations living in other habitats. This gives some insight into the demographic potential of this species, as well as its sensitivity to ecological changes.

Demographic data on great apes are rare in general because the long-term perspective it requires makes such data collection strenuous from several points of view. At present, only a few populations have been followed long enough to provide the necessary information. For chimpanzees, we have so far detailed information on two communities from Tanzania, the Mahale Mountains (Nishida *et al.* 1990) and the Gombe Stream site (Teleki *et al.* 1976; Goodall 1983, 1986). Some information is also available from Bossou in Guinea (Sugiyama 1984; 1994*b*) and Kibale in Uganda (Ghiglieri 1984).

The demographic data we present here are based on sixteen years of observation on the Taï chimpanzee community in Côte d'Ivoire, West Africa. All known chimpanzee communities consist of 'fission–fusion' social groups, because we hardly ever see all members of the community together, and the temporary grouping units have a constantly fluctuating composition (Goodall 1986; Nishida 1990; Boesch 1991*c*; Wrangham *et al.* 1992; Sugiyama 1984). The 'community' or 'unit-group' includes all individuals seen together in various grouping units over a period of many months. Most community members are seen regularly, others much less, and some can be absent for weeks from the groups followed by observers.

Three major problems have to be mentioned in introducing a demographic study on chimpanzees.

1. Prolonged absence of community members can be for purely social reasons, but it can also be due to illness. Sick individuals tend to join the group only when in good health again. Thus, illnesses are probably underestimated in most chimpanzee studies.

2. Definitive disappearance of an individual can be ascertained only with time, because we are rarely sure of the reason why a given individual is absent. As a rule, we started to worry about the fate of an adult male only after ten weeks of absence. For females with infants, this period can be longer, for example one mother was absent for five months without any obvious sign of having been ill when she returned. It is

thus difficult to determine the causes of disappearance of an individual. We noticed only twice that an individual was ill. For example, Schubert, an adult male, lost so much weight that we could see all his muscles under the skin, and we noticed that the social rivals rapidly exploited his declining health. This explains why sick individuals tend to isolate themselves until they have recovered. We only rarely found dead bodies in the forest that were fresh enough to allow some hints of the causes of death. Because corpses decay within a week in the tropical rainforest, the cause of death usually remained a mystery and a frustration to the observers (see mortality below).

3. Determining the permanence of membership of any individual in a chimpanzee community is made difficult by the uncertainty whether an individual's absence is caused by death or by transfer to another community. We knew about the studies on the Gombe and Mahale chimpanzees when we started our project, and we formed our expectations about community transfers on their observations: that is that males remain in their natal community but females transfer. If a certain class of individuals transfers between communities, some should disappear from the study community and some should enter it. Since the full identification of community members in 1982, no males have entered the community, but 13 previously unidentified females did (see Table 2.3, below). These immigrants were without exception adolescent females, as in the Tanzanian studies. We therefore concluded that females transfer between communities when adolescent, whereas males never do. For demographic analysis, whenever a male disappeared definitively we considered him as dead. If an adolescent female disappeared while apparently in good health, we assumed that she had transferred into a new community, but when the same happened with an adult female we assumed that she was dead.

Physical development of wild chimpanzees

For a demographic study we need a method to determine the age of the chimpanzees. The physical development of wild chimpanzees is surprisingly slow. The first data from Gombe chimpanzees (Goodall 1968) were greeted with some scepticism as they presented the image of a much slower developing species than was assumed on the basis of data on captive chimpanzees. Data from Mahale chimpanzees (Nishida 1968) confirmed that the physical development of chimpanzees is comparable to that of human. We based our estimates of individual age on criteria that differ between infants, juveniles, adolescents, and adults. These categories were first proposed by Goodall (1968) and later refined by further subdividing each class (Goodall 1983, 1986). We applied the same classification to the Taï chimpanzees, but in the end used only Goodall's original more simple classification, as individual variations are so important in morphology (Table 2.1).

Taï chimpanzees have a long lifespan, with a few males and females still being very active at an estimated age of 42 to 44 years. Those who had reached that age looked as if they would be able to survive for even longer, and it therefore seems realistic to expect that wild chimpanzees might live until they are 50 (for a similar conclusion see Nishida *et al.* 1990; Goodall 1986). Animals with such a long lifespan present a challenge for those who want to study their demography. As the Taï study is now in its eighteenth

Table 2.1: Development in wild chimpanzees: age classes observed in wild chimpanzees with an indication of the morphological and behavioural criteria considered typical for each class

Age class	Age (years)	Physical criteria	Behavioural criteria
Infant	0–5	Small	Suckles
		White tuft	Dorsal riding
Juvenile	5–10	Half adult size	Constant association with their mother
Adolescent		$\frac{2}{3}$ to $\frac{3}{4}$ of adult size	Decreasing association with their
Female	10–13	Start sexual swellings	mother
Male	10–15	Testis descend	
Adult			
Female	>13	First parturition	
Old	>40	Loss of hairs on head	
Male	>15	Larger rump size	
Prime	25–40	Fully grown	
Old	>40	Reduced body width	Fall in dominance rank

year, we can still only estimate the age of most of the adults. This was a major problem at the start of the project, particularly for subadult individuals, and in 1982 we got help by visiting Gombe. Because the Gombe study was then in its twenty-second year, we could see wild chimpanzees of known ages, and Jane Goodall introduced us to her criteria for estimating the age of the chimpanzees. For males, we used the testis-descent criteria for adolescents, and the rump-size for adults. Young adults continue to grow both at the rump and at the shoulder until they are fully grown (at 25 years). For older males, it was more difficult and we had to compare them amongst themselves and also judge by their appearance; some individuals look older than others. The situation is different for females who barely grow once they are adult. Their differences in morphology are due to individual differences rather than age, except for the older ones (over 40 years old). We decided to estimate their age by using the age of their infants, assuming that they were 13 years old at first parturition. Later we corrected for this, as it was obviously an underestimate of maternal age, given possible miscarriages and early death of infants.

Individuals also vary considerably in their rate of development. When we visited Gombe, a young male, Michaelmas, looked as if he was 6 years old, but Jane Goodall told us that he had broken his leg some years ago and that his physical development was seriously hampered, in fact he was 12! Similarly, one adolescent male in our community, Darwin, had two deformed feet (club-feet) and must have suffered some kind of malnutrition, having barely any teeth, but he was probably young, judging by his favourite companion.

To judge our accuracy in estimating age, we later performed two tests using the demographic data: First, we compared our age estimates for the females immigrating into our study community with those that emigrated, whose estimated age should be more accurate as we had known them for much longer. The estimates proved to be equally accurate (Wilcoxon test: $p > 0.50$) (see Table 3.1, p. 44). Similarly, our estimates of the interbirth interval for siblings that were already born when we started the project with those for which we knew the date of birth are equally accurate, differing by only 3.48 months out of an average interval of 69.53 months (Wilcoxon test: $p > 0.50$) (see Table 3.4, p. 52). Based on these controls we were confident that despite the difficulty in estimating the age of the chimpanzees, we had gained a certain feeling for it.

Fig. 2.1: Baby: a two-week-old female.

Fig. 2.2: Infant: a two-year-old female suckling.

Fig. 2.3: Juvenile: a nine-year-old female eats winged termites after having pounded the termite mound against a root.

Fig. 2.4:   Adolescent: a 12-year-old male collects some nuts while holding his wooden hammer.

Fig. 2.5:   Adult: female looking at her round belly three months before giving birth to her third infant.

Size and composition of the Taï community

The Taï community structure resembles that of other chimpanzee study populations: a multi-male and multi-female society whose composition can be stable for years (Table 2.2). During the six years following the full identification of all the individuals, the size of the community remained stable at about 80. However, during the 16 year span of this study, the community varied from 82 to 29 individuals with an average of over 60. During the whole study, adults represented about 45% of the community.

From 1982 to 1996, we observed 75 births and 188 deaths (Table 2.3). There were on average 5 births and 7.87 deaths per year. The decrease in community size was paralleled by a decrease in the number of births. The number of deaths showed a peak in 1988 and 1989, and was lowest for the first years of the study when the community was the largest and at the end of the study period when the community was the smallest. Births occurred in all the months of the year (Fig. 2.6), without any sign of a 'birth season' in this community ($p > 0.05$)*. This is in agreement with the observations in Gombe and Mahale chimpanzees (Goodall 1986; Nishida *et al.* 1990; Wallis 1995).

There were 20 instances of females transferring in or out of the community. More females immigrated than emigrated. The most likely explanation is that survival from 1988 onwards was low, and most juvenile females died before reaching the age of transfer. No male of any age was seen to immigrate, or emigrate. We took care throughout the study to check for such a possibility. In February 1984, four adult males and three adolescent males

Table 2.2: Composition of the Taï community over a period of 15 years

	Year															
	82	83	84	85	86	87	88	89	90	91	92	93	94	95	96	Mean
Adult male	9	10	10	8	9	8	7	6	7	6	6	6	4	2	2	6.67
Adolescent male	6	5	6	3	2	2	1	2	3	3	4	3	3	1	1	3.00
Juvenile male	4	5	4	4	4	9	9	7	4	2	1	1	1	1	3	3.93
Infant male	9	10	11	10	10	8	9	6	6	6	7	6	6	5	4	7.53
Adult female	21	25	27	27	27	27	27	25	22	19	20	17	16	11	11	21.47
Adolescent female	9	8	3	6	6	7	5	7	3	2	1	2	2	1	1	4.20
Juvenile female	6	6	10	6	5	7	4	3	3	4	5	3	3	3	4	4.80
Infant female	9	9	9	11	13	11	10	9	8	8	7	7	6	5	7	8.60
Unknown	1	3	2	2	2	1	2	1	0	0	0	0	0	0	0	0.93
Total	74	81	82	77	78	80	74	66	56	50	51	45	41	29	33	61.13

'Unknown' are newborns, whose sex was not established before they died. Counts are done yearly on 1 January; individuals that were born and died within a same year are thus not included.

---

* Statistical power is always dependent upon the sample size. With chimpanzees, it is difficult to work with samples as large as could be achieved with animals of smaller size and shorter generation time, such as ants, fruitflies, or mice. This presents us throughout the book with the problem that the statistical tests will have a limited power in detecting relatively weak effects. In other words, positive results will be more reliable than negative ones. If a test is not significant, we will not be able to exclude that our sample size was too small to detect a weak but real effect. Therefore, we shall discuss some tests showing a tendency only (*p*-value between 0.05 and 0.1).

Table 2.3: Birth, death, immigration, and emigration in the Taï community

Year	82	83	84	85	86	87	88	89	90	91	92	93	94	95	96	Total
Births																
Male	5	2	3	0	0	3	3	2	3	2	2	0	2	1	1	29
Female	4	3	1	3	4	3	4	1	1	4	3	2	1	2	1	37
Unknown	0	4	2	1	1	1	0	0	0	0	0	0	0	0	0	9
Total	9	9	6	4	5	7	7	3	4	6	5	2	3	3	2	75
Transfers																
Immigrant	–	–	1	–	1	1(1)	1(1)	4	–	–	–	1	2	–	–	11(2)
Emigrant	–	–	–	1	2	2	–	–	1	–	–	–	–	1	–	7
Deaths																
Male																
Adult	–	–	–	4	–	1	1	1	–	1	–	–	2	2	–	12
Adolescent	–	–	–	3	–	–	2	–	–	–	–	1	–	2	–	8
Juvenile	–	1	–	–	–	1	3	1	–	2	1	–	–	–	1	10
Infant	–	–	–	–	–	–	2	2	3	1	2	1	1	–	–	12
Female																
Adult	–	–	2	2	1	–	2	5	4	4	–	4	–	5	1	30
Adolescent	–	–	–	–	–	–	–	–	1	–	–	–	–	2	–	3
Juvenile	–	–	–	–	–	2	2	1	–	–	4	–	–	–	–	9
Infant	–	2	1	–	–	6	3	2	2	–	2	1	3	–	1	23
Unknown	–	2	3	3	1	1	–	1	–	–	–	–	–	–	–	11
Total	0	5	6	12	2	11	15	13	10	8	9	7	6	11	3	118

In the transfers, 2 females remained in the community for only some weeks and are indicated in brackets.

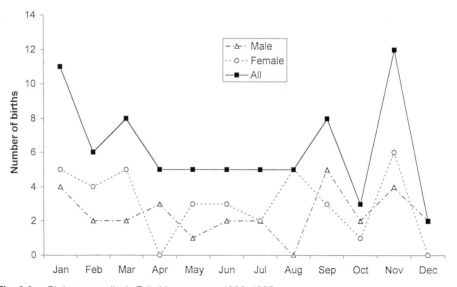

Fig. 2.6:   Birth seasonality in Taï chimpanzees: 1982–1995.

disappeared. This could have been a case of fission, and we controlled in all subsequent territorial encounters to see if any of the stranger males could be one of them. We never found any trace of them and concluded that they had died, probably as a result of poaching (see below).

Sex ratio

Under the most simple assumption, sex ratio at birth is expected to be close to 1:1, because it pays the parents to produce more offspring of the sex that is rarer and that is expected to yield more grandchildren, which has the effect of bringing any sex-ratio bias back to balance (Fisher 1930). If offspring of one sex cost more than offspring of the other sex, that would affect the argument above. In many animal species, a sex ratio differing from 1:1 has been discovered, for example in spotted hyenas (Holekamp and Smale 1995), the red deer (Clutton-Brock *et al.* 1984), some birds (Gowaty 1993) and in some primates (Clark 1978; Altmann 1980; Silk 1988). The most complete explanation of this phenomenon takes into account how the future competitive situation might affect the lifetime reproductive success of the parent and the offspring. If offspring compete with each other and with the parents, more of the sex that disperses should be produced. If the offspring cooperate with each other and with the parents, more offspring of the sex that does not disperse should be produced. This has been clearly illustrated in a group of hyenas that rapidly changed the sex ratio of their offspring in agreement with such a model (Holekamp and Smale 1995).

Taï chimpanzees always had a strongly female-biased adult sex ratio near 1:3 throughout the study period (Table 2.2). It was not affected by the decline in the size of the community. Mahale chimpanzees showed the same tendency in the same direction (Nishida *et al.* 1990), but Gombe chimpanzees had a sex ratio close to 1:1 for most of the time. It is important to remember here that at Mahale and Taï, all females transfer between communities, whereas most Gombe females remain in their natal community (see also Chapter 3).

Does the adult sex ratio reflect the sex ratio at birth? At Taï, the birth sex ratio (1:1.27) did not significantly differ from 1:1 with a slight female skew (Table 2.3), that was also found at Mahale, with a birth sex ratio of 1:1.14 (Nishida *et al.* 1990), and at Gombe of 1:1.5 (Teleki *et al.* 1976). Hence, the birth sex ratios in the three chimpanzee populations are similar, near to unity and slightly female skewed. The slight skewed production of one sex at birth in wild chimpanzees does not explain the large bias in adult sex ratio, and we need to explain the adult sex ratio in terms of differences in survival between the two sexes from birth to adulthood.

At Taï, the sex ratio remained stable for the first ten years of life and the bias in favour of the females appeared only during the transition to adolescence and became especially strong in the adults (adult + adolescent versus infant + birth: $X^2 = 4.34$, $df = 1$, $p < 0.05$). The transfer pattern of females could explain the first effect on the sex ratio, as more females immigrated at adolescence into the community than emigrated (Table 2.3), skewing the sex ratio in favour of the adolescent females. To explain the second increase in the bias toward females (from adolescence to adult), we would expect a lower survival rate of the adult males compared to the adult females. We shall see below that the data confirm this expectation.

Population dynamics

Estimation of the vital statistics of an animal population is done with the help of life tables. By taking a regular census of a population, we can get a precise idea of the number of individuals present in all age categories and of their yearly survival. The most reliable method consists in following a cohort of individuals born within a short period until the last dies (Caughley 1977; Begon *et al.* 1990; Krebs 1994). This allows to estimate the variations in mortality over different age categories. However, for large mammals, it is not easy to have enough individuals to follow through their life, and the time required to do so can be very long. To overcome this problem, alternative methods have been developed to analyze life tables and follow population dynamics. At Taï, we knew all individuals of the population and followed them regularly. We knew the number of individuals entering each age class and the number who survived until the end of that age class. This allowed us to estimate the survival rate for each class, using the instantaneous life table method (Caughley 1977). In doing so, we assume that birth and mortality rates remain constant in time in our study community. This is not a realistic assumption, but it is the best we could make to produce an equivalent of a cohort table. For each age class, we used the proportion of individuals that were still alive at the end of the year class to estimate survival probabilities.

Figure 2.7 gives the survivorship curves for the females and the males in the Taï community for the whole study period. Prospects of survival for this community are not very good, since the population was decreasing by 6% per year. Males and females had a mean life expectancy at birth of 7.2 and 8.2 years respectively. Because they can live up to 50 years in the wild, this reflects their high juvenile mortality. Survival remained low for adults. The expectation of survival once adult was only between 9 to 10 years for both sexes.

The survival rates of the two sexes were similar for the first 10 years of life, as expected from the observed sex ratio. However, starting in the fifteenth year of life, males had a consistently lower survival for the rest of their life than females: 4% less than the females' survival rate from 15 to 23 years of age and 8% less between 28 to 38 years of age. Thus in Taï chimpanzees, the skewed sex ratio we observed in the adults can be explained by the lower survival of adults males. Later in this chapter we discuss which mortality factors affect the males more.

To understand the dynamics of the community, we also tested the validity of the assumption that the mortality rates remained constant throughout the study period. Table 2.2 suggests that this assumption does not hold, for we can define three periods: a stable period from 1982 to 1987, during which the size of the community remained around 80 individuals, followed by a first period of decline from 1988 to 1991, when the size of the community decreased from 80 to 50 individuals, and a second decline from 1992 to 1994 leading to a reduced community of about 30 chimpanzees. To understand the demographic changes within the community, we shall construct a life table for each of these periods.

During the first six years of the study, once identification was completed, the size of the community remained stable with small fluctuations. In other words, deaths and emigrations were compensated for by births and immigrations. We use this period to characterize the normal state of a chimpanzee population in a tropical forest environment. We

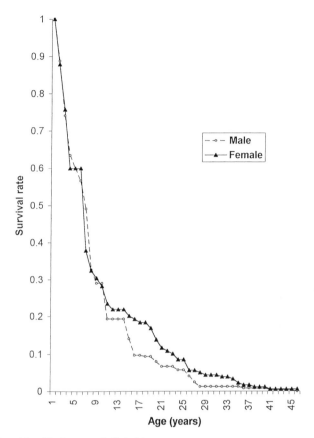

Fig. 2.7: Survivorship of both sexes in Taï chimpanzees.

then aim to discover which parameters affecting the population changed to produce the subsequent decreases.

The stable phase (1982–1987)
Community composition
Community size fluctuated between 74 and 82 individuals with a female-biased sex of about 0.35 for this six-year period. On average, the community included 9 adult males, 26 adult females, and 44 subadults. It had regular encounters with neighbouring communities, but the outcomes did not allow us to conclude whether they dominated the neighbours or not. This stability suggests that such a community size is about normal for chimpanzees living in a tropical rainforest with abundant food. Their territory was about 25 km², and the population density was 4.08 to 2.70 individuals per km² (Chapter 7).

Fluctuations
Although its size was nearly stable, the community nevertheless had to deal with some adverse effects of the environment. Illnesses are always present but apparently did not represent a major source of mortality. Poaching also occurred in this territory at the edge

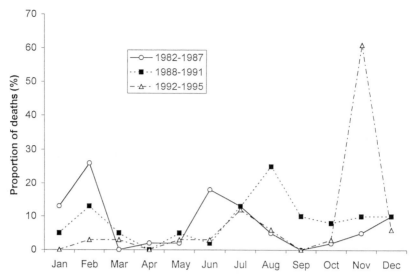

Fig. 2.8:   Monthly mortality rates in Taï chimpanzees.

of the national park and close to some farmers' fields. In February 1984, 4 prime males and 3 adolescent males disappeared all together, probably victims of poachers. This sudden loss reduced the adult sex ratio from 0.42 to 0.32. However, the community was apparently healthy enough to compensate for the loss, and the community remained at 80 individuals for several years. Other individuals probably disappeared because of poaching, but the uncertainty in attributing disappearances did not allow us to assert the importance of poaching. Nevertheless, the loss of 7 males all at once had an impact on the social structure, and we cannot exclude that this important loss made the community more vulnerable in the future.

There was a tendency for deaths to occur during the dry seasons: we registered 80% of all the deaths during the two dry periods (Fig. 2.8). If, however, we exclude the 7 simultaneous disappearances of males we attribute to poaching in February 1984, this tendency disappears, and the distribution of deaths does not deviate from a random distribution. Births and deaths occurred throughout the year with little if any seasonality.

Survivorship

Figure 2.9 depicts the survival of the chimpanzees during the stable period. In striking contrast to the overall results (Fig. 2.7), during this period survival in the Taï chimpanzees was high, with 60% of them reaching age of maturity (15 years old). This is higher than the survival of most other chimpanzee and many primate populations. The population presented during this period a rate of increase very close to zero, each female produced on average 0.899 female infants per generation with a generation time of 19.80 years. Despite the fact that they suffered losses through poaching, the population compensated them and remained nearly stable. This is the important baseline against which we shall compare the other periods and it shows that wild chimpanzee populations have the potential to increase in a stable environment.

The high survival of chimpanzees during this period, with over 30% of the individuals surviving at age 24 is reflected in the mean expectancy of life at birth: 16 years for the two sexes. Mortality rate were moderate, as infant death was seen only in youngsters of less than 2 years, while no infant between 3 to 5 years died during this period (Table 2.4). Subadult mortality rates were generally low.

First period of population decline (1988–1991)
Community composition
The community decreased steadily from 74 to 50 individuals with an average female-biased sex ratio of 0.28 during this 4-year period. On average, the community included 6 adult males, 23 adult females, and 32 subadults. The size of the territory remained about the same as in the previous period, and thus the density of the chimpanzees diminished.

Fluctuation
The decrease in number was constant over this period for most age/sex classes. Whereas births remained dispersed over the year, the mortality pattern changed with about 48% of deaths occurring from July to September (Fig. 2.8). This trend was, however, not statistically significant. As discussed above, it is very difficult to ascertain the cause of the death of a chimpanzee, but we felt that in addition to the causes of mortality they encountered during the first period, predation by leopards became a major factor. During this period, predation was the primary source of mortality in the study community, responsible for 39% of the casualties. Each chimpanzee had an expected probability of being attacked by a leopard every 3.3 years and of being killed by a leopard within 18 years (Boesch 1991c). The previous disappearance of 7 males due to poaching might have diminished the community's ability to counter leopard attacks effectively. Although leopards can kill adult chimpanzees, we had the impression that they concentrated their attacks on older infants and juveniles, and this is reflected in a significant increase in the mortality rate for these two age classes during this period (Table 2.4). For juveniles of both sexes, the mortality rate showed a twofold to twelvefold increase over the previous stable period.

Table 2.4: Mortality rates for the two sexes for different age classes during the three periods of the study on Taï chimpanzees

Age class (years)	1982–1987		1988–1991		1992–1994	
	Male	Female	Male	Female	Male	Female
0–2	0	0.22	0.29	0.21	0.83	0.47
3–5	0	0	0.66	0	0.29	0.21
6–7	0	0.18	0.33	0.65	0	0.67
8–15	0.58	0.06	0.75	0.83	0.92	0
16–24	0.24	0.38	0	0.5	0.70	0.64
25–30	0.33	0	0.5	0.5	0.22	0.4
over 30	1	1	1	1	1	1

Survivorship

Figure 2.9 shows that during this period the survival of the chimpanzees was extremely low compared to the period before: only 4% survived to the age of 15 years. This represents a fifteen-fold decrease in survivorship compared to the stable period. The population was decreasing by 13% each year, females on average produced 0.101 children per generation and the generation time decreased to 17.81 years. Mean life expectancy at birth was only 4.17 years for males and 5.1 years for females. The expectation of further life at 15 years was close to 9 years for both sexes, which is close to the one achieved during the stable period. This reflects the fact that predation was mainly directed against subadults.

This chimpanzee population showed a high sensitivity to an increase in the mortality rate of the subadults. We cannot exclude that the chimpanzee community might have been able to defend itself against increased pressure from predation if it had not previously suffered the loss of 7 males through poaching. However, our observations suggest that chimpanzees are not physically able to challenge an ambush killer like the leopard.

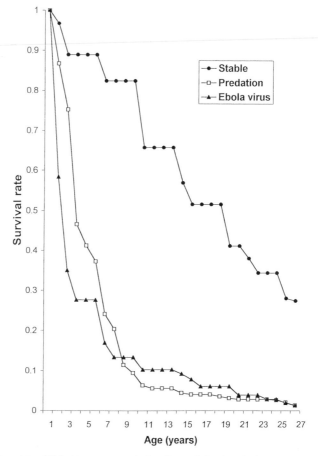

Fig. 2.9:   Survivorship of Taï chimpanzees during three distinct periods.

The victim itself has to be able to fend off the leopard until support arrives, and this is precisely what isolated subadults could not do (Boesch 1991c). In 1991 and 1992, the pressure from leopards seemed to diminish, and the community was starting a new period of stability with the potential for increase. The sudden drop in leopard attacks supports our speculation that the unusually high leopard predation was due to one individual, a 'chimpanzee killer' (Boesch 1991c).

Second period of population decline (1992–1995)

In late October 1992 and again in October 1994, the population decline suddenly worsened as a result of disease epidemics: the chimpanzees fell victims to a virus, a new Ebola virus strain, *Ebola Côte d'Ivoire* (Le Guenno *et al*. 1995). We are still trying to understand why this virus became dangerous for the chimpanzees at this time and how they became infected.

Community composition

The community decreased from 51 to 29 individuals with a female-biased sex ratio close to 0.26. On average, during this period, the community contained 4 adult males, 16 females, and 21 subadults. The decrease was dramatic and threatened the survival of the community as a social unit, since by the end of 1994 only 2 adult males and 1 large adolescent survived. We wondered whether the females would start to leave a community with so few males to offer security and join a larger one, as suggested by the example of the Mahale chimpanzees (Nishida *et al*. 1985). So far this has not happened (see Chapter 7).

Fluctuation

Both epidemics were seasonal, killing chimpanzees in late October and November. In 1992, 8 individuals disappeared within 12 days, and in 1994, 12 individuals disappeared within 3 weeks. The distribution of the casualties differed significantly from a random distribution: 70% of them occurred at the end of the year (Fig. 2.8). In addition, the distribution of the casualties was significantly different to those observed during the two previous periods ($p < 0.05$) suggesting a new mortality factor that was absent during the earlier periods. Mortality rate of young infants increased the most, but adults were also hit hard (Table 2.4). Although we could confirm with certainty the cause of death of only one chimpanzee, the similarity of the symptoms we observed in 4 further individuals and the very sudden disappearance of all of them makes it very likely that they died of the Ebola virus. It is probable that the chimpanzees became a new host of this virus, with 31% of them dying within 3 weeks during the second outbreak. The virus seems to be 100% fatal in chimpanzees, for none of the infected individuals seems to have survived, based on the negative results of the analysis of blood samples of some survivors (Formenty *et al*. 1999). Between these outbreaks, the mortality rate was especially low.

Survivorship

Compared with the first stable period, the survival of the chimpanzees was extremely low in this third period (Fig. 2.9), with only 8% surviving until the age of 15 years. The population decreased by 15% per year, females on average produced 0.07 children per life time and the generation time decreased to 17.82 years. Mean life expectancy at birth was very low: 2.4 years for males and 5.3 years for females. Similarly, life expectancy once adult, which was 15 years in the stable period, dropped to 6 years for both sexes. These results leave little hope for the future survival of this population. The drastically reduced sample size also affects our research, particularly on males (see Chapter 4).

We see from these results that chimpanzee populations are very sensitive to variations in the mortality rate; infants and juveniles more so than adults. It is, however, noteworthy that despite constant but moderate poaching pressure (we are at the edge of a national park), and the sudden disappearance of 7 males, the chimpanzee population was stable during a period of six years.

Causes of mortality

The information presented here was gathered during the entire study and gives an overview of the variety of causes of mortality in the Taï chimpanzee population.

Accidents

Falling from trees

We expected falling from trees to constitute a significant risk for forest chimpanzees, but only a single casualty by such a cause has been inferred in 17 years (see Table 2.5). We most rarely saw a chimpanzee fall from a tree, despite the fact that they often feed on fruit and leaves in the highest trees of the forest. All adults we saw fall did so because of violent fights in trees. In one case, an adult female fell from a height of more than 15 meters without releasing the big *Treculia* fruit she was holding in her hands. We saw

Table 2.5: Causes of mortality in Taï chimpanzees: factors known to have affected the health of the chimpanzees

Factor	Effect	Number of individuals affected	Criteria	Comments
Accident				
Fall from a tree	None	10	Seen	
	Infirmity	1 adult	Seen	Dislocated arm
	Death	1 infant	Inferred	
Leopard attacks	Deep cuts	16 (4 dead)	Seen	
		18	Inferred	
Chimpanzee attacks	Deep wounds	1 adult	Seen	Intercommunity encounter
	Small wounds	30	Seen	Intracommunity aggression
Cannibalism	Death	1 infant	Seen	Eaten by females
Snares	Swelling	9	Seen	Hand/foot in snare
	Death	1 juvenile	Seen	Gangrene
Gun shot	Death	1 adult	Seen	Skull with impact
	Death	6 adults	Inferred	
Illness				
Influenza	Runny nose	80	Seen	Rainy season
Conjunctivitis	Swollen eyes + whitish liquid	25	Seen	Due to caterpillar eating
	White sclera	5	Seen	Due to caterpillar eating
Lameness	Progressive	2	Seen	
Ebola virus	Death	20	Inferred	1 confirmed by test
Monkey pox	Skin eruption with white spots	6	Seen	
	Death	1	Inferred	

only 3 infant falls; one of them we believe to have died as a result of the accident. Gombe chimpanzees fall occasionally from trees and may suffer severe injuries (Goodall 1986; Teleki 1973). This might also be the case at Mahale, although there are no direct observations (Nishida *et al.* 1990).

Leopard predation

Chimpanzees, due to their large size, were long thought to be free from natural predation. However, at Taï, observations of numerous and aggressive interactions with leopards showed that chimpanzees could fall prey to them (Boesch 1991c). We could confirm leopard attacks by the wounds we saw on the chimpanzees that survived, and circumstantial evidence such as leopard calls and fresh footprints, combined with the reaction of the chimpanzees. Leopards can instantly kill both subadults and adults. During a five-year period between 1988 and 1991, pressure from leopards seemed especially important for our study community. The chimpanzees systematically and energetically chased away all

(a)

(b)

Fig. 2.10a,b.   Adult males guarding Tina's body minutes after she has been killed by a leopard. Her neck was broken. The leopard claws have also split opened her belly and some intestines are apparent. The male on the right is grooming her, while the other tests her reaction by gently shaking her leg.

leopards they encountered or heard, possibly to discourage hunting attempts. We never saw them injure a leopard. From marks found near places where chimpanzees had been attacked, we can confirm that leopards would sometimes even attack individuals that were part of large and noisy parties. During the 1988–91 period, leopards constituted the primary mortality cause for the chimpanzees (Boesch 1991c). Predation by leopards, thus, might become a major threat to a chimpanzee population. At times we had the impression that possibly the chimpanzees had always encountered the same individual leopard, rather than there being a general tendency of leopards to hunt on chimpanzees.

Case study, reaction of group members to Falstaff wounded by a leopard: On 18 September 1987 at 10.43, we heard very load screams from a party in the northern part of the territory. Two minutes later as I contacted them, I saw Falstaff, an old adult male, laying on his back on a large tree log surrounded by 3 male chimpanzees grooming him intensively, 8 other individuals were resting in the vicinity. It took me some time to realize that Falstaff was covered with 16 bleeding wounds on his left arm, near his right ear, on his upper right thigh, and on the right side of his back. His right eye was missed by 1 cm. The wound on the back was the biggest, bleeding during the first 2 hours. Falstaff showed no fear of any of the chimpanzees, which would be the case if he had been wounded by one of them, nor did I hear any stranger chimpanzees in the region. However, Falstaff replied with 'ouh' calls to alarm calls of a black-and-white colobus and he was the only one to react with a fearful full open-mouth grin, when hearing a distress 'hooh' of a chimpanzee in the south. This supported our impression that he must have been wounded by a leopard.

Falstaff was attended by several chimpanzees with special care and intensity, but not equally by all the individuals. Snoopy, a young adult male and also a very keen hunter like Falstaff, was caring for his wounds, licking the blood off the wound and the fur and removing any piece of dirt near or in the wounds. Falstaff positioned himself so as to present a wound to Snoopy who would then lick it carefully. I noticed that Snoopy was the only chimpanzee that Falstaff allowed to lick the wound on his side under the right arm. Perla, a newly transferred female, was even more attentive and she remained with Falstaff for almost the entire 3 hours and 15 minutes he was resting. Snoopy groomed him for 35 minutes and tended his wounds for 60 minutes, Perla groomed him for 30 minutes and cared for his wounds for 85 minutes. Other chimpanzees, like Macho, Ulysse, Chanel, and Xérès, were grooming him regularly and tended his wounds as well. He was constantly groomed. After 3 hours and 15 minutes he started to move southwards followed by the others. He was walking slowly and without using his right arm.

On the 19 September, Falstaff was still within the group 1.5 km south of the leopard attack, and he was given regular care by the adult members of the party. Brutus and Darwin licked his wounds and groomed him for 122 minutes. After a successful hunt, Falstaff let the other chimpanzees move away and he remained alone in the south of the territory, two valleys away from the leopard attack. Eleven days later, as the group was entering this same valley, Falstaff joined them. By this time, we knew he was there before we saw him because of his strong smell. All wounds were looking much better, except for the one under his right arm from which a whitish fluid was running. He stayed for the day with the group and we gave him an antibiotic shot with a blow-pipe, convinced he might risk septicaemia. The next day, he remained with the group the whole day as they were not moving far, and his condition seemed slightly improved. Unfortunately, we were unable to give him a second shot as he reacted immediately to the sound of the blow-pipe and avoided the needle with surprising skill. During the next

2 weeks, Falstaff was seen in the group only for 3 days. All wounds had healed except for the one under his arm. On the 4 November, we saw him for the last time.

Between 1992 to 1995, no casualties could be attributed to leopards, which supports the idea of one individual 'chimpanzee killer' responsible for the previous attacks. During that period, we obtained further evidence that leopards have a taste for chimpanzee meat. First, on three occasions leopards scavenged the meat of freshly dead juvenile chimpanzees of the study community (see p. 249, Miriam Behrens and David Jenny, personal communication). Second, while radio-tracking leopards, one adult male spent more time near chimpanzees than the others (Dind 1995), showing an attraction to these apes without actually having attacked them while under study. This shows that potentially some leopards might be dangerous for chimpanzees, but the sudden interruption of attacks attributable to leopards in 1992 suggests that only a few of them take the risk of becoming a 'chimpanzee killer', and face the adult chimpanzees.

The predation episodes by leopards on chimpanzees of the Taï study community were the first observed cases of predation on this species. In 1989, lions entered the territory of the Mahale chimpanzee community and were suspected to be responsible for the death of at least four individuals (Tsukahara 1993). We guess that predation on chimpanzees by big cats probably happens throughout their sympatric range in undisturbed habitats. Human presence can reduce this pressure which might be the case at Gombe (Goodall 1986; Boesch 1991*c*; Isbell and Young 1993), where only one old leopard still existed in 1983 (Goodall, personal communication).

Chimpanzee attacks

Aggressive interactions between members of the same community regularly lead to some minor superficial injuries that usually heal rapidly. During the whole study period, we saw no chimpanzee handicapped for long, although some wounds certainly were quite painful. The most serious injury occurred during a change of the alpha position in which the defeated male lost two toes. Thus, we cannot exclude intracommunity aggression as a cause of mortality, but in Taï chimpanzees it must be very rare. Aggressive interactions between communities (Chapter 7) may result in more serious injuries in adult males or females in surprise attacks by another community. At Gombe, a whole community has been exterminated by its neighbours, and the same has been suspected at Mahale (Goodall *et al.* 1979; Nishida *et al.* 1985).

At Mahale, 10 infants died from injuries inflicted by conspecifics during the 22-year study period (Nishida *et al.* 1990). Infanticide has also been reported from Gombe (Goodall *et al.* 1979). The only case of infanticide and cannibalism we saw at Taï (see Table 2.5) is not very conclusive. The student who witnessed the event, Penny Simpson, arrived at the site when several females were already eating a dead infant. The adult female Poupée kept the dead infant all the time but shared some meat with the other females. We do not know for certain whose baby it could have been, although a young mother appeared 27 minutes later without her newborn baby. Curiously she showed no sign of tension or distress, as we would expect if her own baby was being eaten in front of her. The location of the dramatic event, in the south-west of the territory, does not allow to exclude an intercommunity interaction. It remains a puzzling case, particularly as we never witnessed any sign of aggression aimed at infants in all the intercommunity encounters we observed (see Chapter 7).

Human predation (snares and gun shots)

Cable snares are placed by poachers near the fields and within the forest at ground level mainly to capture forest duikers. Poachers often cannot re-find their snares, and so many animals suffer this agony for nothing. The territory of the study community borders the western edge of the National Park with fields not too far away. Of the 9 chimpanzees we know to have been trapped in such snares, one juvenile female died. We were present when one adult female, Véra, had her right hand caught in a snare. Screaming, she pulled at the cable and spun round wildly on the ground while the other chimpanzees approached silently. After one or two minutes, the cable, totally distorted by her spinning, broke and she quieted down, looking at her wrist where the cable of the snare was already deeply buried in her flesh. She was not bleeding. She pulled unsuccessfully with her teeth at the end of the cable sticking out. Then the adult male Schubert approached her. She panted softly and let him come close while holding her wrist towards him. Schubert was with his back towards us. Bent forwards with his head close to her wrist he must have inserted his canine underneath the cable. We saw him give a strong jerky move and Véra ran away, screaming, but freed of the cable. After that incident, the whole party reversed direction and moved into the park for more than an hour without stopping.

On another occasion, Brutus freed himself rapidly from a snare around his left wrist, but his inner arm was injured. Similarly, Macho was caught around his right ankle, broke the cable rapidly, and succeeded in removing it 7 days later without being injured. Darwin was caught around the third finger of his left hand, but kept the cable for 6 weeks before getting rid of it. The specially thin cable was buried deep in his flesh and the pain might have prevented him from removing the cable earlier. In one instance, a small female infant became trapped in a snare from which her mother probably freed her rapidly.

In contrast, for juvenile chimpanzees, snares can be fatal. Juveniles seem incapable of removing the cable, either because of their lack of intellectual development or because of the pain, and they do not allow their mother or any other adult to touch them. We saw 5 juveniles with a cable embedded in their flesh; 4 kept it for between 8 to 11 months and were badly handicapped in all their movements, while the fifth died, probably from gangrene. Tina, a juvenile female, had the cable deep in her wrist and it seemed that her hand was slowly but surely dying. However, when the cable finally fell off after 11 months, she progressively recovered the full use of her hand. Goma, as a young adolescent in January 1983, had an impressively swollen left wrist due to a snare cable. She was not able to use her hand for nut cracking for two years after she had lost the snare. During this time she learned to crack nuts with the other hand, switching back to use her preferred hand in late 1985. Marius, a juvenile male, had the cable higher up the forearm, not too deep in the flesh and was able to use his hand during the 8 months the cable remained there. Nino and Pitchou had snares around two or three fingers and the delicate skin there was badly injured. Pitchou's finger swelled a lot and started to smell very badly. She disappeared within weeks, most probably dying from the gangrenous infection. In conclusion, snares can have fatal consequences, especially for juvenile chimpanzees.

Fig. 2.11. Nino, an adolescent male, looking at his left hand with the index and the major fingers tightly trapped in an iron cable. The cable will fall off ten months later.

Poachers visited the territory of the chimpanzees at intervals to hunt at night for duikers and during the day for monkeys (mainly red colobus monkeys). Whenever we were with the chimpanzees in the vicinity, we had the impression that they would shoot at the chimpanzees. On six occasions poachers shot in our direction, luckily missing both the chimpanzees and us. However, we found the skull of an adult female (possibly Kiri) with two holes on the left side and one iron bullet still inside the skull. This is our clearest proof that chimpanzees within the park fall victims to poachers. When farmers prepare new fields illegally within virgin forest in the National Park, these pieces of forest are cut within the territory of a chimpanzee community. The chimpanzees may have started to eat and like the cocoa beans that taste quite alike the fruits of the liana *Landolphia dulcis*, on which they feed regularly. To our distress, the study chimpanzees went on various occasions to raid the cocoa fields within their territory, after getting accustomed to this bonanza of new food. The farmers traditionally spare some of the emergent trees in their plantations and the chimpanzees, unaware of the fatal danger, may use these for their night nests. On such occasions, farmers, as they explained to us, surprise them very early in the morning when they still are asleep and can kill up to 9 chimpanzees in a row. We strongly suspect that this is how the 7 males of the study community were killed in February 1984. Thus, direct shooting at the chimpanzees occurs, but it is difficult to quantify its real impact as we so rarely find the bodies of disappeared chimpanzees.

Illness

Illnesses occur in all natural animal populations, but it is difficult to determine their impact on the population. A survey of the parasites of the Taï chimpanzees showed that endoparasites were present in all individuals for at least one period of the year and that the parasitic load was generally higher during the rainy season (Roduit 1999). However, it was not possible to determine their virulence. Parasitism and viral and bacterial infections could be influenced by human presence, as has been observed at Gombe where chimpanzees died of poliomyelitis, influenza, and pneumonia presumably contracted from humans (Goodall 1986). We cannot strictly exclude the possibility that human presence at Taï has caused casualties in the chimpanzees, as we did no post-mortem analysis of fresh dead bodies when we found them before 1994 when the second Ebola outbreak started. However, the parasite study showed that all parasites found in chimpanzees were typical of wild chimpanzees, reducing our fear that despite strict observation rules, we might have affected the chimpanzees' health.

We regularly observed flu-like symptoms, such as a runny nose, during some of the rainy seasons, and we found no evidence that it had a long-lasting effect on the sick individuals. In October 1991, after 3 nights of uninterrupted rain, all the chimpanzees suffered from a runny nose for about 10 days, and they recovered without incident. A year before, about half of the community also had a cold during September but in a less synchronized manner.

Conjunctivitis, which turns the chimpanzee's black sclera of the eye to white, hit about a third of the community members during a period of heavy caterpillar feeding in December 1986 ($N = 25$ chimpanzees). Each collected more than 500 caterpillars (about 2–3 cm long, *Geometritideae sp.*) per day from leaves, feeding on them for hours. We suspected the inflammation was caused by some itchy secretion of the caterpillars that they inadvertently put into their eyes. Two adults had an acute conjunctivitis, a female in both eyes and a male in one, that lasted about 6 weeks. Once they recovered they used their eyes normally, that is we did not have the impression their sight was altered, but the sclera of the infected eyes remained white. Saphir, the female, then had the 'human glance' that was described for the male McGregor at Gombe (Goodall 1970*b*). In humans, too, it is known that certain types of conjunctivitis can lead to the depigmentation of the sclera, but as it is already white, this is less obvious than in chimpanzees. In two youngsters, the conjunctivitis became chronic and a whitish liquid ran from their eyes for years.

Monkey pox seemed to be a problem in the chimpanzee community. The black skin of Chanel, one of the first adult females we recognized in 1983, was covered with small whitish dots, typical for the scars remaining after such an infection. In 1987, the chimpanzees apparently faced an outbreak of this infectious disease. On 3 January 1987, Ali, a juvenile male, started to look weak and was coughing a lot. Two days later, he was covered with numerous pustules on his face, hands, and feet and later on his whole body. He looked weak and could barely walk. Being an orphan, he remained alone for most of the time, although he had at that time already been adopted by the alpha male, Brutus. Brutus led the group back to the area, where Ali was left behind on average every second or third day. On these occasions, Ali was groomed by various group members. Ten days later, he walked better and could follow the group. Most pustules were by now red crusts, while others looked dry. He scratched himself often, removed the crusts and ate them. At the same time, Héra left the group with her two sons. The oldest one, Haschich, presented the typical dots on his body when they came back 4 weeks later. The 3-year-old male Gallus, and the young female Goshu were also covered with pustules and looked emaciated, but both recovered quickly. Things took a much more dramatic turn for Schubert, the beta male. On 28 December 1986, we were alarmed by his emaciated look and slow reactions. We were not the only ones to notice it — he was immediately challenged by Falstaff, the gamma male, and within a day Schubert left the group and was never seen again. We suspect that he died from monkey pox.

The Ebola virus outbreaks

The Ebola virus has gained a sombre publicity as it was at the origin of several fatal outbreaks in 1976 amongst the human populations of Zaïre and Sudan. It led to hundreds of casualties in the villages and stopped before the scientists had found its origin or a cure for the illness. The virus was suspected of coming from monkeys as two more outbreaks had occurred in laboratories housing monkeys in the US and in Italy (Formenty *et al.* 1999). Until 1992, there had been no further sign of the virus in Africa.

In late October 1992, we were extremely worried by the sudden disappearance, within 12 days, of 8 individuals. None of them had looked ill before. The field assistants and students found three of the bodies later, but did not notice anything obvious that could have explained their deaths. We suspected that a virulent disease leading to an almost immediate death must have affected these chimpanzees, but we were technically not prepared to confirm this presumption by virological analysis for which fresh samples are needed. The horror stopped as suddenly as it had appeared.

Two years later, in early November 1994, the sudden disappearances started again, and 12 individuals vanished within 3 weeks. We were alerted immediately and the three students present in the field at that time were able to dissect the two corpses they had found. Samples of many organs were sent to the Institut Pasteur in Paris. The analysis revealed the frightening result that the infant female, Piment, had died of a new Ebola virus strain, *Ebola Côte d'Ivoire* (Le Guenno *et al.* 1995). Three further fresh chimpanzee corpses showed obvious signs of haemorrhagic fever, a typical symptom of a possible Ebola infection. From these indications and the epidemic-like disappearance of most individuals, we inferred that they must all have died of this Ebola Côte d'Ivoire virus (Formenty *et al.* 1999). We collected blood samples from three surviving chimpanzees (Brutus, Macho, and Loukoum); all were seronegative for the Ebola virus. We suspect that the chimpanzees were contaminated by a common source and that those who had contacted the virus died of it (fatality rate = 100%). In 1992, the attack rate on the community was probably of 17% and in 1994 it was of 28%, which is a dramatic blow. If this supposition is correct, the Ebola virus was the major cause of mortality for the chimpanzees during the last few years. If it were to strike again, it has the potential to completely eradicate the community during the next few years.

How do chimpanzees infect themselves with the Ebola virus? In an analysis of the factors that might have contaminated the chimpanzees, we found that the amount of colobus meat eaten by a chimpanzee appeared to play a major role: the more colobus meat an individual had eaten during the hunting season, the more likely it was to die from Ebola (Formenty *et al.* 1999). At present, the most likely scenario is that the chimpanzees infected themselves towards the end of the chimpanzees' hunting season by eating red colobus monkeys, which were healthy carriers of the Ebola virus (healthy, because chimpanzees rarely hunt and eat visibly sick monkeys). Knowing that the chimpanzee hunting season starts in August and that the Ebola virus is suspected of killing its host within one or two weeks, we assume that the red colobus monkeys were contaminated in late October. This scenario suggests that red colobus are an intermediate and temporary host, and eating them might be fatal for the chimpanzee predator only at certain periods of the year. Contamination from other chimpanzees through for example grooming or sexual interaction, cannot explain the 1994 outbreak, as individuals that groomed

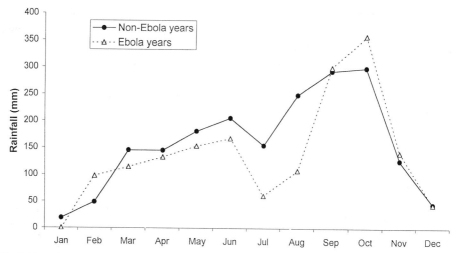

Fig. 2.12:   Rainfall in Ebola and non-Ebola years in the Taï forest.

or copulated with a future Ebola victim were not more likely than others to infect themselves with Ebola (Formenty *et al.* 1999).

What could explain the new emergence of this virus in 1992? Within the fairly stable environment of the rainforest in the Taï National Park, there is one factor that changed in the late 1980s — rainfall. Total annual rainfall decreased from 1800 mm to 1500 mm and, perhaps more significantly, there were clearly marked dry months (with less than 30 mm rain) for the first time at Taï (see Fig. 2.12). It is known from various studies around the world, that ecological changes lead to the emergence of pathogens (Garrett 1995), as they affect the ecological equilibrium — a species might invade a new habitat or increase in density, allowing virus and bacteria to get access to new hosts. Due to such changes, at Taï, fruit-eating bats or rodents could have increased sharply in density and contaminated new hosts, such as the red colobus. Thus, although animal populations are protected within National Parks, they might still be affected by global ecological changes due to human activity.

Relative importance of different mortality factors

During the second period (1988–1991), predation by leopards probably accounted for 42% of the casualties, illness for 34%, cannibalism was seen once explaining 2% of the deaths, one chimpanzee presumably died by falling from a tree (2%), and 20% of the deaths were attributed to unknown causes (Boesch 1991*c*). During the third period, the Ebola virus probably accounted for 66% of the casualties, leopard predation for none, and 34% were of unknown origin. Therefore, chimpanzees were subject to different types of mortality, low during the first period of the study and high for the second and third period. During both high mortality periods, an increase in one mortality factor was followed by a sharp decrease in the population size. In the Taï forest, the chimpanzee population was quite sensitive to variation in mortality factors. Our strong impression is that the vulnerability was increased by the presence of poaching.

Poaching was present throughout the study but was very hard to quantify. Human pressure has increased during the last 10 years of the study, when farmers encroached the park limits and planted fields much closer to the territory boundaries. At the start of the project in 1979 there was only one farmer within the park towards the edge of the park boundaries. By 1987, the fields had multiplied and were much closer to the chimpanzees' territory. This matches the period when the community started to decline.

Survivorship among chimpanzee and human populations

Demographic statistics presently available are still not sufficient to make detailed comparisons for all age classes of chimpanzees. Data exist for infant survival up to 5 years of age, and they show some strong variations, with survival being twice as high in chimpanzees at Kibale or during the stable period of Taï than at Mahale (Table 2.6). For Bossou, Sugiyama (1994*b*) proposed that when young chimpanzees under 10 years of age disappeared, it was because they emigrated. If we used Sugiyama's assumption, the survival rate of all chimpanzee populations would increase. However, no case of independent emigration of infants or juveniles had been witnessed in other chimpanzee populations. Because we doubt this assumption, we recalculated the survival rate of infants up to 5 years of age at Bossou by equating disappearance with death, as has been documented in one instance at Bossou (Matsuzawa *et al.* 1990). As a result, survival of infants at Bossou falls within the value for the Mahale chimpanzees.

With Gombe and Kibale chimpanzees, a more detailed comparison is possible. Whereas at Gombe mean life expectancy at birth is lower than at Taï during the stable period, it is higher at Gombe once they reach adulthood (Teleki *et al.* 1976). In Kibale chimpanzees compared to Taï, mean life expectancy is higher until adulthood and this population enjoys one of the highest survival for all age classes known in wild chimpanzees. As the average fertility of the females is about the same in both populations (see Table 3.7), it appears that the Gombe population was decreasing during the period surveyed. Since then the Gombe population recovered, for its size in 1984 was similar to the one of 1971 (Goodall 1986). Two important causes of mortality in Taï chimpanzees are missing in Gombe — leopards and human predation. At Gombe, a few leopards were present at the start of the project but new sightings are very rare and supposedly there is at most one very old individual in the whole park (Jane Goodall, personal communication). Similarly at Mahale, leopards and poaching were apparently not interfering with chimpanzees, but lions did so for a limited period of time (Nishida *et al.* 1990). In contrast to Taï, at both sites the chimpanzees seemed to have suffered intermittently from human transmitted diseases. At Gombe, and to a smaller extent at Mahale, they were affected by epidemics of human origin, for example influenza, pneumonia, or poliomyelitis which hindered the population increase (Goodall 1986; Nishida *et al.* 1990). At Gombe this was especially the case between 1967 and 1978, but the population was able to recover in the meantime.

What differentiates human from chimpanzee demography? Lovejoy (1981) proposed that during human evolution our ancestors started to reach much higher reproductive rates than great apes and that this gave them a lead that resulted in their expansion from Africa throughout the world. Others suggested that early hominids remained in many respects very ape-like and that their demography was similar to that of the apes. The comparison of Table 2.6 is mixed, because some aspects of survivorship in chimpanzees favourably compare with almost all extinct and living hominids, while others do not. If we compare the average life expectancies at birth for different hominid populations, it remains very stable near 16 to 21 years. This is a remarkable result. Populations of *Homo sapiens*

Table 2.6: Comparison of some demographic parameters in wild chimpanzee populations, extinct hominid species and some human populations

Species or population	Expectation of life at age		Survivorship (%)		
			Age class		
	0	15	5	10	15
Mountain gorilla			62	–	58
Chimpanzees					
Bossou			43(71)		
Gombe[1]	11.9	12.9	55	45	36
Kibale[2]	19.9	15.4	82	68	57
Mahale			40		
Taï (all)	7.7	9.6	58	21	14
(stable)	16.4	8.9	88	66	52
(predation)	4.6	8.3	37	6	4
(Ebola virus)	3.8	6.1	27	10	8
Extinct hominids					
Australopithecus[1]		12.7			65
Homo erectus[1]					32
Modern humans					
Magreb (10 000 BC)	21.1				46
Carlston (3000 BC)	22.4				56
Libben (875 BC)	19.9	18.7	69	60	53
Proto-agriculturalist	19.0	19.8			50
Ache	37.5	40.0	77	69	65
Yanomani[3]	21.5				49
!Kung San	34.6	41.5	65	61	58
Hadza			71	55	47
Xavante[3]					33
USA (1964)[3]	70.0				95
World (1985)[4]	61.1		89.5		

Mountain gorilla: Gerald (1995).
Bossou: recalculated from Sugiyama (1994*b*, Table 2 and Figure 4) (estimation in brackets following Sugiyama's assumption [see page 39]).
[1] Data from Teleki *et al.* (1976).
[2] Wrangham pers. com. in Hill *et al.* in press.
[3] Data from Meindl 1992; Neel and Chagnon (1968).
[4] World Resources 1988–89.
Ache: Hill and Hurtado (1996).
!Kung San: Howell (1979).
Hadza: calculated from Table 3, Blurton Jones *et al.* (1992).
Libben: Lovejoy *et al.* (1977).

living more than 10 000 years ago as well as modern populations living under very different conditions had on average the same life expectancy at birth. Only contemporary humans have a much higher life expectancy at birth, over 60 years. This favours the idea that major demographic changes did not affect this parameter, and that from the point of view of survival and life expectancy hominids were rather ape-like. However, when incorporating the data from mean life expectancy at 15 years, we see that in *Homo sapiens* a change appears for the first time, since life expectancy is much higher than the one observed before. Such a change may reflect the fact that at a given time mortality rate decreased for the adults, and that once individuals reached this stage, their life expectancy was quite high, which in turn gave them potentially more time to reproduce. It is possible that with the emergence of tools used as weapons, adults were able to deter predators effectively and protect themselves better.

Chimpanzees living in the virgin tropical rainforest of West Africa seem able to do well if not affected by more than one negative factor. At Taï, the population was stable for a period of several years. Poaching, a serious problem in most chimpanzee populations, had a negative effect and kept them from increasing. Poaching increased their vulnerability to other natural mortality factors. The Taï chimpanzee community lost 24 individuals during the period with high pressure from leopard predation. This pressure was short-lived and the chimpanzee population could have recovered if poaching had been absent.

The Gombe, Mahale, and Kibale chimpanzees present the image of populations free of poaching. This relates to the special situation of these populations as they live in regions with mostly Islamic human populations who require special hygienic measures for killing animals for food. But the vast majority of chimpanzees live in forest regions of Central and West Africa, regions where the Islamic influence is traditionally less important and where bushmeat is often the main source of animal protein. Therefore, we think the Taï chimpanzees might be representative of large sections of the remaining chimpanzee populations throughout Africa. If true, clearly we have to worry seriously about the survival of this species as, in addition to these factors, the chimpanzees' habitat, the tropical forest, is rapidly disappearing. These studies confirm that the chimpanzee should be considered as highly endangered. Because all detailed studies have been done in national parks, where the protection of the fauna is much better than in other areas, we expect the decline to be even more dramatic in the chimpanzee populations living in unprotected areas. Increased logging, with its accompanying pressure from poaching are at the origin of the disastrous bushmeat crisis that is badly threatening large chimpanzee and gorilla populations living in Central Africa outside of protected areas.

Infant and juvenile mortality is important in all chimpanzee populations with 40 to 60% of them dying within the first 5 years. One response to an increase in the juvenile mortality rate is to mature earlier. We shall discuss this point in the next chapter, and see if there are some constraints that keep chimpanzees from responding in this way. It is human intervention, by providing safe areas, that will allow the chimpanzees to achieve a positive survival rate as observed at Taï during the stable period. The sex ratio has been found to be highly female-biased in some chimpanzees populations. At Taï this relates to the lower survivorship of males. The relative scarcity of males would favour mothers that invest more in sons, as those that survive will have access to a large number of females. We investigate the females' reproductive strategies in the next chapter.

3 *Female life histories*

Forest scene: Salomé, a high-ranking female, was in full oestrus with an impressive pink swelling. All the adult and adolescent males of the community guarded her closely and followed her every move. She was leading this crowd for the whole day. Snoopy, her 16-year-old son, a strong young male normally very interested in oestrus females, followed her like the others, but did not try to mate. Her youngest son, Sartre, now 6 years old, seemed fascinated by the success of his mother and at the same time required her full attention and wanted to be carried for all longer distances. He was clearly one of the oldest infants to still enjoy such favour. There was also Ondine and her offspring in the cortège, the highest-ranking female and very close friend of Salomé. Ondine in contrast to Salomé had only daughters: Chloé, her oldest, was starting to produce some small swellings and tried to gain some attention from all these males (she will soon migrate). Her second daughter had died during the early days of the habituation before we could identify her. The third one, Orée, was now 4 years old and Ondine was again beginning to show some sexual swellings. She is to have another daughter within a year, whereas it will take two more years before Salomé will give birth to a female infant.

Salomé and Ondine, both high-ranking females of about the same age, had a rather different reproductive life. Why do some females conceive so quickly, have shorter birth intervals, or produce more sons than others? What is the optimal solution for a female chimpanzee? We shall attempt to understand such aspects in this chapter.

Females are the determining element in population dynamics; they are the limiting resource for the males, and their life history strategies determine how a given population will evolve. Males have higher reproductive potential, but this is limited by the number of females that are available to them. Therefore, independently of the strategies that males choose, it is the females that will make a population increase or decrease. This makes females of central importance for understanding the biological properties of a species. The aim of this chapter is to understand how Taï chimpanzee females organize their life and what their reproductive strategies are.

In most social mammals, females are the stable sex remaining in the natal group, whereas males transfer between groups before adulthood. This is, with some exceptions, also true for primates (Pusey and Packer 1987). Daughters form a close association with their mothers and often gain ranks adjacent to their mothers (Melnick and Pearl 1987). Exceptionally, females transfer between groups and males stay in their natal groups (for example, Hamadryas baboons, red colobus, gorillas, and chimpanzee). The fact that at least one sex transfers between groups limits incestuous interactions in most cases. In some species, dispersal of one sex appears to be contingent on reproductive success: gorilla and red colobus females change group after having lost a baby. In others, the costs of dispersion play a key role. For example, if predation risk is large, females may not transfer.

From the few chimpanzee communities that have been studied, we know that only females transfer between communities (Goodall 1986; Nishida *et al.* 1990), although the frequency of transfers varies between sites. Data from more communities are needed if we want to understand the factors that affect frequency of transfer. The fact that females transfer makes it difficult to determine their life history, as we know either the first part (before emigration) or the second part (after immigration).

The adult females of the Taï study community, as we saw in Chapter 2, were always about three times more numerous than males. They were very fecund, females without infants being rare. This favourable situation allows insight into many aspects of their life history. We describe here their late adolescent and adult life starting from the time they transfer into the community, through their motherhood, and analyse the different reproductive strategies they adopt.

Immigration and emigration

Soon after the identification of all Taï community members was achieved, we noticed that new adolescent females appeared, migrating into the study community. This suggested that females of the same age that regularly disappeared from the community in fact transferred to other communities. We have not followed any neighbour community for long enough to confirm this. Table 3.1 lists all the females that left or joined the study commu-

Table 3.1: Female transfers from and into the Taï community for a 15-year period

Name	Birth year	Natal group	Immigration date	Emigration date	Age at transfer (years)
Agathe	1977	No	2/89	–	12
Ariane	1983	No	1/94	–	11
Belle	1977	Yes	–	No	
Bijou	1975	No	5/84	–	9
Bonnie	1977	Yes	–	7/90	13
Castor	1976	No	11/88	–	12
Chouchou	1981	Yes	–	5/95	13
Cloé	1975	Yes	–	12/86	11
Coco	1975	Yes	–	7/87	12
Dilly	1978	No	8/89	–	12
Fossey	1979	No	8/89	–	10
Imi	1976	No	4/86	–	10
Perla	1976	No	8/87	–	11
Narcisse	1982	No	8/94	–	12
Nina	1976	Yes	–	6/85	9
Sandra	1975	Yes	–	7/86	11
Vénus	1978	No	8/89	–	11
Zerlina	1982	No	9/93	–	11
Zircé	1976	Yes	–	11/87	11

Total:	All transfers	18	mean = 11.17 y
	Emigrations	7	mean = 11.42 y
	Immigrations	11	mean = 11.00 y

nity. As mentioned in the previous chapter, an adolescent female that disappeared from the group was normally considered to have transferred if she had been visibly in good health before. We found only once a fairly recent skeleton, corresponding in size to that of an adolescent female who was missing, and we assumed then that she had died.

Temporary transfers between communities have been observed in all chimpanzee populations. During aggressive encounters between communities, oestrous females have been seen to transfer to copulate. At Taï, such transfers lasted, however, only a few minutes. Oestrous females sometimes disappear for a few days (see Table 4.6), and it is possible that they made temporary visits to neighbour males. We considered a stranger female as having immigrated only when the transfer was permanent. In all the cases of emigration, the females that had disappeared did so permanently and we never saw them again.

With the exception of Belle, all adolescent females emigrated when about 11 years old ($N = 8$). Emigration seemed to be a one-off event as adolescent females who were at the age to transfer disappeared one day for good. The precise mechanism by which a newcomer is absorbed within a group is not well known because the human observer interferes with the behaviour of the new, non-habituated female. It was, however, striking that in all the cases of immigration at Taï, habituation was achieved very rapidly. Zerlina, seen for the first time in the group on 28 September 1993, was always associated with the adult males. After running away the first few times she saw us, she became very tolerant after the eighth encounter, when she joined males in a high tree despite our presence. It was similar with Narcisse, who we saw for the first time in the community on 17 August 1994. When we met her for the third time 10 days later, she climbed a tree and stayed for a long time fully in view with four other chimpanzees. This rapid habituation to human observers was typical for all immigrants, and contrasts sharply with the five years we needed to habituate the first males at the beginning of the study — not to mention some very shy resident females that remained cautious even after many more years. The process of transfer seems to happen very rapidly. Once we noticed a new immigrant female, we caught sight of her regularly despite her lack of habituation, suggesting strongly that the transfer was done once and for all. For example, Zerlina, Narcisse, Castor, Fossey, Venus, and Imi were seen on an almost daily basis once we had recognized them as immigrants. As a rule, new immigrants stay with the males until they have their first baby. Then they progressively join the females. We saw resident females behave quite aggressively against some of the newcomers (Imi, Zerlina, Agathe, Dilly), who were then supported by the adult males. In other cases, the resident females tolerated them within the first week (Ariane, Narcisse, Fossey, Venus, Perla, Castor) without any visible aggressive interactions. We do not yet have enough detailed observations to generalize, but our impression is that resident Taï females are less aggressive against newcomers than is reported from other populations (Goodall 1986; Nishida 1989).

The transfer process between two Mahale chimpanzee communities differed between individual females and lasted between 6 months to 2 years — a period during which the females were associating and mating with the males of the two communities (Hasegawa 1989). In the new community, the immigrant females associated mainly with males or other immigrant females, whereas their relationship with the resident females were described as aggressive (Nishida 1989). The strong aggression and their difficulty in integrating the new community were thought to cause these females to establish their range at the periphery of the new community's territory. These areas were less safe and they suffered higher attack rates from the neighbouring communities, as well as from the males of their new community once they had a newborn infant (Nishida 1989).

At Taï, failure to integrate the new community was observed twice. Both females were adolescents and they disappeared after having been observed a few times. We cannot exclude the possibility that our presence prevented the integration process. We never saw an adult female with or without an infant that transferred between communities, nor did we see any male of any age join the study community.

So far, only reports from Bossou, in Guinea, suggest that chimpanzees of both sexes might transfer between communities (Sugiyama 1984, 1994*b*). However, the one adult male considered to have immigrated appeared three months after the start of the study and might have returned to its natal community after a longer period of absence, as has been observed for males in other chimpanzees communities (Nishida 1983). The concomitant and sudden disappearance of juveniles of both sexes from this small community (Sugiyama 1984) might unfortunately be related to external factors, such as poaching for food or providing medical centres with young chimpanzees, as this population is surrounded closely by human settlements.

In chimpanzees, except for the ambiguous Bossou case, only females seem to transfer, but each population presents a unique pattern. At Taï, only adolescent females transfer and almost all of them do (95%). At Gombe, only adolescent females transfer, but fewer do so than at Taï (13% in Goodall 1986, or 50% in Pusey *et al.* 1997). At Mahale, females of all age classes have been seen to transfer, with a majority of adolescents, and 13% of them transfer more than once (Nishida *et al.* 1990).

Factors affecting female dispersion in chimpanzees

Individuals are expected to transfer between groups to avoid inbreeding (Pusey and Packer 1987). High levels of inbreeding have been shown in some species to lead to lower fertility with higher rates of miscarriage or lower rates of conceptions, lower survival, or higher frequency of malformation (Keller *et al.* 1995). In other species, inbreeding has no visible effect and is regularly observed. Alternatively, transfer between groups can be viewed as a strategy to increase reproductive success, for the cost of dispersal, especially the risks of mortality, are weighed against the possible gain of increased mating success when changing groups, and this explains differences in male dispersal patterns between populations in savannah baboons (Alberts and Altmann 1995).

The only chimpanzee communities with genetic information available so far are Gombe and Taï: they revealed a moderate level of inbreeding at Gombe, perhaps due to the recent isolation of the population (Morin *et al.* 1994*a*), and a low level of inbreeding at Taï (Gagneux *et al.* 1999). This result mirrors the female transfer patterns. Inbreeding avoidance is incomplete at Gombe: a few incestuous matings have been observed between sons and mothers (Goodall 1986). Why do Gombe chimpanzees not avoid inbreeding more strictly? At Mahale and Taï, the chimpanzees live within large national parks with several neighbour communities, whereas the Gombe main study community inhabits a small area with only two neighbour communities. In this situation, the possibility of avoiding inbreeding is limited. In addition, the two neighbouring communities are situated at the edge of the park, and therefore exposed to greater human influence. From the top of the ridges between valleys at Gombe, the chimpanzees can see human settlements, which might inhibit transfer. In any case, observations at Gombe show that staying in the natal community confers some clear benefits, for example the support of a powerful mother, or the alliance and protection of a brother (Goodall 1986).

The avoidance of inbreeding could be achieved easily by seeking extra-group paternity instead of emigrating permanently. Females could do so during temporary visits to neighbouring communities without incurring the costs associated with a permanent transfer. This solution is widely used in Taï chimpanzees (Chapter 4) and possibly at Gombe, where 13% of the observed copulations were performed with males from other communities (Goodall 1986). In wild chimpanzees, future reproductive success might explain permanent transfer better than the avoidance of inbreeding. At Gombe, a permanent transfer out of the central community would mean joining communities in closer contact with human populations, which might reduce a female's reproductive success. At Mahale, future reproductive success seems to explain the pattern of the observed transfers, as all the secondary transfers observed were from the strongly decreasing K group into the much larger and more stable M group (Nishida *et al.* 1990). At Taï, 10 of the 13 immigrations happened during the nine years when the community included seven or more adult males, whereas only three occurred during the six years when the community included six or fewer adult males (see Tables 2.2 and 2.3). Thus, increasing reproductive success through access to a larger number of males could explain the transfer patterns observed in the three chimpanzee populations. Note that by reproductive success we mean not only finding good fathers (available through extra group paternity), but also living in a safe community with low infant mortality. For this the number of males in the community plays a crucial role (see Chapter 7).

Age of maturity in females

Age at maturity, defined as the age at which individuals first reproduce, is one of the most important life-history parameters, since fitness is often more sensitive to changes in this trait than to changes in any of the others. Natural selection exerts a strong pressure to mature early. However, this is balanced by trade-offs with other factors such as survival and fecundity, which might contribute to delayed maturity, since increasing these factors is normally possible only if maturation slows down (Stearns 1992). Captive chimpanzees are known to mature about three years earlier than wild ones (Tutin 1996). This shows the potentiality for chimpanzees to mature much earlier, but the costs associated with earlier maturation might be very high in the wild. We thus address two questions: is a delayed maturity observed in Taï chimpanzees, and what factors might favour such a delay?

Adolescent sterility in females

The first sexual swellings of the genitals occur before adolescence but are too small to attract males. With time, the size of the swelling increases and progressively attracts young males. Observations suggest that females become attractive to adult males once adolescent. Emigration then follows. Orphan females might differ in this respect, and of the three cases from Taï, one, Belle, did not transfer at all and the two others, Bonnie and Chouchou, transferred later than all the others (Table 3.1). Table 3.2, lists all young immigrant females that were in full oestrus when first observed, and the time span until they gave birth to their first infant.

Given the gestation time in captive chimpanzees (226 +/– 11.8 days, $N = 116$; Martin 1992), we can calculate the length of the adolescent sterility period in Taï female chimpanzees (time from transfer to the first parturition minus the gestation time). The seven immigrant females were sterile on average for 2 years 8 months (average = 2.65 years, median = 2.25, sd = 1.43). The special case of Belle is discussed separately as we suspect that her abnormal history influenced her very late first pregnancy. Such a long period of adolescent sterility has also been found in the Tanzanian chimpanzees (Goodall 1986; Nishida *et al.* 1990; Wallis 1997), and Goodall (1986) suggests that it might allow the females to gain some sexual experience with various males to make the best choice later.

An alternative, but not exclusive, explanation would be that new immigrant females need a sort of 'social passport' to integrate the community and to gain the males' tolerance and support. The sexual swellings they produce could be this 'social passport'. The most striking example of this process at Taï was observed with the tiny female Imi. Imi joined the community in April 1986 as a small 10-year-old female with a large sexual swelling that she had for each cycle for several months. Then, in April 1987, she stopped producing any swelling; by that time she also seemed more at ease with the community members. She started again in August 1988. Having a baby before being accepted by the group members could put immigrant females under strong pressure and these could be at risk of losing the baby. At Gombe, several instances of successful and failed cannibalism of newborn babies, including those of young mothers by resident dominant mothers, have been reported (Goodall *et al.* 1979; Pusey *et al.* 1997). The males of the community themselves can also be dangerous. At Mahale, six infants were stolen from young mothers, killed and partly eaten by the dominant males (Hamai *et al.* 1992). These observations suggest that the social pressure on immigrant females is so strong that it pays them to delay maturity and reproduce once the social environment guarantees a lower infant mortality rate. We shall see below that this argument applies not only to new immigrant females, and that the sexual swelling might be generally used as a 'social passport' in wild chimpanzees.

Table 3.2: Duration of adolescent sterility in Taï female chimpanzees

Name of female	Transfer date	First parturition date	Duration to first parturition (months)	Age at first parturition (years)
Agathe	2/89	11/91	32	14.7
Belle	No	8/94	No	18.5
Bijou	5/84	2/89	69	14.75
Castor	11/88	4/90	17	13.75
Dilly	8/89	11/91	27	14.25
Fossey	8/89	2/92	30	12.5
Perla	8/87	11/89	27	13.25
Venus	8/89	5/91	21	12.75
Total	8			14.3
Immigrants	7		31.8	13.7

Belle was the only female that remained in her natal group, and she first gave birth when 18.5 years old. She had lost her mother at the age of six years. Having no older siblings, she was not adopted, but herself adopted her little 8-month old sister. Despite Belle's efforts, the baby died some three months later. Belle remained another year on her own before being adopted by an adult female of the community. This traumatic childhood might have contributed to her late pregnancy.

Age at first parturition

Age at first parturition varies among individuals and species, and early maturation may lead to higher fertility (Stearns 1992). Some species that suffer a high mortality rate have been observed to reproduce early. There is also a general tendency in many animal species for more rapidly growing individuals to mature earlier (Stearns 1992). Delayed maturity is associated with larger adult body size and long life in many mammals. Primates have a tendency to have a very slow maturation rate that seems to be constrained by the requirement of a larger brain (Charnov and Berrigan 1993). Primates with relatively large neonates have relatively delayed maturity (Martin 1990).

Seven females gave birth for the first time at an average age of 13 years and 8 months (average = 13.7, median = 13.75, sd = 0.91) (Table 3.2). All females tended to give birth at a similar age, despite the varying length of their period of adolescent sterility, which did not correlate with their age at first parturition ($r_s = 0.5, p > 0.05$). This period of sterility following the first oestrus appears to serve females when transferring into a new community. When in oestrus they manipulate males to obtain their tolerance and support. They can apparently influence their physiology so as to produce complete sexual swellings while still young and small. In contrast, age at first parturition and age at maturity are about the same for all females, showing that this trait is less easy to vary, and it is one that is more strongly affected by nutrition. The case of Belle, the only female that has not, so far, emigrated at Taï, shows that additional factors can negatively affect the age of first parturition.

Age at first parturition is very similar for all three chimpanzee populations, despite important differences in the environment and in the availability of food. Mahale and Taï chimpanzees are of a similar weight, whereas Gombe chimpanzees are lighter (Uehara and Nishida 1987; Boesch and Boesch 1989, Morbeck and Zihlmann 1989). Such differences in weight apparently do not affect the age of first parturition. We suggest that age at maturity in wild chimpanzees is constrained not only by food but also by social factors related to the risk of having infants before full social integration in the community. If these constraints are lifted, chimpanzees may mature much quicker, as is seen under captive conditions.

Interbirth interval

A second life-history parameter that can be manipulated to increase reproductive success is the interbirth interval. The females could either reduce the interbirth interval to produce more offspring in a given period (Stearns 1992), or invest a different amount of time in an offspring according to its sex (Trivers and Willard 1973), or according to the social rank or age of the mother (Clutton-Brock 1988). Such variations are expected to be related to the special biology of the population under consideration.

Fig. 3.1:   Castor and her son, Cacao. Chimpanzees care for their young through suckling, providing attention, support, and transport for years.

The total maternal investment in an infant can be measured by the interbirth interval (IBI) (Silk 1988). The IBI, which includes the whole period during which a mother nurses and transports an infant, is the time period between two consecutive births. Since spontaneous abortions and early miscarriages could not be recorded consistently, these events were not taken into account in our measurements of IBI. We shall look at two kinds of interbirth interval: when the first infant dies before it is weaned, and when it is successfully weaned.

Interbirth interval after the first infant died

Following this important loss, it pays the mother to produce another baby as quickly as possible. This could be a measure of the rapidity of the physiological response of the female body to the disappearance of an infant, which might be influenced by her age or health.

   All Taï female chimpanzees resumed their oestrous cycle within a month of the death of an infant (mean = 1.06, median = 0.5), and their fertility remained very high as they became pregnant within 2 months and 11 days after resuming their oestrus. Thus, on average females give birth to another offspring within one year and one month after the death of a dependent infant (mean = 12.9 months, median = 12.0) (Table 3.3), which matches expectation.

Interbirth interval after successfully weaning the first infant

Predicting when mothers should ideally start to invest in a new baby while the previous one is alive, is difficult. Chimpanzee youngsters are dependent on their mother for a very long period of time, and this investment is probably critical to their survival and success in life. Thus, interrupting the investment too early might have damaging effects observable only years later, whereas waiting too long would reduce the reproductive success of the mothers. We shall see how Taï chimpanzees solve this dilemma.

Table 3.3: Interbirth interval in Taï chimpanzees when the dependent baby died

Name of mother/infant	Age of infant at death (months)	Months until resumption of oestrus	Interbirth interval (months)
Bijou/Bambou	25	0	12
Castor/Pollux	5	0	14
Castor/Congo	13	0.5	12
Fossey/Diane	0.3	3	21
Héra/Homère	32	3	10
Loukoum/x	2	?	14
Mystère/Mognié	6	0.5	13
Nana/Nabu	0.5*	?	11.5
Nana/x	2.5	?	10
Poupée/x	0.5	?	11
Poupée/X	13	1	10.5
Poupée/Popeye	12.5	0.5	14
Xérès/x	4	?	15
Mean		1.06	12.9

* Nana's child did not die, nor did the mother, but was adopted when she was only 2 weeks old for unknown reasons by Malibu, Nana's close associate.

At Taï, for the 33 cases in which we could estimate or measure the interbirth interval, the average IBI was of 5 years and 9 months (mean = 69.09 months, median = 65 months, sd = 17.3) (Table 3.4). This high IBI confirms the long maternal investment observed elsewhere in chimpanzees. The IBI in Taï females correlates only slightly with the age of the mother (Boesch 1997). Long IBI is expected in slow maturing animals, and the chimpanzee is a classic example of such a case.

Mothers with living infants, compared with those who had lost their infants, had a period of sterility with sexual swellings 22 months longer (former: mean = 34.5 months, sd = 24.9 [Table 3.4]) (Wilcoxon test: $W_x = 34$, $N_a = 6$, $N_b = 7$, $p < 0.05$). This has been observed in other chimpanzee populations and Goodall (1986) suggests it would allow females to accumulate more knowledge on the males before choosing again. However, this argument would also apply to females whose baby died, so that another explanation is required for the additional 22 months. We suggest that the sexual swellings of mothers with young infants have a social function, that the females postpone an immediate conception while using this 'social passport' to contribute to the socialization of their youngsters. Such mothers could benefit from the oestrous cycle as it increases the tolerance of the adult males whose support they will need more frequently as they are going to be regularly involved in squabbles resulting from their infants' social play behaviour. We shall return to this explanation below.

Maternal investment in sons and daughters

The theory of parental investment predicts that parents should invest more resources into the offspring gender which promises more grandchildren under the prevailing conditions (Hamilton 1967; Trivers and Willard 1973). Skewed sex ratio at birth has been observed in a variety of vertebrate species and differential parental investment in sons and daughters has been documented (Clutton-Brock *et al*. 1984; Holekamp and Smale 1995). These

Table 3.4: Interbirth interval when the mother succeeded in weaning her child

Name of mother/infant	Mother's dominance rank	Estimated(*) or known date of birth	Interbirth interval (months)
Chanel/Coco	high	1975*	72
Chanel/Chouchou	high	3/81	105
Ella/Kendo	high	1969*	72
Ella/Fitz	high	1975*	96
Ella/Gérald	high	1983	60
Fanny/Marious	low	4/82	65
Fanny/Manon	low	9/87	60
Gala/X	low	1977*	72
Gitane/Gipsy	low	10/80	57
Gitane/Goyave	low	7/85	95
Goma/Goshu	low	3/86	66
Héra/Hashich	low	4/78*	51
Héra/Eros	low	11/82	55
Kiri/Kummer	low	8/82	58
Lola/Lorenz	low	5/74*	60
Lola/Lolita	low	5/79*	87
Loukoum/Lychee	high	3/86	67
Momo/Molière	low	2/82	64
Nova/Négus	high	82	83
Ondine/Cloé	high	1975*	48
Ondine/X	high	1979*	51
Ondine/Orée	high	10/83	49
Ondine/Sirène	high	11/87	60
Pokou/x	low	1977*	60
Ricci/Ninette	low	1976*	72
Ricci/Nina	low	1982*	67
Ricci/Nino	low	3/88	78
Salomé/Snoopy	high	1970*	120
Salomé/Sartre	high	1980*	92
Tosca/Sandra	high	1975*	48
Tosca/Tina	high	1979*	55
Xérès/Xindra	low	8/84	75
Zoé/Zircé	low	1978*	60

Summary:
All:	N = 33	IBI = 69.09 months	
Male infant:	N = 15	IBI = 71.40 months	
Female infant:	N = 18	IBI = 67.10 months	

empirical observations are, however, difficult to explain as in some species more males are produced and in others more females. Much theoretical work has been devoted to explain how differential parental investment should evolve (Trivers and Willard 1973; Clark 1978; Silk 1983).

Several factors have been proposed as affecting the parents' investment in one or the other sex in social vertebrates: maternal condition, competition between individuals over local resources, and individual dispersal patterns. Trivers and Willard (1973) predict that mothers in good condition should bias investment toward offspring of the sex that benefits more from the mother's contribution. The basic idea is that physical condition limits the number of offspring a female can produce in a lifetime, whereas the reproductive success of males is constrained less by the impact of physical condition on sperm production then by his ability to find suitable mates. If many mates can be found, a male can potentially produce much more offspring than a female, but he could also fail entirely if he is not competitive (Trivers and Willard 1973). Thus, a male's success is less certain but potentially greater than that of a female. Therefore, if a mother can, through her investment, increase the ability of her sons to gain more mates, she should invest more in sons. If she cannot influence her sons' success, she should play safe and invest in daughters who will produce a limited but more certain number of offspring. In polygynous vertebrates, breeding success in males is more strongly influenced by body size and condition than in females, and we should expect parents to invest more in sons when resources are plentiful and in daughters when they are scarce. As predicted, high maternal rank in red deer, which guarantees access to better foraging places, is associated with more sons produced at birth and with higher reproductive success for sons than for daughters (Clutton-Brock *et al.* 1984).

In many social vertebrates, one sex has a stronger tendency to disperse than the other. Depending upon the conditions, members of one sex might compete or cooperate for food, and mothers might react to this circumstance (Clark 1978; Silk 1983). In cercopithecine primates with female philopatry and intense competition for local resources, dominant females produce an excess of daughters, for daughters benefit them by serving as allies in competition with other matrilines. When the competitive conditions are weak, dominant females produce an excess of sons, since daughters represent a liability to high-ranking mothers competing with them for food (Altmann 1980; van Schaik and Hrdy 1991). In baboons, where females remain in their natal groups, dominant mothers invest more in female offspring, to whose adult social rank they can contribute, which in turn affects the reproductive success of the daughters. In subdominant mothers, the opposite is observed, because the social rank of male baboons is less influenced by the mothers' rank as they transfer between groups (Altmann and Altmann 1970). In chimpanzees, the transfer pattern is the opposite, thus mothers might be expected to invest more in males, but this has not been found in the Tanzanian chimpanzee population (Nishida *et al.* 1990; Goodall 1983, 1986). Do female Taï chimpanzee facultatively adjust their investment in their offspring according to sex?

Wild chimpanzees tend to isolate themselves when they give birth, but in 16 cases we were able to ascertain the date of birth within a few days. In 17 cases, we had to estimate the date of birth either when birth occurred before the study had started or when we had not seen the mother for 3 to 4 months. If the mother or the infant died before weaning, the data were excluded. Mothers were classified as either being dominant or subdominant on the basis of priority of access to feeding places (Table 3.4). This classification in only two dominance categories minimizes the impact of small changes in dominance rank and the impact of changes resulting from important demographic events. Ranks were sufficiently stable during the study period for no individual female to need to be moved between dominance categories. Age and dominance rank did not correlate in the females considered in this study.

The analysis of 33 interbirth intervals in Taï chimpanzees revealed that the length of the maternal investment period did not depend upon the rank of the mother or the sex of the infant considered alone (Boesch 1997). Age of the mother had a small effect if corrected for both sex of the infant and rank of the mother. However, the combined effect of the rank of the mother and the sex of her offspring played a major role: dominant

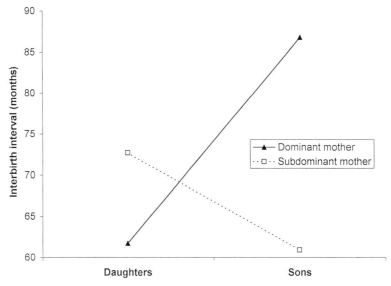

Fig. 3.2:   Interaction of the effect of the sex of the infant and of the rank of the mother on the length of the interbirth interval ($p < 0.0001$).

females invested about 2 years more in sons than in daughters, whereas subdominant females invested 11 months more in daughters than in sons (Fig. 3.2). Dominant mothers invested about the same amount of time in daughters as subdominant mothers invested in sons. The 33 IBI analysed were recorded from 19 females over a period of 25 years. Thus, some females contributed more than one IBI (7 females contributed 2 IBIs, 3 females contributed 3 IBIs, and one 4 IBIs) (Table 3.4). To test whether this affected the result, we reanalysed the data using only one IBI per female, if they contributed to more than one IBI for infants of the same sex. This procedure reduced the sample size to 24 IBI and produced the same significant differences (Boesch 1997).

Higher investment by the mother makes sense only if the favoured offspring benefits more from it than the other (Trivers and Willard 1973). One important fitness component of young chimpanzees is survival until adulthood. Figure 3.3 shows the survival of the two sexes according to the dominance status of their mother. Remember that Taï males, in general, tend to have a lower survivorship than females (see Fig. 2.7). Sons of dominant mothers had higher survival than those from subdominant mothers, and the effect increased from two years onwards, but was not apparent within the first five years (Boesch 1997). However, for daughters no effect of rank-dependent differential investment by the mothers was seen. Dominant mothers invested 13 months more in their sons than subdominant mothers invested in their daughters (Fig. 3.2). These data do not clarify why subdominant mothers invest more time in daughters; it might be that the small sample size did not allow us to document a small but real effect on their survival. It was not possible to test for effects of parental investment on reproductive success of the offspring, because the females leave the community before reproducing and the high mortality rate observed since 1992 reduced our male sample size dramatically.

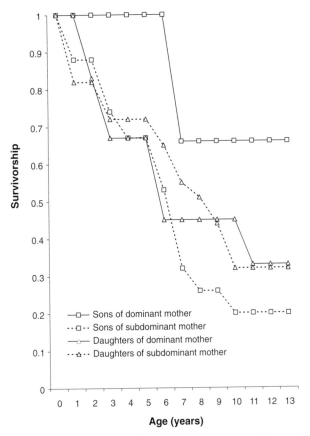

Fig. 3.3: Survivorship curves for the offspring of dominant and subdominant mothers in Taï chimpanzees (comparing son curves: $p < 0.05$).

We would expect young dominant mothers to have more chance of influencing the social status of their sons, since older mothers would most probably die before being able to support their sons when they enter the adult social hierarchy (when 15 years old). Only two dominant females in our sample could help us to test this; both, Ella and Salomé, support it by having shorter interbirth intervals when they were older (20 months shorter for the last IBI compared to the first one) (see Table 3.4). Mothers were seen to support their sons actively in social life; we describe in the next chapter how Ella helped her adult son Kendo to become the alpha male of the community.

Dominant mothers in Taï chimpanzee invest more in sons, and this increases their sons' survival. We cannot explain the differential investment in sons by a higher cost of rearing sons, as chimpanzee males and females have the same birth weight and size, and grow with the same rate during the first 10 years of life (Gavan 1971). The present result mirrors results obtained in cercopithecine primates under conditions of intense competition for local resources (van Schaik and Hrdy 1991). The difference in sex-biased investment is explained by the opposite dispersal patterns: macaques are expected to invest in philopatric daughters, chimpanzees are expected to invest in philopatric sons. All other

Fig. 3.4:   Perla with her two infants, the juvenile son Papot and her baby daughter Pandora. The six years difference between the two siblings is typical for low-ranking mothers.

things being equal, the higher survival rate of sons of dominant mothers would guarantee them more grandchildren compared with lower ranking mothers. At the same time, the longer interbirth interval limits the intensity of the conflicts between siblings. If, in addition, surviving sons of dominant mothers achieve higher social status and higher reproductive success, the longer maternal investment we observed in Taï chimpanzees would be even more adaptive. We investigate this possibility in Chapter 4.

Interbirth interval in other chimpanzee communities

The tendency for Taï mothers to invest more in one sex according to their social rank has not been found in other chimpanzee populations. What is special in Taï chimpanzees that can explain this? Most wild female chimpanzees have been observed to have a very similar reproductive profile to the one we observed at Taï including both an adolescent sterility period and a prolonged IBI (Table 3.5). However, a major difference appears in the tendencies for females to transfer between communities: in Mahale, they transfer more than once and by the second time they are normally already mothers. At Gombe, only a minority of the females transfer. This difference in the female dispersal pattern could provide an explanation as to why East African chimpanzee mothers do not invest differentially in their offspring. At Gombe, the benefit attributed to dominant mothers when investing in sons seems to be compensated for by the important affiliative and supportive interactions observed between mothers and daughters (Goodall 1986), which renders a daughter an ally rather than a competitor. Such a 'local resource enhancement' argument has been used to explain male-biased sex ratios in birds with 'helpers at the nest' (Gowaty and Lennartz 1985) and female-biased sex ratios in dominant mothers among cercopithecine primates with intense local resource competition (van Schaik and Hrdy 1991). In Mahale National Park in Tanzania, all daughters disperse but no difference has been documented in the IBI (Nishida *et al.* 1990; Nishida, personal communication). Maternal investment after infancy may also be required to guarantee higher

Table 3.5: Reproductive profile of wild female chimpanzees and bonobos

Population	Female transfer			Adol. sterility (years)	First parturition (years)	Interbirth interval (months)	Oestrus resumption (months)
	1st transfer	age (years)	2nd transfer				
Chimpanzees							
Bossou						61.2	
Gombe	13–50%		0%	2.4	14.9	66.0	46.3
Kibale						86.4	
Mahale	100%	11.0	13%*	2.9	14.6	72.0	53.2
Taï	95%	11.0	0%	2.6	13.7	69.5	24.5
Bonobos							
Wamba	100%	7–9	0%		14.2	54.0	<12

*At Mahale the average age for the 2nd transfer was 20.8 years.
Bossou: Sugiyama (1994*b*).
Gombe: Goodall (1986), Wallis (1997), Pusey *et al.* (1997).
Kibale: Wrangham *et al.* (1996).
Mahale: Nishida *et al.* (1990).
Wamba: Furuichi (1987); Kuroda (1979); Takahata *et al.* (1996).

reproductive success of the sons. At Mahale, females from the M-group transferred more than once, and in 17% of the cases they left their juvenile infant behind before transferring to a new group and were therefore not able to provide any additional support (Nishida *et al.* 1990). However, this explanation might not be so general since the double transfers were associated with the extinction of the M-group (Nishida *et al.* 1985).

In a new analysis of interbirth intervals at Gombe, high-ranking mothers had smaller IBIs than low-ranking mothers, and the survival of these high-ranking offspring was higher (Pusey *et al.* 1997). The sex of the infant was not found to have an effect. Individual females possess different core areas, and high-ranking females were suggested to have access to the best ranges. At Taï, the females, more social than at Gombe, have not been observed to possess such clear-cut different individual home ranges.

Table 3.5 shows that IBI can vary from 61 to 86 months in wild chimpanzees, while the IBI of captive chimpanzees is about 50 months (Tutin 1994). This difference of 2 years is quite important and shows a certain flexibility in this life-history trait. Bossou chimpanzees possess the shortest IBI of all known wild populations. At the other extreme, the Kibale chimpanzees have an IBI of 2 years longer. This would be expected if they were adapted to different infant mortality rates (Stearns 1992). In support of this idea, we have seen in this chapter that when Taï females increase their interbirth interval, they increase the survival of their infants.

Thus, in chimpanzees, immigration patterns, access to high quality food, and social aspects all seem to affect female interbirth intervals.

Sexual swelling as a social passport?

The females of many Old World primate species present a large swelling of the sexual skin near or at the time of ovulation. These oestrus swellings occur in cercopithecine monkeys like baboons and macaques, in colobines like the red and olive colobus monkeys, and in the chimpanzees (Dixson 1998). However, they are absent in many forest-dwelling cercopithecine monkeys, the gorilla, the orang-utan, and man. The presence of the sexual swelling in those groups is strongly associated with the presence of a multi-male breeding system, especially in the apes and the cercopithecines (Hrdy and Whiten 1987). Four hypotheses have been provided to explain the association of sexual swelling with multi-male systems: (1), in the 'best male' hypothesis, the sexual swelling increases male competition and thus improves the likelihood of the female finding the best male of the group (Clutton-Brock and Harvey 1976); (2), according to the 'many male' hypothesis, the sexual swelling increases the opportunities for the females to mate with several males, which by confusing male paternity would increase the total amount of paternal care received by the offspring and decrease the risk of infanticide (Hrdy 1977); (3), in the 'obvious ovulation' hypothesis, the sexual swelling signals the time of ovulation and increases the likelihood of paternity and paternal care by one or two males (Hamilton 1984); (4), in the 'female quality signalling' hypothesis, the size of the swelling informs the males about the quality of the females in a society in which inter-male competition exists anyway and in which any oestrus is enough to advertise that a female is in oestrus (Pagel 1994). This last hypothesis is a 'best male' hypothesis but viewed from the female/female competition point of view.

All these hypotheses have in common the belief that the sexual swelling is an honest signal that increases female choice and gives reliable information to the male about the proximity of the ovulation. However, this has not proved to be the case. In primates, oestrus is generally less strictly associated with ovulation than it is in other mammals (Martin 1992), with the consequence that males are less accurately informed about the female's state, which gives the females the chance of manipulating the males. In Hamadryas baboons and chimpanzees sexual swellings are observed 1 to 2 years before sexual maturity and the sexual swellings seem to have other functions. In bonobos, mothers with young babies regularly show sexual swellings for years that are not related to ovulation (Kano 1992; Badrian and Badrian 1984). We propose that sexual swellings are used as a 'social passport', so that females obtain the support of the males when facing severe competition with other females. The swellings could be elicited unconsciously by hormone production above a certain level of stress. We can try to test these alternative explanations by analysing the sexual swellings of the adult Taï females.

The oestrous cycle lasts 36 days and is marked by the increasing swelling of the sexual skin around the female genitals: on average, the skin is flat for a week, during the second week it swells, reaches maximum tumescence for 10 days and then deflates. In captivity, ovulation seems to happen during the last 2 or 3 days of the maximum tumescence period. If sexual swelling is an honest signal, female chimpanzees would show the swelling only when mature and fertile. The absence of menstruation (also called amenorrhea), a period of time normally without any sexual swelling, occurs during gestation and lasts for the post natal period when a mother suckles her infant. The end of amenorrhea would coincide with the weaning of the infant or its death.

Gombe and Mahale female chimpanzees typically resume their cycle more than four years after the birth of their infant, and they have a new baby within 20 months (Table 3.5). In this case the classical view that oestrus is bound to fertility is well supported. Is it the general explanation in chimpanzees? And is the bonobo–chimpanzee dichotomy in this aspect a specific difference? The Taï data support the 'social passport' hypothesis and suggest that the bonobo–chimpanzee dichotomy is not a real one. The average anoestrous period after a birth (lactational amenorrhea period = LA) for 26 observed instances in 19 females was 24.54 months (median = 22.5 months, sd = 18.32) (Fig. 3.5). The high standard deviation indicates wide variance in the data and this is also seen in Figure 3.5. In fact, 12 LA periods in 7 females are short, with the females resuming their cycle on average 7.7 months after parturition (median = 8.2 months, sd = 3.6). The others have much longer LA intervals of 38.9 months or 3 years and 3 months (median = 37.5 months, sd = 12.3). For 6 other females the resumption of the oestrus was not observed, either because they disappeared or because they are still under observation, but the tendency was towards long LA intervals. Thus, out of 25 females, 7 showed a surprisingly short LA period and resumed their cycle within the first year after parturition, while their infants were still suckling and grew steadily.

The interbirth interval is not influenced by the duration of the previous LA period ($r_s = 0.27$, $N = 12$, $p = 0.38$). Thus, at Taï sexual swelling is often a dishonest signal that does not provide information about the fertility of the females, and therefore does not support the four first hypotheses on sexual swelling, all of which see it as an honest signal. What factors affect sexual swelling in Taï chimpanzees?

Why do some females resume oestrous cycles so much earlier than others? Age and dominance of the mother and the sex of the infant were statistically tested. Young mothers have clearly shorter LA intervals than older ones (ANOVA: $p < 0.02$), and also inferior mothers have shorter LA intervals than older ones ($p < 0.004$). The sex of the infant, as an independent factor, plays no role. In other words, young low-ranking mothers have a shorter LA interval (mean = 7.6 months) and a longer oestrous period. The more dominant they become, the longer the LA period. Some young females achieved a

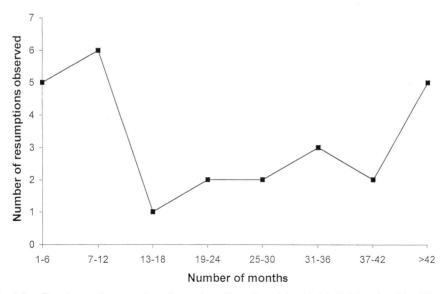

Fig. 3.5: Duration until resumption of sexual swelling after giving birth in Taï females ($N = 26$).

dominant position very rapidly (Lou, Pou) and they immediately had a much longer LA interval (mean = 33 months). Thus, the Taï females' oestrus is notably not always an honest signal for ovulation and seems to be used as a 'social passport'. Young subdominant mothers that are more at risk when in conflict with other group members use it to obtain the support or at least the tolerance of the adult males, while dominant females of whatever age do not need to rely on this subterfuge to cope with intra-community aggression. Thus, both newly immigrant adolescent females as well as young mothers produce oestrus to diminish social stress.

Are the males lured by the sterile sexual swellings? Males copulate with new immigrant females in oestrous, with young mothers in oestrus, and with mothers of young infants who have resumed their cycle. However, the mating frequency is not very high for the sterile oestrous females and we never saw competition between males for them (see Chapter 6). At a later stage, the males start to compete and guard the same females, indicating that they can differentiate between sexual swellings. Nevertheless, although the attraction in the young sterile females is limited, it might give them some male protection. In addition, they are of particular interest to the adolescent males, who make good use of their easy access.

Does the social passport hypothesis explain the varying LA intervals in other chimpanzee populations or the bonobos? The longer LA interval reported from Tanzanian populations may be explained by weaker social stress the young females suffer compared with Taï. At Gombe, most adolescent females remain within their natal community and thus have no integration problems (Goodall 1986). At Mahale, where most females migrate into a new community, the rarity of short LA intervals may be explained by the lower level of intra-group competition due to smaller party sizes (see Table 5.2). Thus, in the two Tanzanian populations, young females have less need for a 'social passport', except during their integration after community transfer, that is until they have had their first baby. This could be tested by checking whether new immigrant females at Gombe and Mahale have more oestrus and shorter LA periods than resident females. There are, however, indications that they do not. Young mothers at Mahale, where party size is intermediate between Gombe and Taï, appear to have difficulties in controlling the aggression of community members (Nishida 1989). Infanticide at Mahale is more frequent than at Taï and is carried out by the males (Nishida 1989). The difference between the two populations in mortality for infants of young mothers under 15 years is important. At Taï, 45% of the infants die during the first 5 years of life (Table 3.6), and 75% at

Table 3.6: Infant mortality and maternal age in Taï chimpanzees: first, we show the number of infants that died during the first 5 years of life (a) and then the number of infants born for each age class (b)

Mother's age	Male infant a/b	Female infant a/b	Sex unknown a/b	Total a/b	%
<15	2/7	1/7	6/6	9/20	45
15–20	2/6	2/6	2/2	6/14	43
21–30	1/4	3/4	–	4/8	50
>30	0/1	2/5	–	2/6	33
Total	5/18	8/22	8/8	21/48	

Mahale (Nishida *et al.* 1992). The 30% higher mortality rate of the infants of young mothers at Mahale indicates that aggression is more difficult to control in this population and that young females suffer higher losses. The difference at Taï is suggestive of the possible profit of a 'sexual passport' at Mahale.

On the other hand, bonobos have a social cohesion similar to Taï or even higher (see Table 5.2), and we would expect intra-group aggression to be at least as high. Kano's observation (1992) that bonobo females have sexual swellings for longer periods than do Tanzanian chimpanzees is consistent with our hypothesis, as is de Waal's (1996) suggestion from a study of captive chimpanzees that intrasexual behaviour such as genital–genital rubbing in bonobos is used to release social tensions. If this applies in natural conditions, it could explain the remarkable homosexual behaviour in female bonobos as a means of controlling the high level of intrasexual aggression. This could be tested by comparing intrasexual behaviour and oestrus periods across communities of different sizes: we predict that females in smaller communities have longer LA intervals and a lower level of homosexual behaviour.

Thus, it seems that the 'social passport' hypothesis is consistent with most of the sexual swelling differences we observe in wild populations in chimpanzees and bonobos. It explains the occurrence of sexual swellings both in sterile newly immigrating adolescent females and in young mothers.

Fertility and senescence in female chimpanzees

The female lifetime reproductive potential measures the number of offspring produced during the lifetime. With such a long-living species as the chimpanzee, it is difficult to follow females through their complete reproductive lifespan. We can, however, use the information we collected on the female reproductive profiles to calculate their potential. On average, mature females produced 0.2 infants per year, and this did not vary with age. At Taï, a female chimpanzee produces her first infant when 14 years old (Table 3.2) and can give birth every 5.75 years (Table 3.4). As she can live up to 46 years with no decline in her fertility, she could rear 5 to 6 offspring before dying (5.39 infants on average). Such a maximum reproductive performance is rarely achieved; only Héra and Ondine produced five infants, and most died before reaching adolescence. Infant survival is low in general and mothers rarely survive long enough to reproduce for 31 years. On average during the stable period of the study (1982–7), a female actually produced 1.98 infants surviving to adulthood, which just prevented the community from decreasing. Thus, under good conditions, Taï female chimpanzees reached 31% of their potential. But during the later periods, females survived less than a tenth of the reproductive span of the stable period, which means they reached less than 3% of their reproductive potential. Females like Agathe or Marlène died after producing just one infant which disappeared with them, and prolific females like Héra and Poupée produced four infants within ten years, but most of them died quite young. Ella was more successful, producing four sons, three reaching adolescence.

Taï females can adopt two different reproductive strategies, whose success does not have to be the same. Dominant females invested relatively more in sons than subdominant females did in daughters, so that the average IBI for dominant females was seven months longer. Thus, dominant females would bear 5, and subdominant females 5.6 infants during their lifetime (assuming they have the same survival). This disadvantage was compensated for by the fact that sons of dominant females survived at least twice as well as the others. The strategy of dominant females would pay even more if their sons achieved a higher fitness than the others (see next chapter).

Senescence, observed in many animal species, is thought to result from weak selection pressure in old age, making it possible for costly mutations to accumulate and express themselves leading to the characteristics of aging (Stearns 1992), including reduced agility, physical deterioration, and reduced fertility. Reproduction may cease entirely, giving way to the menopause. Age-specific fertility does not diminish in Taï chimpanzees, showing that no senescence is observed in this trait (Fig. 3.6). All the 11 females reaching 30 years of age continued to reproduce normally and had young infants until their death. The two oldest, 45 and 46-years old, gave birth during the two previous years. The absence of menopause in Taï chimpanzees concurs with what has been found at Gombe (Goodall 1983), but contrasts with the observations of menopause in 9 out of 11 old Mahale females that neither gave birth nor had a cycle for 2.5 to 9.5 years before their death (Nishida *et al.* 1990). In Mahale females, the reproductive period ceased on average at 39 years, about seven years earlier than at Taï. The physical costs of a pregnancy are high, requiring an abundant food supply. In some environments old females may be undernourished. We have the impression that the Taï forest is a rich environment for chimpanzees (van Rompaey 1993), and the relative richness of different environments may help to explain why only Mahale chimpanzees experience menopause. The data from Table 3.6 show that mortality rates of infants from Taï mothers of different ages do not differ. This contributes to the absence of senescence in Taï chimpanzee reproductive abilities. In contrast, Mahale infants of young mothers (under 15 years old) and old

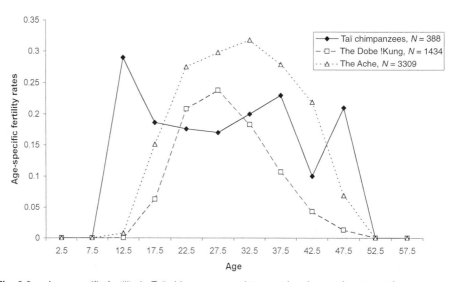

Fig. 3.6:   Age-specific fertility in Taï chimpanzee and two modern human hunter–gatherer societies.

mothers (over 30 years old) suffered high mortality, that is 75 resp. 60% (Nishida *et al.* 1990). This suggests that old age in Mahale chimpanzees is associated with low infant survival, possibly leading to menopause.

How frequently are chimpanzee females sterile? Sterility has been suspected in many females at Gombe but has been confirmed only in Gigi, who, when over 30 years old was still having regular cycles but never had an infant. Similarly, at Mahale, two females did not show any signs of pregnancy, despite the fact that they had cycles for 12 to 16 years (Nishida *et al.* 1990). At Taï, all females surviving to maturity had at least one infant ($N = 40$). One female, Gala, had three infants and was not pregnant during the seven years before her death. We estimated her then to be 30, too early an age to qualify her absence of pregnancy as a sign of menopause. Thus, in Taï females, we did not observe menopause, but there were some problems with fertility.

Female reproductive profiles among great apes and humans

The lifetime reproductive profile of females can at present be contrasted between five populations of wild chimpanzees living in different environments (Table 3.7). Age of maturity (first pregnancy) is reached later in all wild populations than in captivity. This shows that some important constraints on the maturation of the females have been removed in captivity and one price of living in the wild is a two-year maturation delay. However, the wide range observed in all wild populations indicates that chimpanzees have the potential of maturing earlier, and we should ask why most of them do not.

Two factors directly affect the maturation process: female transfer and infant survival. Integration in the new community takes time, and the survival of a new infant is guaranteed only when a certain level of integration is achieved. Data from Mahale chimpanzees show higher infant mortality during the first period of the reproductive life (Nishida *et al.* 1990). Delaying reproduction might improve infant survival. The two East African populations (Gombe and Mahale) reach maturity about a year later than the Bossou and Taï populations, this despite the fact that most females at Gombe do not transfer between communities and do not need to wait for full integration before reproducing. Another factor might explain this difference. Both East African populations live in open woodland habitats, whereas both West African populations live in the forest. Are West African forests richer, allowing for quicker development?

The interbirth interval is one year shorter in captive chimpanzees than the shortest wild IBI at Bossou. The analysis of IBI in Taï suggest that female reproductive strategies may vary with rank and offspring gender, making use of the average IBI misleading. The females adopting the quickest strategy at Taï have IBIs of 61 months, similar to Bossou. Kibale chimpanzees have a surprisingly slow reproductive rate with IBIs of 86 months, which is within the range of the dominant females with sons at Taï (Fig. 3.2). Since at Kibale the mortality of infants seems to be especially low (Hill *et al.* in press), it is possible that the long IBI in the Kibale forest contributes to high infant survival.

Gorillas start to reproduce a year earlier than at Bossou and is captive chimpanzees and have the same IBI as these chimpanzees. Intriguingly, unlike in chimpanzees, the gorillas' performance is not improved by captive conditions. A possible explanation is that

Table 3.7: Female reproductive profile in great apes and humans

Population	Age of 1st pregnancy (range)	IBI (months)	Length of reproductive span	Total fertility rate
Chimpanzees				
Bossou	13.0	61		5.29*
Gombe	14.9	66	32	4.56*
Kibale		86		
Mahale	14.6	72	25	4.23*
Taï	13.7	69	32	7.44 (5.39*)
Captive	11.0	50		
Bonobos				
Wamba		54		
Gorillas				
Karisoke	10.2	47	22	5.55*
Captive	9.5	51		
Hominids				
Australopithecus		69		
Homo erectus		48		
Homo sapiens		42		15.1
Modern hunter–gatherers				
Ache	19.5	37	24	8.03
Yanomani	18.4	34	19	6.86
!Kung San	18.8	49	15	4.69
Gainj	20.9	42		4.31
E. Hazda	30.9			6.15
Gidjiingali	15.9	44		6.40
Batak	18	28		3.94

Total fertility rate (TFR) is the sum of all age specific fertility across the reproductive span, when not available we gave the estimated number of births within the maximum lifespan marked *.
Chimpanzees: Captivity from Tutin (1996).
Gorillas: Tutin (1996), Gerald (1995).
Hominids: Teleki *et al.* (1976), Wood *et al.* (1985).
Hunter–gatherers: some data from Teleki *et al.* (1976).
Ache: Hill and Hurtado (1996).
!Kung: Howell (1979).
Gainj: Wood *et al.* (1985).

wild gorillas do not face as high a level of intragroup competition as chimpanzees. They live in smaller social groups with often only one sexually active adult male, making social interactions less tense. They also have a folivorous diet within an environment very rich in undergrowth foliage, so that feeding competition is probably minimal (Harcourt *et al.* 1980; Watts 1996). Therefore, female gorillas can start reproducing as they mature; delaying reproduction might not increase the survival of their offspring.

Modern human beings (*Homo sapiens sapiens*) have a wide variety of life styles, ranging from tradtional hunter–gatherer to post-modern industrial societies. With the varied use of medicine, hygiene, and contraceptive methods, these societies have produced very different reproductive profiles in women. However, the populations living under conditions similar to those of great apes, with no or minimal access to medicine, and no contraceptive methods, could be compared. Care is needed since the chimpanzee data in Table 3.7 represent the maximum reproductive performance, while data on humans represent the average reproductive performance. Figure 3.6 compares age-specific fertility between Taï chimpanzees and two hunter–gatherer populations showing that chimpanzees start to reproduce about 5 years earlier than humans and have then a comparable age-specific fertility for about 30 years. Since human live about 25 years longer than chimpanzees, the decrease of fertility with age is more striking in humans than in chimpanzees. Such a comparison makes sense only if we take in account that only 7% of the Taï chimpanzee females reaching menarche survive to age 32, while 87% of the Ache female do so. Thus fertility in humans is dramatically higher due to higher adult survivorship, despite a potentially similar length of reproductive span than the chimpanzee.

The female reproductive profile of Taï chimpanzees is characterized by a transfer between communities at about 11 years, followed by a period of adolescent sterility of 2 years and 8 months. They have a first baby when 13 years and 8 months old, and have an interbirth interval of 1 year and 1 month if their infant dies, or one of 5 years and 9 months if their infant survives. They have the same number of sons and daughters and have a fertile period of about 28 years (from 13.8 to 43 years).

The sexual swelling in Taï females seems to function primarily as a 'social passport' allowing females to control the aggression of other community members in critical periods; the sterile sexual swellings are observed during the integration in a new community, during most of the nursing time in young and subdominant mothers, as well as during the weaning process of older infants. Such sexual swellings in young females have been mainly observed in Taï chimpanzees and bonobos. This seems to reflect the social tensions within different social communities. At Mahale, young mothers could need such a 'social passport', but for unknown reasons have failed to adopt it.

We observed two reproductive strategies at Taï. The dominant females had interbirth intervals 26 months longer for male than for female offspring and by this additional investment give their sons a higher survival rate until maturity. Subdominant females invested 12 months more in their daughters than in their sons, not enough to give them a higher survival. In other chimpanzee populations, females use only one reproductive strategy, and we think that the different dispersal patterns of the females in those populations might explain this. The dominant female strategy at Taï pays only if the contact between mother and son remains uninterrupted until he has reached the peak of his social potential (when about 20 years old). We shall see in the next chapter how females may support their sons and how males increase their reproductive success.

4 Reproductive strategies of male chimpanzees

Forest scene: Salomé, one of the highest ranking females, in full oestrus with a large glowing pink swelling, was followed closely by all the males of the community all day long. She could not move without the whole crowd moving with her. The tension between the males was perceptible, although for long periods of time none of them mated, and conflicts exploded whenever Salomé and her friend Ondine neared a fruiting tree. The alpha male, Brutus, then wanted to be the first behind her, but his young challenger Schubert was not always willing to let him do so. The ensuing conflicts between the two were intensely followed by all the other males. Schubert relied on the support of his young ally Macho. Falstaff, the oldest male of the community, was making the best of the situation, either supporting Brutus or attempting to sneak away with Salomé while the other males were too busy threatening one another. Regularly, despite the difficult situation, the big males tried to mate with Salomé by silently and inconspicuously leading her away. She only rarely followed one of them and they came back looking unconcerned.

At the same time, Goma, a young newly immigrated female, was also in full oestrus. She, however, barely aroused any interest. It was Goma who tried to remain in contact and regularly presented her swelling to the males, with some success only with the adolescent males. While, whenever one of these adolescent males as much as neared Salomé, his move would be immediately discouraged by aggressive barks from most of the dominant males. Why are the males competing so fiercely for some females and hardly at all for others? How do rank and possible allies affect the males' reproductive success?

Male reproductive strategies are constrained by the access to suitable mates, and many male traits are expected to be aimed at finding and securing a large number of mates (Andersson 1994; Krebs and Davies 1993). Sperm production is less costly than egg production, allowing males to produce many sperms and, in theory, to inseminate a large number of females. This, however, is limited by female receptivity. If females are not always receptive, males need to keep them under guard until they can inseminate them; their ability to do so will determine the number of females they can monopolize (Dunbar 1988; Krebs and Davies 1993). In species with a harem-like social structure, like many monkeys and the red deer, one male can potentially monopolize all the females in a group. In others, like the gorilla, the alpha male has to tolerate some males in his group, but succeeds in suppressing at least part of their sexual activity. Multi-male groups are seen in other species, such as in chimpanzees, baboons, and lions, where the competition between males is so strong that they gain by associating with other males to control females. In lions, coalitions of more than four males control a pride of females for a longer period of time and succeed in producing more offspring than coalitions of two males. The cost for the alpha male is having to allow the other males some copulations, but this apparently remains the best solution for him (Packer *et al.* 1991). In chimpanzees, larger coalitions of males occur in larger communities, they control larger numbers of females and are better able to resist the coalitions of other males (Goodall *et al.* 1979; Nishida *et al.* 1985; Chapter 7). The major question remains: how do males share the paternity of the offspring, and what affects their success?

In chimpanzees, males remain all their lives in their natal community, whereas females may transfer between communities. The males have two ways of increasing their reproductive success. First, they can increase their success within the community by monopolizing the resident females, which has been described as being the strategy of the alpha male at Gombe and Mahale (Goodall 1986; Hasegawa and Hiraiwa-Haegawa 1990). Second, they could try to find additional females outside the community, which has been described in the Gombe chimpanzees and baboons (Collins 1981; Goodall 1983). This could be achieved either by attracting females from neighbouring communities or by moving into the territory of other communities to find mates. The latter strategy entails, however, the risk of having to face the other resident males, who can be very aggressive (Chapter 7). Here we compare the strategies used by the males in the different populations under study.

How can males share paternity in a multi-male group? Two factors have been proposed to play a role. First, male dominance rank may decide priority of access to mating partners. Dominance, if it limits aggressive interactions, has a mixed value as a reliable predicator of reproductive success (Dunbar and Colishaw 1992; Bercovitch 1995). Indirect measures for reproductive success, such as copulation rates, time with oestrous females, or time spent in consortship, do not always correlate with dominance rank in primates. One problem is that the advantage from higher dominance status may be countered by coalitions of other males in primates (Noé 1992; Bercovitch 1988; Chapter 6). Also subdominants may use alternative strategies to have access to females. Male baboons form long-term bonds with certain females who then prefer to copulate with them (Altmann 1980). Subdominant male chimpanzees may leave the group with a female before she is in oestrus and of interest to other males, and remain away until the end of her oestrous cycle (Goodall 1983, 1986). Second, kin selection can influence paternity in multi-male groups. In lions, males in large prides are all related and leave most of the mating to the dominant individuals, whereas in small prides of unrelated males, matings are equally shared (Packer *et al.* 1991).

Female mammals do not have the option of multiplying sexual partners in order to increase the number of their offspring, but they can increase their quality by selecting high-quality mates. Female mate choice is expected, and may be based on the result of male competition, for example males winning the most contests against other males being preferred by females (Andersson 1994; Eberhart 1996). Female mate choice that is not correlated with the outcome of male competition would indicate its independent role. Recently new evidence of female mate choice has accumulated. This evidence suggests both that female choice increases genetic benefit for their offspring (Kempenaers *et al.*

1992), and that females can choose between the sperms of the different males that mated with them (Olsson *et al.* 1996). The importance of female choice in different species remains unclear and it is considered limited in chimpanzees. Females at Gombe and Mahale are thought not to have the opportunity of choosing their male partners (Goodall 1986; Hasegawa and Hiraiwa-Hasegawa 1990).

Here we must distinguish between two levels of analysis. The first consists of indirect behavioural measures of reproductive success, such as copulation rates and association time with oestrous females, that might be useful indicators both for chimpanzees and for human observers. However, these are indirect measures, and the assumption that higher copulation frequency with a given female increases the likelihood of paternity must be tested. The second level consists of the direct measure of reproductive success as the number of offspring sired by an individual during its lifetime, which can only be ascertained by genetic analysis. We recall here the surprising results of genetic analysis of monogamous birds. In some species, up to 30% of the offspring were not sired by the male that was feeding them (Andersson 1994). This result should serve as a warning, since the male birds seemed unaware of the extra-pair copulation, as most fed the chicks in their nest just as actively as did the progenitor (Lifjeld *et al.* 1993), and since the human observers also failed to notice it. In other words, males may base their behaviour on inaccurate indicators that females are apparently able to manipulate.

Genetic studies are now a fairly reliable tool for ascertaining how successful behavioural strategies used by males are. For this purpose the genotypes of all possible fathers of a given infant have to be established and compared with the genotype of both the infant and the mother. A non-invasive method based on nuclear DNA microsatellites has been developed recently and allows to study paternity in wild chimpanzees (Morin *et al.* 1994*a*). The first tests were done on the Gombe chimpanzees, with only two paternities determined for the Kasakela community (Morin *et al.* 1994*b*). A similar study is now in progress on the Taï community (Gagneux *et al.* 1997, 1999).

Chimpanzees use behavioural strategies to improve their reproductive success. Our aim is to analyse a multi-male chimpanzee community in the Taï forest to see which strategies males and females use to increase their reproductive success, how successful they are, and who takes the decisions.

Male dominance rank order

In Taï, dominant mothers invest longer in sons than in daughters (Chapter 3). Do these sons achieve higher reproductive success than sons of low-ranking mothers? Two prerequisites might be needed: first, a larger proportion of sons of dominant mothers need to reach adulthood and have the time needed for many successful matings. This is the case, for compared with low-ranking mothers, 30% more sons of dominant mothers reached adulthood (see Fig. 3.3). Second, they may need to achieve higher social status, for this has been suggested as leading to higher reproductive success in chimpanzees (Hasegawa and Hiraiwa-Hasegawa 1990). We now take a closer look at this aspect.

Dominance rank has classically been decided by placing a preferred object between two individuals (Sade 1967). The individual that obtained it was considered as dominant over the other one. If this is done for all possible pairs, a dominance matrix could be produced out of these encounters. At Gombe, it has been shown that submissive behavioural elements like the 'pant-grunts' (Goodall 1968) were always given to the dominant individual, and aggressive behaviours such as attack charge or display were directed by the dominant to the lower ranking one (Simpson 1973). The unidirectionality of these behavioural elements was confirmed at Taï (Chapter 6), and we used them to establish the dominance relationship within the community. Table 4.1 presents the yearly dominance rank order of the males in the community. Dominance rank tends to follow for each male a ∩-shaped curve, as males ascend in rank from maturity onwards but descend again after a certain age.

Not all individuals reach the highest ranks, for example, at Taï, Darwin, Rousseau, and Ulysse were always low-ranking. Some males behaved as if they were not interested in improving their position in the social rank, an impression also reported from Gombe (Goodall 1986), but it is difficult to say if this lack of interest is genuine or if it is the result of the fact that these individuals know or feel that they would not be able to improve their status and therefore do not try. In support of the second alternative, we gained the impression that Ulysse was not interested in increasing his social rank until Brutus had lost his alpha position, and then, to our surprise, we saw Ulysse display against Brutus until he pant-grunted towards him. Therefore, we shall look for physical or historical factors that might explain the success of males in raising the social ladder.

Darwin was so named because of the malformations he had on both feet that exemplify the struggle to survive. His left foot looked like a club-foot with only one toe clearly visible and facing backwards, the right foot looked as if it had been broken in the middle, and he could not use its toes. Thus, Darwin was unable to grasp anything with his feet. He could not run at a normal speed and climbed trees much more slowly than other chimpanzees. He probably remained low-ranking because of these serious physical problems. But when young Fitz started to show some interest in the alpha position, Darwin became his ally and started to display with him. He thereby gained a certain status that allowed him for a short

Table 4.1: Dominance rank of the adult males in the Taï community

	82	83	84	85	86	87	88	89	90	91	92	93	94	95
Balzac	5	5	–											
Brutus	1	1	1	1	1	1	4	5	4	5	5	5	4	2
Darwin	+	+	+	+	9	8	6	6	6	6	6	6	–	
Falstaff	4	4	2	4	4	7	–							
Fitz	+	+	+	+	+	+	+	+	7	1	1	1	1	–
Kendo	+	+	+	7	5	3	2	2	1	2	2	2	2	–
Le Chinois	6	6	–											
Macho	8	8	4	3	3	2	1	1	2	3	3	3	3	1
Pistache	3	3	–											
Schubert	7	7	3	2	2	–								
Snoopy	+	+	+	8	7	4	–							
Rousseau	9	9	5	6	6	6	5	4	5	4	4	4	–	
Ulysse	+	10	6	5	8	5	3	3	3	-				
Wotan	2	2	–											

1 is the highest rank
+: sub-adult
–: died during the year

period to rise his rank. He was not able to respond to the challenges of the other males on his own, and Fitz, once established as alpha, did not support him any more.

Rousseau did not suffer from any physical handicap to start with. He, Falstaff, and Pistache were the three tallest males, with particularly long legs and thorax that gave them a very characteristic build and way of walking. Rousseau was the tallest of the three. However, he remained low-ranking throughout his life and rarely displayed against other males. He was a poor hunter, specializing in cheating roles (Chapter 8). In 1985, Rousseau dislocated his right shoulder and could not use his right arm at all until, two weeks later, it was somehow repositioned correctly. For the rest of his life he was unable to apply much strength with this shoulder and no longer directly climbed large tree trunks, waiting on the ground for the others when there was no alternative route to join them. The consequences of this accident probably simply accentuated his already low social ambitions, although he still participated in wild scrums to gain access to meat and copulated with females. In fact, he did father one of the infants (see Table 4.6).

Whereas physical handicaps probably contributed to Darwin's and Rousseau's lack of interest in social rank, Ulysse was never physically handicapped. He developed as a strong male and became one of the most gifted hunters in the community (Chapter 8). His apparent lack of interest in social rank was probably for social reasons. As long as Brutus was the alpha male, Ulysse remained uninterested in improving his social position and remained a lazy hunter for most of this period. When Macho, his close friend, became the beta male in 1986, Ulysse suddenly revealed himself as one of the best hunters. When Brutus lost his alpha position in 1987, Ulysse changed his indifferent attitude and started to challenge him successfully.

Thus, all the males we followed regularly since 1983 eventually reached one of the three top ranks, except those with a physical handicap.

Coalitions and dominance in male chimpanzees

It has become generally accepted that male chimpanzees increase their social rank through the coalitions they form with one another (Foley 1989). Male chimpanzees are known to form coalitions in social contexts in all populations studied (Nishida and Hosaka 1996; Goodall 1986; Chapter 6). These coalitions are frequent and opportunistic, in the sense that two or more males (whatever their relationship) temporarily join forces against a common rival. Thus, one male may have many different coalition partners, and most coalition partners have supported many males in the community. In this context, it is difficult to prove that a given male would have achieved his present position had he not been supported by another male at a given time. Coalitions tend to increase the confidence of a male and might help him to attack a rival more quickly than he would if alone. The situation is more straightforward in an alliance between two males that repeatedly support one another over long periods of time. But alliances are comparatively rare in chimpanzees. The only convincing example comes from two brothers at Gombe, Figan and Faben, where the younger Figan was unable to dominate his rival, Evered, until Faben supported him systematically and helped him to defeat Evered three times in a row (Goodall 1986, 1995). Even in this case, Jane Goodall had the feeling that Figan would eventually have done so on his own.

At Taï, we have one clear example of increasing social rank thanks only to a coalition partner. In February 1983, the study community lost four adult males, possibly to poaching, changing the social relations between the surviving males. Schubert, a young prime male, found himself in the gamma position within a month, and in March of the same year he started to challenge Brutus, the alpha male, who in turn had been affected by the disappearance of the beta male, Wotan, probably his brother. In addition, Brutus was not at his full strength; his wrist had just been tightly strangled by an iron cable. Schubert took advantage of Brutus' weakened state, and Macho was Schubert's ally, constantly supporting the older male and displaying in parallel. We cannot be sure that Macho and Schubert were not brothers, but we thought their alliance was not a family affair but based simply on common interests. We estimated Macho to be 21 years old and he was not a very large male. Both were clearly younger than Brutus, and, perhaps because of that, they challenged Brutus for the first time some 50 m up in a tree above the canopy. The pair displayed spectacularly for 20 minutes at Brutus. He replied, in what looked to us, like a crazy and highly risky aerial ballet. For each of these challenges, Brutus was the first to retreat, his face twisted with fear-grin, typically screaming in a high pitched voice.

Then the two also started to challenge him constantly on the ground. Brutus started to seek support from Falstaff, the beta male, who like all the other males was attracted to these scenes of high tension. Falstaff hesitated to become involved against the two allies he was still dominating. Then, Brutus, screaming with fear, positioned himself so that the pair that was threatening him was also facing Falstaff. Falstaff then joined forces with Brutus and put the two to flight. But Falstaff was an unreliable ally, whereas Macho constantly supported Schubert, so that by the end of April Brutus was clearly frightened of facing the two of them. Macho was the weak point of their alliance; he was not very strong and still submissive to Brutus when facing him alone. At this point, Brutus changed his strategy and, when facing the alliance, would wait for Macho to be about five metres away from Schubert, or displaying in another direction,

Fig. 4.1: Brutus groomed by Kendo. Both Kendo and Fitz, sons of powerful Ella, became alpha males of the community.

and then immediately charge at him, hitting and biting him rapidly before Schubert could come to his rescue. After three such attacks, Macho started to diminish his support to Schubert. By mid-May, Brutus was seen for the first time to attack Schubert successfully. Schubert and Macho changed targets, attacked Falstaff, and quickly dominated him. They reached rank two and three respectively, as Macho on his own could defeat Falstaff.

The alliance between Schubert and Macho was not successful against Brutus because neither was able at any time to defeat him alone. Looser alliances, in which both partners are lower ranking than the individual challenged, can be successful on some occasions, but the partners have to remain together almost permanently, otherwise they are exposed to immediate revenge. Chimpanzees are intelligent enough to judge the weaker member of a coalition and concentrate the attacks on him. This might explain why, except for brothers, alliances between males against dominants tend to be short-lived (Schubert–Macho remained allies until Schubert's death, but did not challenge Brutus any more).

Darwin, as mentioned before, supported Fitz, whose success he tried to use to his own advantage. But when he became the target of higher-ranking males while Fitz was out of range, he rapidly stopped the attempts to improve his social position. The alliance between Ella and Kendo described in the following section is an example of another type of coalition aimed at increasing the rank of a male. However, in all struggles to climb the social ladder that we saw at Taï, actual gain in rank was ultimately achieved by direct physical confrontation between two males. Coalitions did not seem to alter the outcome, although they might have made it happen faster.

Mothers' contribution to their sons' social status

In chimpanzee societies, males remain all their lives within their natal group and can as adults still have their mother around. Vivid descriptions are provided from Gombe of adult males maintaining strong relationships with their mothers, remaining with her when they are wounded; and the mother, from a distance, sometimes providing vocal support to an adult son (Goodall 1986, 1995). In bonobos, mothers are described as being direct and active supporters of their presumed sons, and, when dominant, as having dominant sons (Kano 1992). Determining mother–son relationships solely on physical similarities, as was done in the bonobo study, is risky, especially when the males are already independent of their putative mothers. Nevertheless, we may wonder why, in chimpanzees, mothers do not take a more active role in promoting their sons' social status. When sons reach an age for showing social ambitions, the mothers are usually either dead or too old to be of any serious help. Moreover, when a son enters adulthood, mothers may have several younger offspring, and too much investment in one might be at the cost of another. In the end, such an investment might be non-beneficial to the mother. It would, however, make sense for mothers with more than one son to be of active support, for the younger ones might profit from their big brother's social status. Owing to the variety of possibilities that could explain an absence of maternal involvement in their adult sons' social life, more observations are needed to clarify this point.

At Taï, only five males grew old enough to enable us to observe their mother's reactions. Three were sons of high-ranking mothers (Kendo, Fitz, and Snoopy), and all progressed rapidly in the hierarchy and seemed very keen to progress. Two were sons of low-ranking mothers (Darwin and Rosseau), and they showed no real interest in progressing in the social rank order and barely improved their rank in the hierarchy. We have already seen that for the two of them a physical handicap might have contributed to this lack of ambition.

Salomé's son, Snoopy, was one of the best hunters in this community. He showed very early on a keen sense of his goal and the investment needed to achieve a capture. During his late adolescence, the hunting success of the community rose to nearly 100%, mainly as a result of his contribution, and the frequency of hunting was very high for most of the year (Boesch and Boesch 1989). His mother was the second-highest-ranking female of the community. His direct rival was Kendo, who dominated him in many fights owing to his more powerful build. Snoopy was, as a young male, already a powerful and expert drummer contributing to the social cohesion of the group. He was also the one male who seemed able to communicate information about movements to other group members with his drumming, as Brutus did for years. Macho, alpha male at the time, considered him as his main rival, but Snoopy disappeared suddenly for unknown reasons and we could not determine if his mother was ready to invest actively in her adult son, while still nursing a daughter.

Fitz and Kendo, both infants of Ella, provided the best observations we have of the maternal investment in adult sons and we report them here in more detail.

A case study: Ella, Fitz, Gérald and Kendo

Ella was one of the shyest females of the study community. We saw her quite irregularly and initially gave her a peripheral status. In fact, Ella was a dominant female, and it was our presence that intimidated her. It took many years and extremely cautious approaches until she lost her timid attitude. Ella was always with her two sons, Fitz and Gérald, born in 1975 and 1983 respectively. Family ties were strong. Fitz showed no desire to leave his mother until he was 13 years old. As he grew older, we progressively saw her more regularly in the group of the adult males. Then we rapidly realized that Kendo, born around 1969, must also be her son: he was surprisingly tolerant, letting her take large pieces of meat and fruit from his hands, and he was very relaxed with her other two sons. Kendo, too, had a strong physical appearance and was certainly the fastest walker in the community. As an adolescent, we noticed how very playful he was with the two other adolescents of the community, Darwin and Snoopy.

In striking contrast to Darwin, even as an adolescent Kendo already showed a keen interest in climbing the social ladder. He often displayed violently against the adult females, regularly challenging the most dominant ones. As a young adult, in 1985 and 1986, he rapidly dominated Snoopy, Rousseau, and Ulysse. We even saw him twice challenge Brutus, who was still the alpha male at the time, biting off a bit more than he could chew. In late 1986, it first became obvious that Ella was interested in helping Kendo to improve his social status. Kendo was trying to dominate old Falstaff and each time that Kendo screamed in front of Falstaff, Ella rushed to support Kendo, followed by her two younger sons, all three of them barking. The sight of this rather impressive trio must have had an effect, as Kendo rapidly dominated Falstaff. Beginning in 1987, Ella started to be a regular member of the males' group, and her association with Kendo increased, as did the association of the three brothers, Kendo, Fitz, and Gérald. Kendo was by that time number three in the male dominance rank order. He was dominated only by the ageing alpha male Brutus and Macho. Fitz was clearly very interested in the males' social relationship and the five-year-old Gérald also was a keen watcher of these interactions.

In October 1987, the number of conflicts between Macho and the Kendo–Ella team increased markedly. Three times the conflicts were clearly initiated by Ella herself. Kendo rushed to rescue his mother, and together they chased Macho away. It became apparent that Macho felt uneasy with Ella, uncertain about how to handle this female always accompanied by Fitz and Gérald, a sturdy team, and an individual who always obtained the support of Kendo whenever she cried long enough. Now, for the first time, Kendo was able to retain an adult red colobus prey in front of Macho when his mother was present. In early summer 1988, Macho became the alpha male of the community. Brutus gave up his position without much resistance, at about 36 years of age. Now Macho used the strategy he had suffered himself from Brutus some years ago: he leaned very hard on Kendo whenever his mother was absent. Kendo clearly did not feel strong enough to challenge Macho, and we gained the impression that he was uncertain about his mother's ambitions for him. He sharply increased his association with her during the summer 1988, both of them keeping away from the group.

Beginning in 1989, Kendo tried to disrupt the bond between Macho and Ulysse by being more associated with Ulysse, without much success — Macho pressed Kendo very hard and slapped him regularly. During the summer, the whole Ella family was again absent from the group and she gave birth to a fourth son, Louis, in August. In September 1989, the whole family reappeared in the group. In November, despite her tiny baby, Ella started to support Kendo actively in his challenges against Macho. This time, Macho clearly feared Ella's support, as we heard him scream whenever he heard her supportive barks, even if she was out of sight. In December, Kendo defeated Macho in a violent fight in which Macho lost a toe and a phalanx of a finger. Kendo became one of the most powerful and aggressive alpha males of the community we had seen. His mother, Ella, died during the summer 1990.

Why did Ella play such an active role in her son's social career? That all her infants were males might be important. In fact, Kendo remained alpha for only 18 months. Without any fights, he gave up his position to his younger brother Fitz, who became alpha when only 15 years old. We considered Fitz adult in December 1990. Since 1988, Fitz regularly left his mother, spending most of the time with his older brother Kendo. In December 1990, he started to display with Kendo and enjoyed his constant support whenever he was screaming in any social squabbles. In December 1990, he dominated Darwin and Rousseau without much trouble. In February 1991, he dominated Brutus, and Macho started to pant-grunt to Fitz in March 1991. In May 1991, Kendo started to pant-grunt to Fitz without any fighting. Thus with the help of his then alpha brother, Fitz became alpha himself in less than six months. Kendo was obviously physically much more powerful than his younger brother, and we have to consider the possibility that he voluntarily gave up his position. Fitz and Kendo then occupied the two highest positions of the social hierarchy, and their younger brother, Gérald, enjoyed one of the most privileged situations imaginable. We will never know whether Ella had anticipated the outcome, but it seems good foresight for a mother of two or three sons to help her first one reach the alpha position. Although the two Ebola outbreaks of 1992 and 1994 killed her three sons, within this short period of time her two eldest sons had been able to sire three infants (see Table 4.6), indicating how successful her strategy might have been if they had survived.

Is longer maternal investment in sons than daughters adaptive? We saw in Chapter 3 that high-ranking mothers have much longer interbirth intervals when they have sons than other mothers (Fig. 3.2) and that this was followed by better survival of those sons compared to all the other infants (Fig. 3.3). If all other things were equal, it would pay dominant mothers to invest more in sons, for these might achieve higher social ranks as adults. The sample is small, five males reached adulthood during our study, but neither of the two sons of low-ranking mothers showed interest in progressing in rank, whereas the three sons of high-ranking mothers did, and two lived long enough to reach the alpha position. This supports the idea that sons of high-ranking mothers not only survive better but also achieve higher social ranks than those of low-ranking mothers.

Leadership: an additional dominance system

A special aspect of the male dominance system at Taï is that beside the social dominance system, two additional systems of dominance, the leader and the meat-sharing rank order, can be distinguished (Table 4.2). Individual males do not achieve the same rank in all of them, except for Brutus, the long-term alpha male who held rank 1 in all contexts. Brutus eventually lost his alpha position in the social context, but retained the top position as a leader and in meat access. His privileged meat access lasted up to 1994, despite the fact that by then he had not been social alpha for more than seven years (see Chapter 8 for factors affecting meat access in males). Dominance for meat access occurred always in a situation with extremely strong social tensions between the individuals, and it was mediated by coalitions between both males and females. On a very regular basis, in 1984–5, Falstaff, Brutus, Ondine, and Salomé supported one another against other aggressive

Table 4.2: Dominance order in 1984–1985 for all adult males and the dominant females of the Taï chimpanzee community in three different contexts: social, leadership, and meat-access

Name	Sex	Social rank	Leader	Meat access
Brutus	M	1	1	1
Schubert	M	2	–	4
Falstaff	M	3	–	2
Macho	M	4	–	6
Ulysse	M	5*	–	7
Rousseau	M	6*	–	8
Kendo	M	7*	–	9
Darwin	M	8*	–	10
Snoopy	M	9*	–	11
Ondine	F	10	–	3
Salomé	F	11	–	5

*Rank order of these males was unclear.
Social rank is determined by the direction of pant-grunting (see Chapter 6).
Meat-access is determined by the time an individual was seen eating meat.

males (mainly Schubert and Macho at the time). Some females reached very high status in the meat-access order.

The Taï community had for years a male, Brutus, who was considered by the other group members as a leader, in the sense that his drumming conveyed not simply information about the location of the drummer in the forest, but also on the direction and speed of the group movement (Boesch 1996*b* and Chapter 10). During the first period of the study, the drumming even seemed to convey symbolic information. During most of the study period, Brutus' drumming was used by other community members as a guide for travel directions. Only with the decrease in the size of the community did he reduce the use of this means of information transfer.

The leader in a dense tropical rainforest seems to act to preserve the auditory contact between different parties moving out of sight of each other. It requires a male able to drum clearly, that is, one that does not drum too frequently, but often enough for others to stay in auditory contact. This role does not seem to require a dominant male, nor do we have the impression the position yields any reproductive benefit.

Sexual behaviour in male chimpanzees

Through careful observations of the Gombe and Mahale males, three different mating strategies were distinguished: promiscuity, possessiveness, and consortship mating (Tutin 1979; Hasegawa and Hiraiwa-Hasegawa 1990). Promiscuous mating takes place within a party without overt signs of conflict between the males that are present. Possessive mating is shown by the dominant males who attempt to monopolize an oestrus female and prevent other males from copulating. Consortship mating is observed when a couple isolates themselves from others for a certain time, thereby ensuring that only the consort male copulates with the female. At Gombe, for 46 conceptions from 1968 to 1984, 35% are estimated to have resulted from copulation in group settings, 28% in consortship, 15% could have taken place in group settings as well as in consortship, and 13% with neighbouring males. Between 1972 and 1983, females went on consortship on average every fourth oestrus (Goodall 1986). In a larger sample of 56 conceptions in Gombe chimpanzees from 1975 to 1994, 25% were again estimated to have occurred during a consortship (Wallis 1997). At Mahale, out of 12 conceptions only one (8.3%) could have taken place during a consortship; over 80% occurred in group settings (Hasegawa and Hiraiwa-Hasegawa 1990).

Following reproductive strategies in the wild is difficult, for individuals may conceal mating from the more dominant members of the group or leave the party inconspicuously. If a male wants to lead an oestrous female away from the other males, they both need to be silent and quick, because the others will search for the female and rush towards any suspicious calls. Thus, a certain willingness is required to make such departures possible. In all wild chimpanzees studies, if a couple leaves the group before the female is in oestrus, observers would usually not specifically follow them and just note that two individuals are absent. Therefore, consortship is only indirectly ascertained by the fact that one male and one female are absent from the group at the same time for a certain number of days. We tried to observe whether they came back together but this was not always possible. In addition, as we did not follow the consorting pairs, we have no idea of how frequently they mated and cannot reliably compare the mating frequency of this strategy with others. The uncertainties associated with determining whether two chimpanzees are together when they are absent makes it speculative to use absence as a proof that that male was the only to have mated with that female, and should, therefore, be the father of the offspring. We therefore use here the results of genetic analysis to test the success of different reproductive strategies used by males.

Genetic analyses that determine the father of the infants born within a community have been used to test the efficiency of the different male reproductive strategies. Of the two Gombe chimpanzee infants whose paternity was proved (Morin *et al.* 1994*b*), one was sired by an adolescent male of the community and the other by a stranger male, suggesting that dominant males might have problems controlling the females, as was apparent in the behavioural studies. The genotypes of all individuals of the Taï community living between 1991 and 1995 have been established (Gagneux *et al.* 1997). This allowed a first analysis of paternity for the 13 infants for which we were sure to have the genotype of all living males above nine years of age at the time of conception.

In Table 4.3 we list all consortships from 1987 to 1994 in the study community. For 13 infants, 30 consortships were recorded, which means that a female went on consortship

Table 4.3: Consortship in Taï chimpanzees from January 1987 to December 1995: we indicate the length of the consort, whether it resulted in a possible conception and the conception date of the infant (determined by subtracting 229 +/– 30 days to the birth date [Martin 1992])

Male	Male social rank	Female absence	Consort	Infant	Consort success	Conception date
Kendo	3	Fanny	12/22-2-87	Manon	No	1.2.87
Kendo	3	Bijou	17/29-8-87	Bambou	No	
Kendo	3	Bijou	5/8-9-87	Bambou	No	
Kendo	3	Bijou	11/19-9-87	Bambou	No	25.6.88
Kendo	2	Zoé	6/29-10-88	x	No	?
Kendo	2	Mystère	11/23-11-89	Mognié	No	20.1.90
Macho	3	Fanny	24-4/10-5-91	Foutou	No	
Macho	3	Fanny	12-8/6-10-91	Foutou	No	
Macho	3	Fanny	6/21-11-91	Foutou	No	14.2.92
Kendo	3	Bijou	3/13-10-91	Baloo	?	27.10.91
Brutus	5	Fossey	13/27-10-92	Fedora	No	
Kendo	2	Fossey	13/22-1-93	Fedora	No	
Fitz	1	Fossey	14/22-2-93	Fedora	Yes	1.4.93
Kendo	2	Belle	17/30-3-93	Bagheera	No	
Kendo	2	Belle	3/27-5-93	Bagheera	No	
Kendo	2	Ricci	10-7/13-8-93	Roxane	No	
Kendo	2	Ricci	16-8/5-9-93	Roxane	No	
Macho	3	Ricci	7/21-9-93	Roxane	No	16.1.94
Macho	3	Belle	9/19-11-93	Bagheera	No	1.2.94
Fitz	1	Castor	16/22-8-93	Cacao	No	
Brutus	5	Castor	31-8/10-9-93	Cacao	No	3.4.94
Macho	3	Héra	20-8/9-9-94	Hélène	No	
Brutus	4	Héra	12/29-9-94	Hélène	No	
Macho	3	Héra	1/12-10-94	Hélène	No	
Macho	3	Héra	17/31-10-94	Hélène	No	
Brutus	4	Héra	10/17-11-94	Hélène	No	
Brutus	4	Héra	17/21-12-94	Hélène	No	23.12.94
Kendo	2	Perla	4/12-10-94	Pandora	No	
Brutus	4	Perla	22/30-10-94	Pandora	No	21.1.95
Fitz	1	Mystère	7/12-10-94	Mozart	No	7.4.95

2.3 times per infant (range = 1 to 6). Four of the consortships happened within a month of the conception. If we used only this criteria, which was used at Mahale and Gombe, 31% of the infants would have been conceived during consortship. This is similar to what has been observed at Gombe and at Mahale. However, of the 10 infants for which we have genetic data as well as observations on consortship, only one was conceived during the consortship. Before conceiving these 10 infants, females had about 208 oestrus periods (one after a previous infant died, the other nine after successful weaning), and they went on 17 consortships. Thus, the females went on consortship in 8% of the oestrous periods and only one of them was successful (6%). Consortship seems to be a strategy with low success. And this is not due to the fact that males did not remain in courtship long enough, as 73% of them included the period near the ovulation (peri-ovulation period).

From another perspective, consortship seems to be a better strategy. Fertile oestrous females, when in the group, were seen to mate about 4 times per day, which means about 40 times for a 10-day full swelling period. Thus, the 10 females with genetic data mated about 7640 times during the 191 oestrous periods for which they remained in the group and conceived 9 infants. Thus, if we adopt the conservative figure of males mating as frequently with females when in consortship as they do when in a group, possible conception success per mating during consortship and in groups both equal 0.001. However, at Gombe, it has been observed that males mate less when in consortship (Tutin 1979; Goodall 1986). This does not consider the fact that the likelihood of mating in group settings differs between males (Table 4.4). Thus, consortship-mating might be a better strategy for some males than group-mating and is certainly associated with less intra-male competition.

A striking feature of consortship at Taï is that half of the males and 56% of the females were never seen to go on consortship (Table 4.5). We might have missed consortships for some females that were not regularly observed in the group, but we doubt that we missed any for the males. Thus, consortship is performed by only about half of the chimpanzees at Taï. How can we explain that? One trait shared by all consorting males and only by them is that they all reached the alpha position during some period of their life. Possessive-mating has been proposed as a strategy of alpha males at Gombe (Tutin 1979), whereas at Taï it is not a very efficient strategy. Indeed, in a low visibility environment it is almost impossible to prevent sneaky copulation by other males, unless you

Table 4.4: Mating occurrences with females of different reproductive status during 2 months in autumn 1993

	Age (years)	Social rank	Fertile female	Sterile female	Total
Brutus	42	4	19	12	31
Darwin	24	5	9	6	15
Fitz	18	1	12	8	20
Kendo	24	2	17	3	20
Macho	29	3	10	14	24
Sartre	13	6	0	1	1
Marius	11	7	0	1	1
Gipsy	13	8	0	1	1
Total			67	46	113

Table 4.5: Individual participation in consortship of all adult members of the Taï chimpanzee community during a 9-year period (1987–1995)

Male	(N = 7)		Female	(N = 16)			
Kendo	13		Belle	3		Agathe	0
Fitz	3		Bijou	3+1		Dilly	0
Macho	8		Castor	2+0		Gitane	0
Brutus	6		Fanny	3		Goma	0
Ulysse	0		Fossey	3+0		Loukoum	0
Rousseau	0		Héra	6		Ondine	0
Darwin	0		Mystère	1		Venus	0
			Perla	2+0			
			Ricci	3			

For females who had more than one infant, we provide the number of consortships for each of them.

guard the female for all the important oestrus days. Thus, consortship is a strategy used by high-ranking males, and females only agree to follow those males that have, are going to reach, or had once reached the alpha position. This reflects the females' willingness to follow males that have superior social potentialities than others. Thus, not all males at Taï can use all the strategies that might contribute to their reproductive success.

At Gombe, all males were reported to leave the group on consortship one to three times per year. It has been suggested that males are able to coerce females into following them, females not having much choice (Goodall 1986). Thus, two strategies are available to Gombe males in controlling a possible paternity, possessiveness for the alpha male and consortship for all of them.

At Taï, 44% of the females never went on consortship, and three mothers that did go went for only one of the two children they conceived (Table 4.5). We have been puzzled by this fact, but have not been able to find an obvious factor to explain this difference, as both young and old females went on consortship, and neither parity nor dominance of the females seems to explain the difference.

Promiscuous mating has been described as a situation in which no competition for mating is observed. However, such within-group mating does not mean free access to any female by all males. Within-group mating was followed in some detail during the autumn of 1993 (Table 4.4). All males mated with oestrous females. However, access was strongly affected by age: all late adolescent males in our sample, Sartre, Marius, and Gipsy, had very limited access to the adult and late adolescent females, independently of whether the females were having fertile or sterile oestrus. This contrasts with the Gombe situation, and indicates stronger competition between males for females. In Mahale chimpanzees, younger males copulate at higher rates than older males with all females, but dominant males have a clear advantage over both younger and older males for fertile females (Hasegawa and Hiraiwa-Hasegawa 1990; Nishida 1997). Thus, at Mahale and Taï, there are indications for high intra-male competition for fertile females.

How is within-group mating distributed among males? Table 4.4 presents data according to the reproductive state of the females. For example, Castor lost her son in March 1993, had her cycle again within one month, and must have conceived in July 1993, since she gave birth in March 1994. Castor, like most female chimpanzees, continued to have genital swellings after conception, and the males continued to mate with her and lead her on consortship as did the alpha male. Gala was having sterility problems (see Chapter 3), and Gitane had just lost her infant and the mating occurred during her first cycle after the event. Oestrus observed when the infant was less than 40 months old were considered as sterile, for the shortest interbirth interval observed at Taï was of 48 months (see Table 3.4).

The males seemed able to detect the reproductive state of the females and tended to copulate more with fertile females than with sterile ones. They mated 8.37 times per male with fertile females, but only 5.75 times with sterile ones (Wilcoxon signed rank test: $T+ = 13$, $p = 0.09$). Adolescent males were allowed to copulate only with sterile females. Direct aggression between males over oestrous females was not observed during this period, but has been seen regularly in other periods. Some female choice was seen as they refused some males access in some situations. Young Dilly was very afraid of Kendo and tended to avoid him. This made him more aggressive, and he was not seen to copulate with her at all. Ricci, the most popular female during this period, refused to copulate with Macho four times and his aggressive display did not change the issue. But overall, females accepted the copulation offers of the males in 113 out of 123 propositions (92%). Nevertheless, the adult males' access to females was not free of competition, some being clearly more successful than others. What determines the adult males' success? Neither age nor dominance rank nor coalition involvement (see Table 6.12, p. 121) explains mating success, suggesting the males' access to females as being free. However, for fertile females, adult males that are more involved in coalitions succeeded in securing more matings ($r_s = 1.0$, $N = 5$, $p < 0.005$). In this sample, coalition-active males were all current or former alpha males.

In conclusion, Taï male reproductive strategies resemble those observed at Gombe and Mahale in that both within-group and consortship-mating strategies were observed. But important differences exist between the three populations. High male competition for fertile females strongly restricts access to the females for young males at Mahale and Taï. Consortship patterns vary the most, from being used by all males at Gombe, to being used at Taï only by males that are, will be, or were alpha, to being very rare at Mahale. At Mahale, alpha males are more or less able to monopolize fertile females (Hasegawa and Hiraiwa-Hasegawa 1990; Nishida 1997), and they might feel that consorting the female is not necessary. In contrast, at Taï, alpha males cannot prevent former- or future alpha males from mating frequently with fertile females, so alpha males consort more regularly with them to secure a higher proportion of matings. At Gombe, all males go on consortship, which suggests less competition between the males.

Reproductive success in male chimpanzees

Based on behavioural observations, it is proposed that high rank at Mahale and consortship at Gombe are the two most efficient ways for adult males to increase reproductive success (Goodall 1986; Hasegawa and Hiraiwa-Hasegawa 1990). Adult females are thought to have limited possibilities of choosing consort partners at Gombe, but propose copulation more frequently to mature males when ovulation is approaching. In contrast, we suggest that Taï males rely both on consortship- and within-group mating, and that Taï females can choose males according to their potential for achieving alpha status. We shall see whether these conclusions based on behavioural observations are sustained by the genetic data.

Theoretically, we should expect males to attempt to mate with as many different fertile females as possible. Males should also attempt to mate with females from other social units than their own, as their reproductive success should increase with the number of partners. Females, on the other hand, are expected to be much more choosy and to select males providing them with more benefits. These benefits may be direct, in the form of social support or foraging gain, or indirect, in the sense that it is the infant that would profit from the female's choice, for example higher likelihood of acquiring genes for strong physical build, for intelligence, or for higher resistance to health hazards. Selecting the 'best males' could lead to such a benefit, but simply choosing different males could be favoured when genetic diversity within the offspring is favourable (Andersson 1994). Since traits important in male–male competition for females may not always be the same as those that are important to females selecting mate, a conflict of interest between the sexes might exist. We now look at paternity in Taï chimpanzees to see which males have higher success and whether females can choose their partners.

A paternity analysis requires that we have not only the genotype of the infant and its mother, but also that of all possible fathers that the mother could have mated with at the time of conception. The high mortality observed in the study community restricted the testing of all infants. Table 4.6 presents the data for 13 infants that lived from 1990 to 1995 for which the conditions for a paternity analysis were fullfilled. For a fourteenth infant, we could identify a possible father although we did not have all potential fathers (Lychee). Fifteen more infants were alive but we could either not genotype the infant (five cases), not genotype the mother (three cases), or not have all possible fathers (seven cases) (Gagneux *et al.* 1997). The analysis revealed the father of six of the 13 infants. As we had all the possible fathers in the community for all of them, we have to conclude that

Table 4.6: Reproductive success of the males in the Taï community

Name of mother	Name of infant	Birth date	Absence from group (seen/possible)	Number of males present at conception time	Father
Dilly	Dorry	24/11/91	3/6	8	Kendo
Fossey	Fedora	12/11/93	0/0	9	Fitz
Goma	Gargantua	21/9/91	0/0	9	Brutus
Vénus	Vanille	30/5/91	7/7	9	Ali
Castor	Congo	28/1/92	2/8	10	Stranger
	Cacao	14/3/94	15/25	9	Fitz
Loukoum	Lychee	1/3/86	?	10	Macho
	Lefkas	7/10/91	1/1	9	Stranger
Perla	Papot	19/11/89	0/11	7	Rousseau
	Pandora	2/5/95	0/2	5	Stranger
Belle	Bagheera	18/9/94	16/16	7	Stranger
Héra	Hector	10/12/80	0/45	9	Stranger
	Hélène	4/8/95	38/38	7	Stranger
Mystère	Mognié	31/7/90	2/4	9	Stranger

Under 'absence from group' are indicated the number of days during the 3 oestrous periods around conception time that the female was not seen in the group, or was possibly absent, which includes the days the chimpanzees were not followed by observers (days on consortship are excluded here).

the missing fathers are to be found in neighbouring communities: 55% of the infants were apparently fathered by males living outside the resident community of the mother.

Could this result be influenced by special circumstances prevailing within the study community? The data include the period of the Ebola epidemics when many individuals died. Might this have affected the female reproductive strategy? This seems unlikely, as four of the seven infants produced by extra group fathers were born before the first Ebola outbreak. In addition, the relatedness among the adult males in the Taï community is the same as the relatedness among the females, showing that gene flow between communities was the same for the two sexes for a long period of time, since adults of the community were aged between 15 to 40 years (Gagneux *et al.* 1999). Thus, the mating strategy of the females at Taï as revealed by the genetic analysis has prevailed for a long time, and was not affected by the recent decrease in the community.

This result cannot be compared with other chimpanzee populations for which such results are not yet available. Preliminary results from Gombe show that for the two infants whose father was identified, one was from another community (Morin *et al.* 1994*b*), and at Bossou, in a community with only one male, one infant out of four analysed was not fathered by this male (Sugiyama *et al.* 1993). If one keeps in mind that 13% of matings are known to occur with neighbouring males at Gombe, we can expect extra-group paternity to be generally widespread in chimpanzee populations.

Hence, male chimpanzee have problems controlling the oestrous females of their community. They cannot use the absence of the females to evaluate extra-group paternity, as females who conceived with males of the community were absent from the group during the oestrus around conception date for a similar amount of time (Wilcoxon sign rank text: $p = 0.67$) (Table 4.6). For example, three mothers were absent for less than five days during the three oestrus periods around conception, but conceived during their absence. In addition, with the exception of Héra and Belle, all were seen to mate with males from the study community during their oestrus periods. This result came as a surprise as we never saw stranger adult females visit the males of the study community (see Chapter 3) except during the territorial encounters and for a very short duration. This also stresses the point that we cannot attribute paternity on the basis of natural observations only. Females seem to be careful when engaged in extra-group copulation, as they might risk losing necessary male support if their behaviour is noticed. Similarly, in some species of monogamous birds, no extra-pair copulation was witnessed, although the frequency of illegitimate offspring was as high as 40% (Andersson 1994).

Can males judge the paternity of the infants? This is tricky to answer as adult males' affiliative behaviour with infants is rather limited. We know that some male birds with illegitimate infants in their nest worked as hard to feed them as males with only legitimate infants (Lifjeld *et al.* 1993). Grooming and play behaviour would be the two most obvious candidates, but this behaviour tends to be observed mainly in older males. This has been explained by the increased willingness of old males to engage in social behaviour because they no longer suffer from the social stress associated with an ambition to attain or keep high rank (Goodall 1986). Adoption of infants by males could be considered as the highest form of paternal behaviour. We saw it three times at Taï: Brutus adopted Ali and Tarzan when they became orphans as five-year-old infants, and Ulysse adopted Brando. For Ali and Brando, the genetic data clearly exclude Brutus and Ulysse as fathers (Gagneux *et al.* 1999). Therefore, it is possible that Brutus and Ulysse mistakenly undertook the heavy investment of adoption. Brutus was seen to support and share food with Ali for eight years. We cannot, however, exclude that adoption of male infants is a long-term investment to enlist their support once adolescent and adult.

Male reproductive success is complex to evaluate, as males may sire infants with females of four to five adjacent neighbouring communities. Therefore we need to differentiate between male reproductive success within and outside the natal community. What factors affect reproductive success of males within their natal community? Of the seven infants born within the study community that survived long enough to be genotyped, five were born to males that were, had been, or were still to become alpha. These were also the males that were followed by females on consortship. The two exceptions were Ali, 12 years old when he fathered Vanille, and low-ranking Rousseau, father of Papot. Whereas Fitz was actually the alpha male when he sired Cacao and Fedora, the others had been alpha but had lost their status by the time they sired the infant. The case of Brutus is striking, as he was alpha for so long and sired only one of the youngsters that we could test ($N = 24$), and he did that when he was as low as number five in the hierarchy. From this sample, neither mating frequency ($r_s = 0.44$, $N = 5$, $p = 0.37$), nor dominance rank ($r_s = 0.74$, $N + 6$, $p > 0.10$), nor frequency in consortship participation ($r_s = 0.28$, $N = 6$, $p > 0.5$) correlates with reproductive success within the community. Due to the small sample size we cannot exclude dominance as having an effect on reproductive success, and this corroborates with the observation that the ability to reach the alpha position does. At Gombe, dominance does not directly relate to reproductive success: for example Figan, a long-term alpha male, is suspected to have sired none or at most two infants, while at the same time Evered, a peripheral male, is supposed to have sired five (measured by consortship participation at the supposed time of conception, Goodall 1986).

What factors affect the reproductive success of males outside their natal community? So far data are lacking, but we can suggest the factors that might influence it. First, we need to know who decides the sexual interactions between individuals not familiar with each other. Males may coerce females during territorial encounters. However, when we saw sexual interactions in this context, it was always the oestrus females that walked towards the stranger males to copulate and not the other way round. With this in mind, the fact that Loukoum, Perla, or Mystère left the community for one or two days during their fertile period emphasizes that females may visit areas where they expect to find stranger males, Second, it seems likely that during inter-community encounters males can impress females from other communities by their powerful appearance and by their fighting abilities. It would thus pay for males to be active during such encounters as well as during patrols, for by doing so they contribute to their future reproductive success. It might be relevant here that at Taï the most active males in such encounters are neither adolescent males nor the older ones. In other words, the dominant prime males are also the most active inter-community fighters.

Female choice in chimpanzees

At Gombe, female choice was presented as limited, mainly since all males engage in consortship, which indicates their ability to impose their will on females (Goodall

1986). However, Gombe females could select males from other communities, especially because they associate less with resident males than do Taï females: two males have been seen consorting with females from neighbouring communities, and both were high ranking males (Goodall 1986). In addition, some female preference in mating with different male partners has been documented at Gombe and Mahale, although we don't know yet the effect it had on paternity (Tutin 1979; Nishida 1997).

At Taï, there are two strong arguments in favour of female choice. First, consortship is not a strategy successfully used by all the males, but only by those that were, are, or will be alpha. Although all the males tried to initiate a consortship by leading females away, only a few were followed. Without a certain willingness on the part of the females, a consortship initiation can easily be turned down by screaming that attracts all the males within auditory distance, putting an end to the initiative. An alternative to female choice would be that only potential alpha males are strong enough to impose a consortship, that is one independent of the willingness of the females and the presence or absence of other males. From Gombe it is reported that if there is no male within auditory distance, even low-ranking males can impose a consortship on a female (Goodall 1986). Hence, as apparently imposition of consortship seems possible by all the males, it cannot explain the observed pattern at Taï. Revealing in this context are the six consortships performed by Kendo between 1987 and 1989, when he was 18 years old and had not yet made it to the alpha position. It looks as if females chose him correctly, betting that he would make it to the top. Female choice seems to be an important aspect of consortship at Taï. Females would then be responsible for the fact that dominance status of the males at a given time is a weak predictor of reproductive success, as they judge the males' potentialities rather than their present status.

Second, the important proportion of extra-group paternities in the community supports the importance of female choice at Taï. Females leave the group for short periods of time in order to conceive with extra-group males and come back quickly. The short absence from the group suggests that they mate much more frequently within the group (Table 4.6). Nevertheless, the success of those extra-group copulations is high, which suggests biased sperm competition or selection in favour of the stranger male as has been shown in birds and humans (Baker and Bellis 1993; Olsson *et al.* 1996). The alternative explanation would be that stranger males are able to force females into copulation and that there is no female choice involved. This seems unlikely since the females' absence from the group was sometimes very short and mainly near ovulation time, a pattern more compatible with female choice than male coercion. Furthermore, the females' absence in our sample did not correspond to periods during which stranger chimpanzees were encountered, situations favourable for extra-group copulation. As most females transfer this leads to the intriguing possibility that by visiting neighbouring males, they may also come back to their natal community and mate with males they know very well, that is with whom they may even be closely related (father or brothers).

Two criteria seem to be used by females when choosing their mates. First, by selecting abilities found in alpha males (future, present, and past), they select potentially healthy, strongly built males with social skills. These males can provide both direct and indirect benefit to the females: direct in terms of support during intra-group competition for resources and indirect in terms of genes coding for healthy build and social gifts. Second, females also select for higher genetic diversity within their offspring as they visit other communities to search for more diverse mates. This mixed choice of mates was exemplified by Castor, Loukoum, and Perla (Table 4.6), who had their first infant with a resident male and their second with a stranger. Héra had her fourth and fifth infant with stranger males; the first three disappeared before we had access to genetic testing. Thus, females seem to select their mates so as to have the benefit of 'good' males from their community, as well as to achieve a high genetic diversity for their offspring. We can relate this point to the discussion about female transfer between communities (Chapter 3), where we suggest that females would transfer between communities to join the one with the largest number of male partners. At Taï, when the number of males decreased within a community, fewer females immigrated into that community, but the proportion of extra group paternity remained the same, which supports our interpretation of females following a mixed strategy in mate choice.

To conclude, the behavioural strategies used by males to increase their reproductive success, such as dominance rank, mating, or consortship frequency, are not reliable indicators of success as revealed by genetic analysis. Females, through active mate choice and possibly sperm selection, seem able to select males to increase both direct and indirect benefit. This includes selecting males from other communities. The high incidence of extra-group paternity may have important effects on the males' behaviour pattern in the territorial context (see Chapter 7), and explain why infanticide by adult males at Taï is absent.

Population differences in chimpanzee reproductive strategies

Table 4.7 summarizes the main differences we have noted between the three chimpanzee populations in terms of reproductive strategies. Within the community, the three populations differ in the males' access to females and in the frequency of use of the consortship. All aspects of consortship differ, including its general frequency, and the number of males and females involved. Between communities, the data are still fragmentary but indicate a possible population difference between Gombe and Taï. Supplementary analysis from all three populations is needed to understand the real importance of extra-group paternity.

Table 4.7: Reproductive strategies in three chimpanzee populations

	Gombe	Mahale	Taï
Within group behaviour			
Fertile female mating partner	All	Alpha	Coalition males
Consortship (frequency)	28%	8%	31%
% male in consortship	100%	10%	50%
% female in consortship	100%	3%	44%
Extra group behaviour			
Mating	13%	–	~1%
Paternity	1 case	–	54%

Two major differences emerge from this comparison: first, female chimpanzees of different populations use different strategies when selecting a mate within the community, and second, male–male competition for females differs between populations. Female choice through extra-group paternity cannot be discussed yet on a comparable level, as we do not know how frequent it is in other populations. From Taï, we know that females may choose extra-group fathers without the resident males being able to prevent them. This could be even easier for females at Gombe and Mahale where there is lower social cohesion than at Taï. However, within-group female choice seems to be affected by male–male competition: the higher the male–male competition, the fewer males are available to be chosen. At Mahale, inter-male competition seems to be very high, so high that within-group infanticide by males is apparently related to the possibility that a female has mated frequently with low-ranking males (Hamai *et al.* 1992). In this situation, females can only choose the alpha male. At Taï, the male–male competition is weaker, so that males that will be, are, or were alpha male achieve higher frequency of mating and consorting. In this situation, females have more partners to choose from and make use of the opportunity. At Gombe, male–male competition seems to be the lowest, but females do not seem to use all possibilities and Gombe males are able to impose their own choice on the females. The reason why male choice dominates reproduction at Gombe is unclear. One possibility is that the greater involvement of Taï females in the males' social conflicts through coalitions (see Chapter 6), compared with Gombe, could explain why Taï males cannot coerce females.

The main difference that remains to be discussed is that of male–male competition for females. At first, this seem puzzling, as the adult sex ratio is 1:1 at Gombe, while at Taï and Mahale there are three times more adult females than males. Since, in addition, Taï females are in oestrus for a longer period than females at Mahale or Gombe (see Table 11), we can understand that inter-male competition for females is lower at Taï than at Mahale. However, it should be higher at Gombe. One possibility is that there is inter-male competition not only for females but also for social rank. Within a stable community, male–male competition for social rank will increase with the number of males. Thus, Gombe males might invest proportionally more in the contest for social rank than Taï or Mahale males. Goodall (1986) suggests exactly this for some males at Gombe, in whom conflicts between the two types of competition were visible. Only after Evered was defeated by Figan for the alpha rank, did he concentrate on sexual success and was apparently able to become the most successful reproductive male of the community, while Figan, the alpha male, was able to sire at most only two infants.

In conclusion, at Taï the males' life history seems to be influenced from very early on by their mother's investment. High-ranking mothers invest more in male infants, who survive better, and who later on seem to be more interested in acquiring a high social rank. The mothers may help them actively in this quest. As a consequence, the sons of high-ranking mothers may have greater reproductive success than low-ranking males. The reproductive strategies used by males are influenced both by male–male competition and female choice. Because the levels of both differ in the three chimpanzee populations considered, different strategies are used. Taï females seem to have enough social power to choose partners, and this may affect males reproductive success decisively. Some males that have been observed to copulate regularly did not succeed in producing any off-spring. More data are needed from more individuals to strengthen these hypotheses. At the moment, they are compatible with the evidence obtained from behavioural and genetic analysis.

The males' dominance rank order seems to have only an indirect connection with reproductive success, for females choose males who have the potential for the alpha position, but they do not base their judgement on the actual position occupied by a male. The conflict between the males' strategy for securing paternity and that of females for choosing the sire of their offspring became obvious thanks to genetic analysis. Ovulation in females is hidden by prolonged sexual swellings, and copulations with stranger males at the decisive time are cryptic, making it difficult for males to control the females at the crucial time. However, we have the feeling that at the population level the dominant males may still be more successful than low-ranking ones, because of a possible greater success in extra-group paternity. A planned analysis of about twelve neighbouring communities will increase our knowledge on the outcome of this conflict of interest between the sexes.

The social unit and the reproductive unit do not overlap fully in wild chimpanzees, and this has consequences for our understanding of the evolution of reproductive strategies, as well as of many social behaviours, since the outcome of any strategy has to be considered for the reproductive unit rather then for the social one. For example, the benefit of being a dominant male in chimpanzees has not always been clear (Goodall 1986). As we have seen, the advantages are at best weak within the social unit, the chimpanzee community. However, from the point of view of evolution it makes more sense to consider the reproductive level, for which data are badly needed to answer this question. Given the growing evidence of extra-group paternity in birds and other vertrebrates, extra-group paternity could change our understanding of many aspects of the evolution of sociality.

5 *Social structure of the Taï chimpanzees*

Forest scene: On 3 January 1986, we followed Falstaff, the oldest male of the community. He was on his own, eating the young leaves of saplings while he moved on slowly. It had not been clear what direction the chimpanzees would take. Now at 6.30, we heard pant-hoots from the west, while in the south we heard the young adult male Snoopy drum loudly. Falstaff sat and listened to his group mates. A third party was vocalizing farther away north. Loukoum, Gauloise, and Goma passed, heading northwards with their youngsters Lychee and Gallus wildly chasing one another. As they disappeared, Mystère passed by, running after them. We heard Schubert, the beta male, drum loudly to the north and Macho bark in support. Snoopy replied by drumming three more times. To the west, someone was pant-grunting intensively at a dominant individual. Falstaff started slowly to move in this direction, 10 minutes later silently joining Ondine and Salomé, the two highest-ranking females, who were feeding on fruits of Dacryodes.

At 8.00, there was a short drumming quite a way farther west. By the loud response of many chimpanzees dispersed around us, we guess it was from Brutus, the alpha male and leader. Within three minutes, Falstaff, Ondine, and Salomé with their infants Orée and Sartre, moved up the crest westwards. Soon the sound of nut-cracking became audible. We were in an area rich in Coula trees and the forest resounded as if carpenters were at work. Orée and Sartre speeded towards a group of nut-crackers and greeted them enthusiastically. We saw Gauloise, Goma, and Tosca at work in the trees, and could hear five more individuals. Ondine and Salomé joined them, both were already carrying hammers. Some of them stayed in this Coula region for four hours. At 12.10, Brutus drummed clearly further south-west, and Ondine and Salomé with six other females moved there. We joined Falstaff, Schubert, Rousseau, and Macho who groomed each other on a large log. The females settled and at once the infants started an animated play session. Suddenly, all the chimpanzees got up and looked southwards, the youngsters rushed there and we heard greeting. In a silent and impressive display, Brutus rushed through all the chimpanzees, who hurried out of his way. Brutus then sat, hair bristled, in the middle of the log and Falstaff and Schubert came to greet him and they started to groom. The females Tosca and Saphir had been with Brutus and joined now the group of females.

At 15.45, Brutus moved southwards and everybody followed rapidly. Soon Brutus, Falstaff, Tosca, Ondine, and Salomé climbed a fruiting tree, while some others started again to crack nuts nearby.

The above is a description of a typical day for the Taï forest chimpanzees. Varying parties fuse and disperse as they forage along, maintaining cohesion by loud calls and drumming. Brutus, leader and alpha at the time, mostly moved ahead and indicated the general direction. In this chapter we aim to analyse these grouping patterns, to find what affects them and how they differ between chimpanzees populations.

The social structure of animals is the result of the attempts of individuals to optimize their survival and reproduction. As group living incurs obvious costs, including higher intraspecific competition for resources and higher risks of parasite transmission, discussion has centred around the benefits that favour the evolution of sociality. Three main benefits have been proposed: decreased predator pressure, better exploitation and protection of food resources, and increased cooperative behaviour, like hunting or cooperative breeding. For specific situations, other factors have been mentioned, for example rearing strategies (Dunbar 1988). Predation pressure has been shown to favour larger group size in many animals including fishes, birds, grazing mammals, and primates (van Schaik 1983; Dunbar 1988). Similarly, seasonal variation in food supply has been shown to have a marked influence on group size in Hamadryas, gelada and anubis baboons, and spider monkeys. However most observers study animals that already live in groups, making it difficult to investigate the factors that originally led to the appearance of social groups. What we can attempt is to study the factors that affect the existing social groups observed in the wild.

All chimpanzee populations live in a 'fission–fusion' social structure (Kummer 1971). This means that group members gather in unstable, temporary groups that usually include only a small subset of the whole community. Goodall (1968) named a 'community' the entity of all individuals seen for long periods of time in various temporary subgroups called 'parties'. The community is equivalent to the 'unit-group' used for the Mahale and Bossou chimpanzees (Nishida 1968; Sugiyama and Koman 1979a). The fission–fusion system has also been observed in eight large arboreal primates and is thought to allow more flexibility in exploiting resource patches of different sizes in a species free of predation (Dunbar 1988). That a species is free of predation is hard to confirm: for predation is a rare event and must be so for the prey to survive, and the presence of human observers may directly interfere with predation when the prey, but not the predator, is habituated. For example, in a study on vervet monkeys, disappearances attributed to leopards were 3.6 times more frequent when human observers were absent than when they were following the monkeys (Isbell and Young 1993). Evaluating the importance of human presence is difficult, but its effect is to underestimate the importance of predation in primates (Cheney and Wrangham 1987). At Taï, we noticed only after some years that predation by leopards on chimpanzees existed and was the main cause of mortality during a certain period (Boesch 1991c; Chapter 2). Thus, species with a fission–fusion system are not guaranteed to be free of predation, and we still need to explain its advantage.

Many attempts have been made to single out the factors responsible for social organization in chimpanzees. Goodall (1986), after analysing two years of data from Gombe, proposed that sex was the single most significant factor in shaping grouping patterns. Wrangham (1986) suggested that the greater dependence of bonobos upon terrestrial herbal vegetation in contrast to chimpanzees explained their larger party sizes, based on comparisons between Gombe chimpanzees and Lomako bonobos living in two very different habitats. Subsequent data showed that neither population relies much on this vegetation (Malenky and Stiles 1991). Then, it was proposed that the size of food patches and their distribution might explain the differences (White and Wrangham 1988). Comparisons between bonobos of Lomako forest and chimpanzees of the Kibale forest did not support this hypothesis (Chapman *et al.* 1994). It was then proposed that social organization reflects the need to adapt to extended periods of food shortage when individuals are forced to be in small parties (Malenky 1990 in Chapman *et al.* 1994). Food shortage is supposed to be less of a problem in forest environments, and chimpanzees living in forests are expected to occur in larger parties. In the wild, animals have to deal simultaneously with many problems in order to survive and reproduce. We should thus expect that the solutions they adopt reflect the balance between several needs. With this in mind we now analyse the social structure of the Taï chimpanzees and compare it with that of other populations.

Fission–fusion in Taï chimpanzees

Party size and composition in a fission–fusion system change frequently over time, and many different methods have been used to study them. Ideally, target individuals, that is, a representative sample of individuals of different age/sex classes would be followed simultaneously from the morning nest to their evening nest, and all parties they join recorded. This has never been done, for only in a few studies is it possible to follow targets all day long, and also the sample size is limited, since one observer can follow only one individual per day. Therefore, methods vary from opportunistic sampling of parties, whenever one was observed in good condition (Nishida 1968; Goodall 1968), to short 10-minute focal animal sampling (Wrangham *et al.* 1996). Different methods may produce different results. At Taï, a comparison between continuous recording and 15-minute-interval samples from the same day with the same target produced significantly different results ($p < 0.001$). The second method underestimated small party sizes and overestimated mixed party types. Thus, we used continuous sampling of the parties containing the target individual and recorded when they joined or left parties. A party includes all the individuals in visible contact with one another.

In 1987, party size and composition were recorded with fully identified individuals. For travelling periods, in which chimpanzees normally progress more or less in a line, and the observer is at the rear, we checked party composition whenever possible. If a change occurred we entered it for the minute following the previous count. Joining and leaving parties were given an arbitrary one minute duration. We followed individuals for hours or days and changed targets according to our interest in specific behaviours, such as hunting in the wet season and nut-cracking in the dry season. Previously, we had analysed only initial party counts, plus a new one for each change of party (Boesch 1991c, used also by White (1988) and Wrangham *et al.* (1992)), but here we include the time a given party remained constant. Small parties last for a shorter time than large ones (Fig. 5.1), and excluding the time overestimates the importance of small parties. This method gives a precise image of the parties that are seen in a population as well as the duration of each party size.

Chimpanzee grouping patterns

We first provide a general description of the social grouping patterns of the Taï chimpanzees for comparison with other chimpanzee populations. Then we analyse the groupings at Taï, to find out what influences them and to test some theories on the evolution of sociality in chimpanzees.

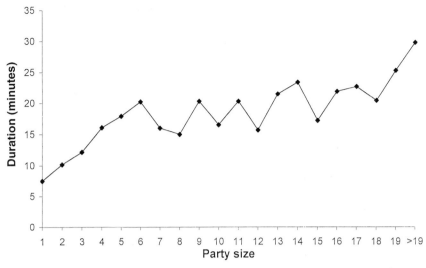

Fig. 5.1: Duration of parties of different sizes in Taï chimpanzees ($r_s = 0.81$, $p < 0.001$).

The fission–fusion nature of Taï chimpanzee society is presented in Table 5.1. Parties last on average 24 minutes. The fluidity of parties is observed for all activities, party type, and size, although there are variations that we discuss later. This fluidity might be explained in part by the fact that many parties forage in auditory contact with each other. The group containing all the parties foraging in auditory contact usually includes more than 80% of the community members (Boesch 1991c). Figure 5.2 shows the frequency of all party sizes we followed from August 1987 to February 1989. The community then included 79 individuals, mean party size was 10 individuals (median = 8, mode = 6) and the largest party seen contained 48 chimpanzees. This reflects the basic structure of the

Table 5.1: Mean duration of parties in different chimpanzee populations

Population	Duration of parties (minutes)
Chimpanzees	
Taï	24
Gombe	69
Bossou	126
Budongo	14
Bonobos	
Wamba	86
Lomako	102

Bossou: Sugiyama (1984, 1988)
Gombe: Halperin (1979)
Lomako: White (1988)
Wamba: Kuroda (1979)
Budongo: Reynolds (in press)

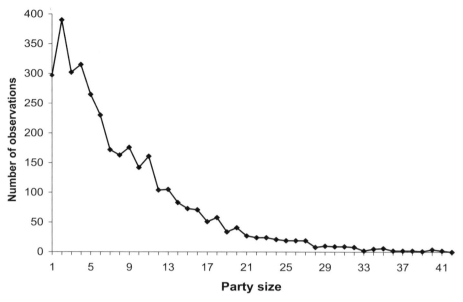

Fig. 5.2:   Party size in Taï chimpanzees: 1987–1989, *N* = 3532.

Taï chimpanzee community: high fluidity, parties rarely including more than a third of the community members, and an average party size of 10 chimpanzees.

Party size

Other chimpanzee populations have average parties of about five individuals, smaller than has been observed for Taï chimpanzees (Table 5.2), and bonobos tend to have larger parties than chimpanzees. This important difference has been explained by some authors as being species specific and requiring a special explanation (White 1988; Wrangham 1986; Kano 1992). However, such a comparison is strongly influenced by the size of the community, and we have to expect that smaller communities form smaller parties. For example, the Bossou community has 20 group members and cannot form parties larger than 20 individuals, whereas in the Taï community with 80 individuals, such party sizes account for 7% of the observations. When comparing party size between communities of different sizes, using the relative size of the parties (mean party size divided by the community size) might thus be more accurate. Chimpanzees have a relative mean party size between 9 and 21%, whereas bonobos have one between 21 to 89% (Table 5.2). This difference is explained primarily by community size: the smaller the community, the larger the mean relative party size (Fig. 5.4). A similar comparison with mean party size reveals no correlation with community size, confirming that party size is a poorer predicator of community size than relative party size. If we exclude Wamba, an exponential fit describes 73% of the variance in relative mean party size. The bonobo community of Wamba is special, as its mean relative party size increased from 29% in 1975 (Kuroda 1979) to 92% in 1984 (Furuichi 1989). The heavy provisioning that started in 1977 may be the most important factor in explaining such a change in social grouping, and as it is a

Table 5.2: Party size for different chimpanzee populations

Population	Mean party size (a)	Community size (b)	Relative mean party size (a/b*100)	Proportion of parties containing <25% of the community members
Chimpanzees				
Bossou	4.0	20	20	–
Budongo	6.0	50	12	–
Gombe	5.6	60	9	95%
Kibale	5.1	27	19	–
Mahale	6.1	29	21	58%
Taï	10.0	76	13	77%
Bonobos				
Lomako	5.4			
Blobs	4.3	10	43	–
Hedons	7.1	22	32	–
Rangers	9.7	21	46	–
Lomako (93)	5.8	36	16	–
Wamba				
Kuroda	16.9	58	29	50%
Furuichi		35	92	–

Bossou: Sakura (1994).
Budongo: Reynolds (in press).
Gombe: Goodall (1968).
Kibale: Chapman *et al.* (1994) (dependent infants and juveniles were not considered by the authors, thus we excluded them from the community size count).
Lomako: Data are from three different communities, White (1988).
Lomako 93: Fruth and Hohmann (pers. comm.), who followed the community previously called the Rangers.
Mahale: Nishida (1968).
Wamba: Kuroda (1979), Furuichi (1989).

Fig. 5.3: A mothers' party with four adult females resting on the ground, while some of the youngsters play together.

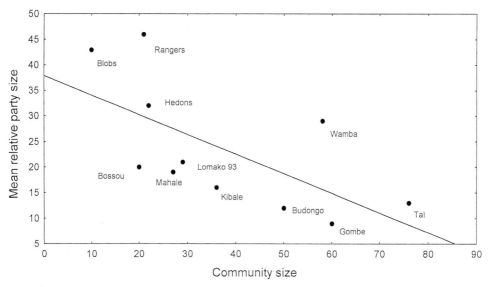

Fig. 5.4:   Comparison of relative party size for communities of different sizes in chimpanzees and bonobos.

human artefact we should avoid generalizing from such data. For the other populations, it seems that the smaller a community is, the more frequently individuals are found in the same parties. In other words, the smaller the community, the higher the cohesion.

However, if we compare the proportion of parties containing less than 25% of the community members, Mahale and Taï chimpanzees, and Wamba bonobos before artificial provisioning (Kuroda 1979) tend to have similar proportions (Table 5.2). Gombe chimpanzees fall at the lowest extreme with 95% of the parties having less than 25% of the community members, thus presenting a more skewed distribution towards the small end in the distribution of party size than other populations.

It seems that small communities in chimpanzees and bonobos retain a fission–fusion structure, but it loses much of its flexibility and the parties remain stable for much longer periods of time than for large communities (Table 5.1). We lack data for some populations, but there is a trend in this direction. At Wamba, once provisioning was successful, the fission–fusion structure tended to disappear; parties lasted for weeks and included 91% of all community members (Furuichi 1989).

Community size affects the fluidity and cohesiveness of the community in both chimpanzee and bonobo populations. This tendency for smaller communities to become less fluid indicates that competition for food is not the only factor affecting fission–fusion structure.

Party composition

The second basic aspect of social structure is the composition of the chimpanzee parties. Taï chimpanzees are found mainly in mixed parties containing adult males, females with their offspring, infantless and subadult females (Table 5.3). Unisexual parties account for only one third of those observed, stressing the cohesion between the sexes in this

Table 5.3: Composition of the parties in Taï chimpanzees from August 1987 to February 1989

Party type	Frequency (%)	Duration (%)
Mixed	50.9	60.3
Females	17.7	12.6
Males	20.0	21.5
Adults	0.3	0.3
Lone males	6.5	3.3
Lone females	4.5	1.8
Total	3532	59 236 mins

community. In addition, the low frequency of adult parties made up of males and only those females who are without dependent offspring can be explained by the high number of females with babies (on average 24 of the 27 adult females). Among adults, females were observed 82% of the time in association with males, and males 74% of the time with females. If individuals were associating at random, we would expect that in the Taï community with 27 adult females and 7 adult males, female parties would be the most common, followed by mixed parties, and male parties would be rare. At Taï females associate less than expected with one another, and more with males, whereas males associate more with females than with males (Boesch 1996c). Thus both sexes prefer to associate with each other. Taï chimpanzees appear to live in a social system in which both females and males associate with each other most of the time.

The proportion of females in mixed parties increases as party size increases (Table 5.4), tending towards the adult sex ratio of the community. Thus, at Taï, females are mostly found with males, and males have a stronger tendency than females to be in mixed parties.

In comparing party type between chimpanzee communities, we found that mixed parties are always the most frequent ones observed (Table 5.5). The only exception is the Bossou community which included only one adult male and 10 to 12 females, explaining the low frequency of mixed parties. Taï chimpanzees have the same proportion of mixed parties as the Lomako bonobos, but more than were observed in Gombe and Mahale chimpanzees (Boesch 1991c). This difference is in reality even larger with Gombe chimpanzees and the Lomako bonobos, as these two populations have an adult sex ratio near 1:1, whereas in Taï and Mahale chimpanzees it is strongly biased towards females. We can correct for this by including the sex ratio in the measure of the frequency of occurrences of mixed parties (mixed party corrected, Table 5.5). From this we see that Taï

Table 5.4: Adult sex ratio of mixed parties of different sizes in Täi chimpanzees between 1987 and 1989. The adult sex ratio of the community was 0.25.

Party size	Sex ratio (male/female)
1–10	1.34
11–20	0.90
>20	0.54

Table 5.5: Party types in different chimpanzee communities

Population	Adult sex ratio (a)	Mixed parties (b)	Mixed parties corrected (b/a × 100)	Adults	Males	Mothers	Lone
Chimpanzees							
Bossou	0.12	42	3.50	–	–	49	–
Budongo	1.09	41	0.37	2	17	23	6
Gombe	0.80	30	0.37	18	10	24	18
Kibale	0.66	52	0.78	–	–	5	23
Mahale	0.66	52	0.78	4	11	13	21
Taï	0.25	61	2.44	0.3	22	12	5
Bonobos							
Lomako 87	0.62	68	1.09	8	5	5	14
Wamba	0.66	74	1.12	3	3	5	6

The proportion of mixed parties correlates with the sex ratio of the population (chimpanzees: $r_s = -0.98$, $N = 6$, $p < 0.001$; chimpanzees and bonobos: $r_s = -0.92$, $N = 8$, $p < 0.001$).

males most actively seek contact with other males and females; that Mahale and Kibale males are intermediate; and the Gombe and Budongo males seek the least contact with females. In chimpanzees, sex ratio of the population explains the proportion of mixed parties observed: the more females are present (that is, the smaller the sex ratio), the less frequent are mixed parties. At Bossou, only one male was present within the community, compared with Taï with six males, emphasizing that it is the relative proportion of the two sexes and not the absolute number that is important.

In conclusion, this comparison of party size and type in chimpanzees and bonobos indicates that community size and sex ratio are important factors in determining the differences observed between populations. Community size explains most of the differences seen in party size, whereas the sex ratio prevailing within a community affects the fre-

Fig. 5.5:   Part of a large mixed party grooming on a large fallen tree.

quency of mixed parties. The more females are present in the community, the more males associate with them. Although our knowledge of the bonobos remains limited, these conclusions apply to both known populations in this species and, therefore, stress the point that both ecological and social factors affect party composition in the two species in a similar way.

Factors affecting party size in Taï chimpanzees

The factors most often considered as influencing the grouping patterns in chimpanzees are the availability of food, predation pressure, and sex. To assess the importance of each of these factors for party sizes at Taï, we first look at the activity budget of the community, then at the other factors affecting party size in Taï chimpanzees.

The Taï chimpanzee time budget varies with the seasons and over the years (Table 5.6). In general, they spend less time feeding on plants and eat more meat in the wet season than in the dry season. Meat eating is, however, an important activity all year round, taking up to 9% of the time. Taï chimpanzees spend the same amount of time feeding as the Gombe chimpanzees, but less than those of the Kibale forest (Chapman *et al.* 1994).

For this analysis, we compare two seasons over a two-year period (August 1987 to February 1989): the wet season from August to October, which is the big rainy season, and the dry season from January to February/March, which is the long dry season in the Taï forest (see Fig. 1.3). The rainy season is characterized by heavy rainfall almost daily, by a slow increase in the general availability of food, and by low temperature, and it is the major hunting season (Chapter 8). The dry season has become very pronounced since 1983 with sometimes no rain at all for up to a month, high temperatures (except for about two to four weeks when the 'Harmattan', the cold desert wind, reaches the forest), high general availability of fruit, including the abundant nuts of *Coula edulis* that the chimpanzees pound with tools. By 1987 the chimpanzees were well habituated and not disturbed by our presence in their foraging activities. Solitary individuals were under-represented in our sample because during the rainy seasons we concentrated on the males in order to follow hunting behaviour, and during the dry seasons we concentrated on the mothers to study nut-cracking behaviour (both mothers and hunters can spend time alone).

For statistical analysis, independence of samples is required. This is difficult with such data. To improve it, we used only one data point per activity, choosing the largest party size during the time the target animal spent for a given activity. For example, if the target joined a party at a tree where three other parties later joined and the target left the tree only after two parties had left before him, we took as the duration the total time the target was in the tree, and as size the largest count in that tree during the time the target was

Table 5.6: Activity budget of Taï chimpanzees between August 1987 to February 1989, 'rest' includes social time

	All	Wet 1987	Dry 1988	Wet 1988	Dry 1989
Feed	45%	35%	56%	47%	45%
Rest	22%	25%	16%	20%	31%
Travel	22%	23%	17%	23%	22%
Meat	9%	14%	10%	10%	5%

recorded. For feeding, resting, and meat eating this greatly reduced the sample size. In addition, all parties joining and leaving the target's party were excluded from the analysis, as they directly resulted in an increase or decrease of the party they left or joined. In this way, the sample size was reduced from 3464 to 1335 parties. This method (method A) might overestimate party size, and we tested this effect by comparing it with method B) where a data point was entered for each change observed in the party of the target followed (Tables 5.7 and 5.8). We see that, although method A might bias the data towards larger party size, this effect is small, except for meat eating episodes (less than 5% increase on average party size).

When corrected for interactions, year, and season, each taken alone had no significant influence on party size, whereas the duration of the party, and the chimpanzees' activity had a highly significant role on the party size (see Boesch 1996c for the statistical results of this ANOVA test). As already shown in Figure 5.1, the larger the party, the longer it remains stable. Meat-eating parties are the largest of all and remain so all year round.

Seasonal variations in party size differ with the year. In 1987–88, the parties were smaller in the dry season than in the wet season, whereas the opposite was true the next year (Fig. 5.6). The dry season of 1988 was one of the worst in twelve years with almost no Coula nuts and barely any other kind of fruit production. The chimpanzees scattered in small parties to feed on leaves and the little fruit available. The difference observed in the two wet seasons is more difficult to understand as fruit production was very similar and we could find no obvious cause. Seasonal variations in party size were also found in the

Table 5.7: **Mean party size of the Taï chimpanzees when engaged in different activities for the period August 1987 to February 1989 using two different methods of data collection (see text for further explanation)**

Activity	A Size (N)	B Size (N)
Travel	9.0 (487)	8.2 (805)
Feed	10.1 (489)	9.9 (895)
Hunt attempt	10.2 (36)	9.1 (45)
Rest	12.2 (281)	11.9 (468)
Hunt	20.8 (42)	15.3 (141)
All	10.48 (1335)	10.0 (2354)

Table 5.8: **Mean party size in Taï chimpanzees when feeding in trees of different sizes for the period between August 1987 to February 1989 using two different methods of data collection (see explanation in text)**

Tree size	A Size (N)	B Size (N)
Small	8.2 (176)	8.27 (217)
Medium	10.8 (113)	10.55 (128)
Large	11.4 (199)	10.19 (276)

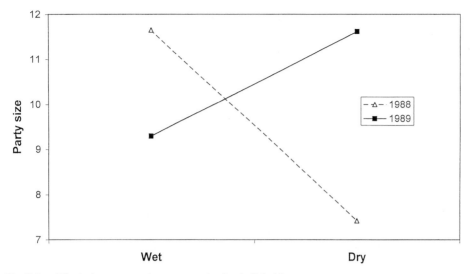

Fig. 5.6: Effect of season and year on party size in Taï chimpanzees.

activities. Resting and travelling parties were larger in the dry seasons, whereas the opposite held for the other activities (Fig. 5.7). If we remember that the activity budget is irregular and that feeding tends to occur for longer periods of time in the wet season, then we can speculate that food is less abundant in the rainy season and that once the chimpanzees find a patch, they feed for longer periods and in larger parties. But this does not explain the trends observed for the other activities. It appears that food availability influences party size, especially for feeding parties, but other factors also have a strong influence.

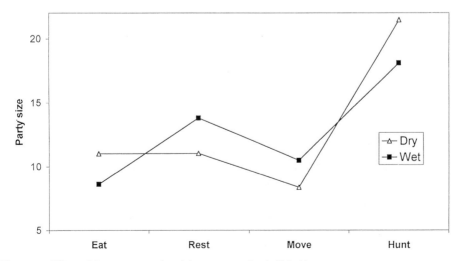

Fig. 5.7: Effect of the season and activity on party size in Taï chimpanzees.

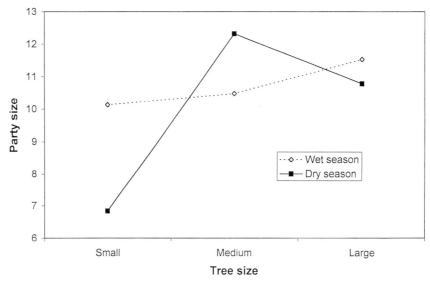

Fig. 5.8:   Effect of season and tree size on feeding party size in Taï chimpanzees.

Because food availability influences some party sizes, it is important to see if it is particularly tree size that does so. If larger trees are producing fruit or leaves eaten by the chimpanzees in the wet season and the chimpanzees feed in larger parties in larger trees, this would partly explain the previous results. We classified trees in three different categories (Boesch 1996c). The larger the tree, the larger the party that fed in it (Table 5.8) (for statistical results see Boesch 1996c). In addition, parties were larger in the wet season than the dry season for small and large tree sizes. Whereas, parties were smaller in medium trees during the wet seasons than in the dry seasons (Fig. 5.8). The year also affects party size, since in the first year all parties were smaller in the dry season, whereas in the second year parties were larger for medium and large trees in the dry season (Fig. 5.9). This illustrates that tree size only partially determines feeding-party size. The dry season in 1989 had a much higher availability of fruit than it did in 1988, and this affected feeding-party size, especially for larger trees, suggesting that party size is influenced more by general fruit availability than by tree size in the dry seasons. For the wet seasons, neither tree size nor fruit availability, which was very similar in the two years, seems to influence party size. Factors other than tree size certainly influenced party size in our sample periods.

Sex has also been proposed to influence grouping pattern in chimpanzees (Goodall 1986). Figure 5.10 shows the number of oestrous females observed in the community for all months of the year. The average number of oestrus females is five per month from September to April, fewer between May and August. Although for the two years considered, more females were in oestrus in the wet than the dry seasons, no seasonal differences in party size could be explained by this trend (Fig. 5.6). However, when we combined females in oestrus with the difference in food availability, we observed that wet seasons have more oestrous females and tend to have larger parties, and that a higher

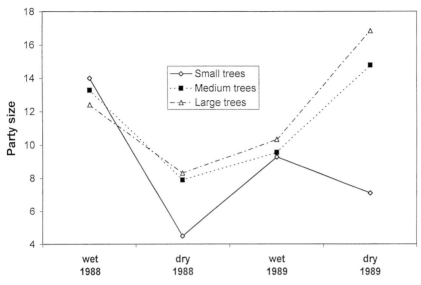

Fig. 5.9: Effect of the year, the season, and the tree size on feeding party size in Taï chimpanzees.

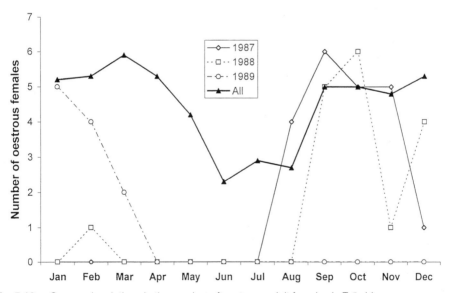

Fig. 5.10: Seasonal variations in the number of oestrous adult females in Taï chimpanzees.

fruit availability in the dry season tends to compensate for the low number of oestrous females. In the dry season 1989, we observed the same party size as in the wet season 1988, despite a strong difference in the number of oestrous females (Fig. 5.6). In support of this conclusion, very small party size was observed during May–July 1988 (5.2 individuals: Doran 1997), when both the general availability of food and the number of oestrous females were very low. However, party size differed significantly between the two wet seasons, and this is not explained by the number of oestrous females. Therefore, in addition to general food availability and the presence of oestrous females, which explain part of the variation in party size in Taï chimpanzees, another factor must be found.

During the wet season, Taï chimpanzees hunt nearly every day (Chapter 8). Successful hunts result in prolonged meat-eating episodes that last much longer than any other activity in chimpanzee social life and lead to especially large assemblies (Table 5.7). Although the main interest of these parties consists in meat-eating, these parties are also the occasion for many social interactions between individuals that do or do not eat meat. Hunting might help to explain the large party sizes during the wet seasons: the average hunting rate during the wet season for the whole study period was 0.89 hunt per day, whereas it was 0.25 for dry seasons. The hunting rate varies with the year for it was 1.32 hunts per day in 1988 and only 0.78 in 1989. This may cause the last difference in party size we need to explain between the two wet seasons.

In conclusion, party size in Taï chimpanzees is directly influenced by fruit availability, sexual opportunities, and hunting rate. These factors explain most of the variations we observed. When all these factors are low, party size can become quite small and the duration of social interactions decreases sharply; this happens mostly during June and July. Throughout the rest of the year, one or more of these factors are higher and parties remain large. Thus, the fission–fusion system gives the chimpanzees the flexibility needed to react precisely to variations in their environment in such a way that party size remains quite high except when two or more factors do not favour sociality. When all three factors are unusually low, as in the first part of 1988 (see also Doran 1997), grouping patterns in Taï look similar to those reported for eastern chimpanzees, but most of the time conditions are such that party size remains larger and allows a close association between the sexes.

Hunting rate and sexual opportunities seem to compensate for a lower availability of fruit in the wet season, whereas high fruit availability can compensate for low hunting rate and limited sexual opportunities in the dry season. Factors that have been proposed in other populations to play a role in social grouping, such as food-patch size (measured by the tree size), or proportion of feeding on terrestrial herbaceous vegetation (that represents only 3% of the feeding activity at Taï), have a very limited role in explaining party size at Taï, because they are directly related to general food availability. A mono-factorial approach is of little value in explaining grouping patterns in Taï chimpanzees, for ecological parameters (fruit availability and hunting rate) and social parameters (activities and sexual opportunities) interact to explain grouping patterns.

Analyses of grouping patterns in other chimpanzee populations do not always consider the impact of such factors. However, important differences in party size have been reported between the seasons at Kibale and Gombe, suggesting that general food availability is an important factor as well as sexual opportunity that has also been proposed to explain yearly and seasonal variations in party size at Gombe (Goodall 1986; Stanford *et al.* 1994a). Bonobo and Taï chimpanzee females are both characterized by having a longer period with oestrous swellings after giving birth (Chapter 3), giving males in those populations more sexual opportunities than in other populations, which should lead to large party sizes. From Table 5.2, we see that bonobo communities of 20 individuals do tend to have larger party sizes than chimpanzee communities of the same size (Bossou, Kibale, and Mahale). A recent analysis of party size in Gombe chimpanzees suggests that high fruit availability, high hunting rate, and a high number of females in oestrus coincide in the autumn when party sizes are the largest (Stanford *et al.* 1994a). Thus, preliminary information suggests that the four factors affecting party size at Taï are also important in explaining variations of party size in other chimpanzee populations.

Association in wild chimpanzees

Owing to the flexibility of the fusion–fission system, chimpanzees can choose with whom they want to spend their time. Association between two individuals is therefore an important aspect of their sociality. For example, coalition partners are expected to be preferred associates, so they are together whenever a situation arises that requires the coalition to work, and individuals will associate with those from whom they benefit the most or with whom they are the most at ease.

We present data based on the dyadic association index (DAI) that measures the time that individual A is seen in the same party as individual B:

$$DAI_{AB} = \frac{Time_{\text{A+B seen together}}}{Time_{\text{A seen without B}} + Time_{\text{B seen without A}} + Time_{\text{A+B seen together}}}$$

A and B are considered as being together when both can see each other, that is when they are members of the same party. Some researchers have used shorter distances between the two individuals, measuring proximity rather than association. We used the association index because this criterion has often been used in other studies and proximity does not add much to it. The data from Bossou, where only close proximity was considered, are excluded here.

Between August 1987 and February 1989, associations between males were about three times stronger than associations between females or between females and males (Table 5.9). Most male DAIs were larger than 24%, while most DAIs between females were smaller than 24% (Table 5.10).

At Taï, as in other chimpanzee populations, males are more strongly associated with one another than are females with other females and males with females (Table 5.9). Because males remain in their natal community and females transfer between groups, males are more intimate with one another. They share similar interests in hunting, in being around oestrous females, and in cooperating to protect the territory against macro-coalitions of neighbouring males (see Chapter 7). The higher DAIs observed between males thus reflect their common interest in many aspects of their social life.

Detailed data exist from three chimpanzee communities on the intensity of DAIs and allow to investigate how the different dyads associate. Eighty-nine per cent of the Taï male dyads are seen together for at least 25% of the time, whereas this proportion falls to 24% of the dyads at Gombe, and none are so frequently associated at Mahale (Table 5.10). Similarly 52% of the Taï female dyads are seen at least 10% of the time together, but only 8% at Gombe (data for females are not available from Mahale). This corroborates

Table 5.9:   Average dyadic association indexes (DAI) for individuals of the same sex or between the two sexes observed in different chimpanzee populations

Population	Male–Male	Male–Female	Female–Female
Gombe	0.24	0.07	0.05
Kibale	0.18	0.1	0.08
Mahale 1968	0.76	0.38	0.39
1996	0.10	≈0.05	≈0.05
Taï	0.35	0.12	0.11

Gombe: DAIs are calculated from data of the 3 following years; 1978, 1979, and 1981 (Goodall 1986, Appendix d).
Kibale: Data from July 1988 to September 1989 (Wrangham *et al.* 1992).
Mahale: Nishida's data (1968) should be considered with caution because they come all from the artificial feeding area, where association is strongly affected by this artificial situation. Nishida and Hosaka (1996) provide more reliable data on the DAI for the adult males during long follows (calculate from Table 9-1 and 9-12a: range = 0.23–0.04) with a figure 7 times lower than the 1968 estimates. Assuming that the early figures for the females were equally inflated, we estimate the female DAIs with a "≈" (No data on female associations are yet available).
Taï: Data from August 1987 to February 1989.

Table 5.10: Intensity of dyadic relationship for three chimpanzee populations classified according to the time the individuals spent associated: figures for each population are the number of dyads with DAI of the given value (e.g. the 7 males at Taï form 28 different dyads).

Time spent together	Male/Male			Female/Female		Male/Female	
	Taï	Gombe	Mahale	Taï	Gombe	Taï	Gombe
0–9%	0	0	22	132	140	88	102
10–24%	3	16	14	133	10	96	23
25% or +	25	5	0	11	3[1]	6	1
Total	28	21	36	276	153	192	126
Statistics:							
Gombe/Taï	$X^2 = 21.67***$			$X^2 = 80.4***$		$X^2 = 39.0***$	
Mahale/Taï	$X^2 = 53.96***$						
Gombe/Mahale	$X^2 = 24.91***$						

[1] Two of the three dyads are mother/daughter pairs, a situation absent at Taï as the mother of the only female that did not transfer died before 1987.

the conclusion from the analysis of party size and types: Taï chimpanzees have larger party sizes, including more members of both sexes (Tables 5.2 and 5.5).

What causes these differences in grouping patterns? We have seen that four ecological parameters affect grouping patterns at Tai and possibly other chimpanzee populations as well: predation pressure (Boesch 1991*c*), general food availability, hunting frequency, and sexual opportunities. In the tropical rainforest of Taï, most or some of these factors favour a party size larger than those at Bossou, Kibale, Gombe, and Mahale. In the forest, leopards are more abundant than in more open habitats (Myers 1976; Dind 1995) and a more serious threat to chimpanzees, and in the forest male chimpanzees hunt more in groups than elsewhere (Chapter 8). Lastly, the more strongly biased sex ratio in Taï chimpanzees indicates that sexual opportunities for males are generally higher at Taï than at Gombe.

Do individuals living in a community with larger party sizes associate more intensively with one another? The comparison of Tables 5.9 and 5.10 suggest a positive answer; Taï chimpanzee society appears to be very cohesive. However, the correlation between party size and association can vary, since more recent data from Taï show that DAIs can increase even when party size does not and community size decreases. Thus, DAIs are affected by the relative party size.

Given the high social cohesion of Taï chimpanzees, social interactions are probably more frequent in larger and more cohesive parties. This should be associated with frequent conflicts between individuals and increase the benefit of having coalition partners. If so, it would pay females to have male and female associates on which they could rely when tension increases. We can see in Table 5.10 that Taï females associate more with males than Gombe females. But do Taï females have preferred female associates?

Female friendships in Taï chimpanzees

The highest of all DAIs in the Taï chimpanzee community were found between adult females: Ondine and Salomé were seen together 66% of the time, Malibu and Poupée 71% of the time, and Loukoum and Gauloise 79% of the time during the 18-month period between August 1987 to February 1989. Such values are within the range of mothers and their adolescent offspring (Ella/Fitz = 70%, Gitane/Gipsy = 52%, Xérès/Bonnie = 77%, and for Brutus/Ali [Ali was adopted by Brutus] = 68%). These close female associations lasted for years and remained very stable; each of these females had the same female associate for the whole observation period. The association between Loukoum and Gauloise lasted for four years, between Ondine and Salomé for five years, and between Malibu and Poupée for five years. These associations were only disrupted by the death of one of the partners. Ondine, Loukoum, and Poupée survived their partners but were never seen in such a close association again. We called the partners of such associations 'friends' because their relationship was not only characterized by high DAIs but also by frequent food-sharing and support in conflict situations. Whenever Loukoum or Gauloise lost visual contact with their partner they whimpered or cried, and the friend would respond. Ondine was the highest ranking female and the most successful one in obtaining meat (see Table 4.2). In the many instances she was observed to possess meat, she usually shared some of it with Salomé and some males. The same was observed for Malibu, who shared with Poupée, although she did not manage to secure meat as well as Ondine. Food sharing was also observed for the large *Treculia* fruit, and whenever one of these females had a large piece, she would share pieces with her friend.

The female friends were estimated to be of a very similar age, making it quite unlikely for them to be sisters, although we could not measure their genetic relatedness. Likewise, the friends of each dyad seemed to share a similar social rank. The two most strongly associated dyads were formed by the highest ranking females of the community, and we observed many fights between dyads over meat and for other reasons. In one dramatic instance, Loukoum and Gauloise challenged Ondine and Salomé for more than twenty minutes, and the dominant males tried to calm the situation by charging against them. But the coalitions were too strong and after some fruitless attacks the males simply screamed without doing anything, while Ondine and Salomé were pursued by the younger Loukoum and Gauloise. In this study community with more than 25 adult females, the highest rank in the female hierarchy was held by members of long-lasting alliances. Ondine retained her alpha position for some time after the death of Salomé, but lost it eventually to Poupée, who was associated at the time regularly with Mystère. Ondine, Salomé, Loukoum, Gauloise, Poupée, and Ella, the highest-ranking females of the community, were all involved in long-term friendships. While female friendships were characterized by strong association, frequent food sharing, and alliances to support in conflict situations, they were not associated with a higher tendency to groom one another (see Table 6.6).

More such friendships may have existed, but we did not observe them. The timid attitude of several females interfered with the association data when one member of a friendship was less habituated than the other. For example, Fanny and Gala were fully habituated to human observers only by the end of 1991. In 1987, many females were still very shy and would leave potential friends if they joined parties that we were following. We could confirm that some females were regularly seen together when they were not with the main group, but as we were mainly interested in being with the main group of parties, we usually let them alone. Therefore, the DAIs of the females underestimate the time some females spent together when away from the main group. For example, Héra and Nova* were always seen together when they were on their own (names of unhabituated females are followed by an asterisk). Ella and Ricci formed a tight sub-group with Fanny*, Pokou*, and Gala*, the five regularly spending days together in the south of the territory. Ella often joined the main group, where she helped Kendo, her oldest son, to gain high social status (see Chapter 4), whereas the other females of her sub-group remained at the periphery. We also had the impression that Kiri and Momo* were often together, but we were less confident about this dyad. To conclude, 25 to 50% of the 24 adult females of the Taï community had, or might have had, one or more friends with whom they preferred to associate.

The importance of friendship in the female association patterns in Taï chimpanzees contrasts with observations from other chimpanzee populations. As a rule, female association is described as lower than male association (Tables 5.9 and 5.10); females are described as usually being solitary and as rarely interacting (Goodall 1986; Wrangham *et al.* 1992; Nishida 1989). Alliances (long-term coalitions) between females in these populations are notoriously absent and coalitions are rare. Mahale is the exception as alliances are formed by the secondary immigrants that transfer together to counter the gang attacks of resident females against them (Nishida 1989). The friendships of Taï females suggest a society more female bonded than other chimpanzee populations and prompts questions about the reasons for and consequences of higher female association and friendship. For males, higher predation pressure, hunting frequency, and general food availability seem to have led to higher degrees of association at Taï. With higher association, intraspecific competition also increases, compared with other chimpanzee populations (Table 5.11). As with males, this puts females in a situation in which it pays to form long-term coalitions, even more so as they live in large parties.

The high mortality rate we observed at Taï (Chapter 2) may serve to test this hypothesis. Salomé (September 90), Gauloise (August 88), Malibu (July 90) died when the community was suffering one of its largest decreases. This decreased the density of chimpanzees living in the same territory, and the surviving females probably faced less female–female competition (see Table 7.4 for density measures). Loukoum and Ondine were not seen to associate closely again with other females. Poupée found an associate in

Table 5.11: Potential intra-specific competition in female chimpanzees: adult female competition index = female density × average DAI of female/female × 100. The higher the index, the more intense the competition

Population	Community size (Nb female)	Territory size (km²)	Density for all (Female)	Female Competition Index
Gombe	57 (19)	17.0	3.35 (1.12)	5.6
Kibale	41 (14)	18.6*	2.20 (0.75)	6.0
Mahale	90 (33)	21	4.28 (1.57)	7.8
Taï	66 (25)	23.7	2.77 (1.05)	11.5

* Kibale: The 14.9 km² published in 1992 was later reported as clearly underestimated; we corrected it by increasing this value by 25% (Chapman *et al.*1994).

Mystère, but with a much lower degree of association than with Malibu (DAI = 30–40% instead of 71%). Generally, DAI between the closer female associates decreased by about 20–30% in 1994 (Steiner *et al.* in preparation), supporting our hypothesis that the benefits of association increase with intra-group competition.

Thus, both higher intra-sexual competition and higher involvement in the social interactions of the males makes it profitable for females at Taï to develop long-term friendships with other females and to form stable alliances. Some of these strongly associated females were mothers who played an important role in the social life of the group and of their sons in particular (see Chapter 4), for example, Salomé and Ella were at the time the only dominant mothers of adolescent males. Their participation in coalitions may have improved their ability to help their sons.

The Taï chimpanzees appear to have a bi-sexually bonded society: Large mixed parties are frequently observed, and the two sexes are more frequently associated than in other chimpanzee populations. The higher cohesion and larger parties for most of the year that seem to distinguish Taï chimpanzees from other populations have also been reported from two bonobo populations. Because all bonobos live in tropical African rainforests, the same features of the environment may have led to the same general sociality. But a greater sociality brings with it more intra-community conflicts, and we should expect mechanisms to evolve that control competition. We have seen in Chapter 3 that Taï chimpanzees, and possibly bonobo females, vulnerable to such competition, use sexual swellings as a social passport, not only when they transfer between communities as adolescents and when their infants become socially more active, but also for years as young low-ranking mothers. We have seen that high-ranking females in Taï chimpanzees have developed very close ties with other females, leading to close friendships and reliable alliances in social contests. Bonobo females are also said to be more affiliate than females of some East African chimpanzee populations, but the data so far do not show how this affiliative behaviour is distributed among the bonobo females. We have seen in Chapter 4 that some mothers at Taï develop strong associations with their sons and actively support them in their social life. This has also been reported from bonobos (Kano 1992). It appears that a suite of social and physiological traits adapt the females to the higher competition that results from the higher social cohesion observed in forest environments. This suite of adaptations appears to be convergent in Taï chimpanzees and in bonobos.

At Gombe, Mahale, and possibly Kibale, female chimpanzees are less social, which decreases female competition within the community, and they do not have preferred associations with one another. At Gombe, the situation is further complicated by the fact that daughters normally stay with their mothers and intimate mother–daughter bonds are reported, in contrast to the generally observed emigration of adolescent females in other chimpanzee populations (Chapter 3). Thus, female sociality seems to be influenced in chimpanzees both by the ecological and the sociological conditions prevailing within one population, and they can produce rather different social systems within the frame of fission–fusion.

Our analysis of the fission–fusion system in chimpanzee and bonobo shows that ultimately two demographic factors seem to explain most of the inter-population variations that have been observed: community size and adult sex ratio. The smaller the community, the larger the proportion of community members to be found within the parties. The lower the sex ratio, the more males seek contact with females. If these rules hold true, we should expect fission–fusion social systems to change flexibly with demographic changes. Bonobos have until now been observed in smaller communities, and they tend to have larger party sizes than some larger chimpanzee communities. Similarly, bonobo sex ratios were lower than most chimpanzees, and mixed parties were more frequently observed. With more information on chimpanzee populations, we see that this is not a species difference but part of the flexibility of the fission–fusion system. This system allows group members to adapt themselves flexibly to the prevailing situation and environment. These solutions can vary within a same population over time, and among populations living under different conditions. We should expect to find a gradient from male bonded to bi-sexually bonded societies.

If our hypothesis about the factors affecting party size and type is true, the next question is what affects community size and sex ratio in chimpanzee and bonobo. As we have seen in Chapter 2, sex ratio is primarily affected by mortality rate. The two populations possibly suffering the most from predation and human poaching, Taï and Bossou, had the most skewed sex ratio and the largest proportion of mixed parties. This suggests that bonobos at Lomako and Wamba might have suffered from similar pressures. The determinants of community size are not well understood, but we saw at Taï that both predation and illness reduce the community size dramatically. And an indication that food availability in the tropical rainforest might support more chimpanzees comes from the high density of chimpanzees found at Taï compared with other sites located in more open environments.

We shall see in the next chapter how social interactions are distributed within a community with strong associations between its members. Under such conditions it should pay to seek more and more complex coalitions.

6 *Social relationships within the community*

Forest scene: One day in March 83, Malibu and Poupée, two friends and both in oestrus, attracted all the males and led the whole crowd on their foraging trip. The desirable females regularly climbed in trees to rest, thus provoking high tension between the males. At 10.30, as they approached another tree, hostility arose again between alpha Brutus and Schubert. Macho immediately joined forces with Schubert. Both young males, their hair standing on end, faced Brutus and screamed at the top of their voices. Brutus had to choose between chasing them or remaining close to the females. Rousseau, another young male, bobbed and hoohed ten meters away without making it clear whom he was supporting. Falstaff, the senior male, approached Brutus in what looked like an attempt to quieten him. Brutus stretched his hand towards Falstaff, and Schubert had now to face both. Falstaff barked at him and Brutus chased Schubert and Macho up the big aerial roots of a Uapaca tree. Macho, farther away from Brutus, displayed above him. Schubert waited for Brutus to return to the females that had not moved and looked indifferent.

Schubert and Macho were back on the spot, hoohing loudly. The young adolescent Snoopy approached Malibu from above and was chased away by loud barks from Schubert and Brutus. Then, Schubert and Macho threatened Brutus again, who then approached Falstaff sitting nearby. As the ill-tempered team approached, Falstaff wanted to leave, but Brutus, upright, stretched his hand towards him, looking alternately at Falstaff and Schubert. Schubert, undecided, hoohed for a while. Nothing happened. Then Schubert threatened Brutus with a big arm wave. Falstaff made a reassuring move towards Brutus. And once more, the two oldest chased the younger team away. Schubert in a wild display rushed through the forest and chased all chimpanzees up the trees, while Brutus quickly returned to the two oestrus females.

Who dominated whom? Who finally gained the females' favour? How do such intense interactions influence the relation between the males, and how can they still remain within a same social group?

Living in groups incurs costs in terms of feeding and sexual competition. One way to diminish these costs is by organizing social life so that competition between group members for food, mates, or other resources does not escalate. This allows the benefits of sociality to be realized in terms of better exploitation and control of resources, cooperative defence of the territory, and control and access to sexual partners. The control of conflicts of interest within a group is classically done by establishing a dominance rank order between group members to decide on the priority of access to resources. Once established, dominance rank order can influence all social interactions: low-ranking individuals will seek tolerance or support from higher-ranking ones, and high-ranking individuals will try to control those with low rank (Seyfarth 1977; Kummer 1971; de Waal 1989, 1996). Dominance style may vary. Dominant individuals may exploit inferiors with relative impunity (this is called 'despotic dominance', and is seen, for example in rhesus macaques), or they may have a low level of aggression against inferiors ('tolerant dominance', for example in stumptail macaques), or may even be challenged by inferiors ('egalitarian dominance', for example in chimpanzees) (de Waal 1996). In primates, social grooming is generally used as a tool to gain some social advantage from higher-ranking individuals (Dunbar 1988; Seyfarth 1977) and can be observed in all dominance styles. Egalitarian dominance permits inferiors to use many sets of behaviour to circumvent some of the limits imposed by the dominance rank order: the most important are coalitions and alliances that dominants use to confirm their position and inferiors use to gain some advantage (Harcourt and de Waal 1992; Noé 1992; Nishida and Hosaka 1996).

Human social organization, with its extended capacity for cooperation, reciprocal interactions between and within the sexes, and high degree of coalition behaviour has been proposed to be unique (Alexander and Noonan 1979; Tooby and DeVore 1987). However, some of the so-called 'human social characteristics' are observed in other primates, especially in chimpanzees: both short- and long-term cooperation and alliance are regularly seen in primate social systems, and more regularly than in non-primates (Harcourt 1989; Dunbar 1988). The sophistication of primates can be seen in their sensitivity to qualitative differences between their partners (Harcourt 1989), while, in addition, only chimpanzees make use of revenge (de Waal 1989; de Waal and Luttrell 1988). Coalitions are of special interest, for they enable manipulation of social partners and may offer a way to counter a hierarchical order.

The fission–fusion social structure of the chimpanzees requires a subtle evaluation of social relations from each group member, as members belong to parties of constantly changing size and composition (Chapter 5). Dominance relations have to be re-evaluated for each party, and the behaviour of the individuals adapted accordingly. This social system probably favours social flexibility. At Taï, we saw in addition that the chimpanzees tend to live in larger parties with greater cohesion between individuals than is known in other populations. To evaluate the impact of greater cohesion on social life is the aim of the following analysis, with particular attention to how grooming and coalition give inferior group members access to dominants and to resources.

Social interactions and dominance in Taï males

Conflicts within a social group are expected because resources are limited. The intensity of such conflicts will depend on both the value the individual attributes to the resource and the number of individuals competing for it. For high-value resources we expect more fights or the use of alternative strategies to attain them. A dominance rank order might diminish the number of conflicts, as it allows an individual to predict the priority of access to resources without having to rely on direct confrontation. The chimpanzee fission–fusion structure, in which all the males of a community are rarely together, makes a direct measure of dominance between them difficult. In one of the first studies of males at Gombe, Bygott (1974) suggested that the direction of the pant-grunts, the soft and repeatedly emitted vocalizations used when a individual greets another one, gives a reli-

able measure of dominance. They have been shown to correlate with aggressive and submissive interactions, and this has been confirmed in the Mahale chimpanzees (Nishida
and Hosaka 1996; Hayaki *et al.* 1989).

At Taï, during the period from mid-August to mid-October 1993, the pant-grunts were
clearly unidirectional, with one exception resulting from a coalition (Table 6.1). Based on
the pant-grunts, the dominance order between the males is linear, except for the equivocal
situation for the two middle-ranking males, Macho and Brutus. They did not pant-grunt at
each other, but both pant-grunted the two highest-ranking males, and Darwin pant-
grunted Macho. It is noteworthy that Brutus, former alpha for many years, pant-grunted
least at the alpha male. In general, 96% of the pant-grunts were directed toward the two
highest-ranking males, Fitz and Kendo, and the alpha received about twice as many as the
beta. Kendo received 77% of them from Macho, the gamma male, who had a particular
reason for doing this. Macho was alpha when he was supplanted by Kendo in one of the
most violent fights we ever witnessed in the community (Chapter 4). Since then Macho
has always greeted Kendo with overemphasis and without being asked. If we exclude
these, 88% of the pant-grunts between the adult males were directed towards the alpha.

The analysis of submissive behaviour given by one male to others confirm the pant-
grunt analysis, since we arrive at the same rank order (Table 6.2). When we combined

**Table 6.1: Direction of pant-grunts between the 5 adult and 3 adolescent males in the Taï
community in autumn 1993 (*N* = 137) (males are listed in decreasing rank order in all tables)**

	Pant-grunt to:							
	Fitz	Kendo	Macho	Brutus	Darwin	Sartre	Marius	Gipsy
Fitz	–	0	0	0	0	0	0	0
Kendo	24	–	0	1	0	0	0	0
Macho	13	33	–	0	0	0	0	0
Brutus	5	5	0	–	0	0	0	0
Darwin	22	5	1	0	–	0	0	0
Sartre	13	1	2	0	0	–	0	0
Marius	8	1	0	1	0	0	–	0
Gipsy	3	0	0	0	0	0	0	–
Total	88	45	3	2	0	0	0	0

**Table 6.2: Submissive behaviours between males of the Taï chimpanzee community in 1993
(*N* = 112).**

	Submissive to:							
	Fitz	Kendo	Macho	Brutus	Darwin	Sartre	Marius	Gipsy
Fitz	–	5	2	4	0	0	0	0
Kendo	3	–	0	1	3	0	0	0
Macho	7	11	–	1	0	0	0	0
Brutus	3	3	1	–	2	0	0	0
Darwin	11	14	4	2	–	0	0	0
Sartre	3	5	0	2	0	–	0	0
Marius	1	7	3	4	0	0	–	0
Gipsy	4	4	1	1	0	0	0	–

Submissive behaviours include intensive greeting, retreat, whimpering and screaming.

pant-grunts with submissive behaviour, the unidirectionality was statistically supported ($Kr = -10$, $p = 0.05$ [Hemelrijk 1990]); individuals directed most submissive behaviours to those males from whom they received the least. When we tested for pant-grunts and submissive behaviour separately, the test was not significant owing to the small sample size and the many ties in the matrices. The position between Macho and Brutus remains unclear, for we saw them each once show submissive behaviour to the other. In fact, the unidirectionality of submissive behaviour is less marked than for pant-grunts. The dominant males show some signs of submissiveness to inferiors, but all of them result from coalition interactions and show how effective coalitions can be. They do not alter the dominance order but force dominants to produce submissive signals to low-ranking individuals under specific circumstances.

Agonistic confrontations were regularly observed, and the matrices indicate the same kind of rank order between the males: the males received more agonistic behaviours from those to whom they showed more submissive behaviour, and did so proportionally to the amount of aggressive behaviour received (for relative reciprocity: $Kr = 19$, $N = 5$, $p < 0.05$, for absolute reciprocity: $z = 3554$, $p < 0.01$) (Table 6.3). As with the pant-grunts, agonistic confrontations tend to be unidirectional, and attacks on dominant individuals involved coalitions. Thus, all three measures of dominance confirmed a linear dominance rank order between the five adult males of the community. But here again we cannot clearly separate the two middle-ranking ones, Brutus and Macho, as each attacked the other twice. If we used the intensity of the interactions of each of these measures, in all of them Macho had much more tense interactions with the two highest-ranking males than Brutus did, which indicates that they both felt Macho was more of a threat than Brutus.

The two highest-ranking males were the most aggressive individuals, performing 80% of the agonistic displays. They were also the only ones making separating interventions, thereby often succeeding in interrupting a close contact between two lower-ranking individuals (Table 6.4). Thus, submissive and aggressive behaviours involved almost exclusively one of the two highest-ranking males.

Some regularly described behaviour patterns in the Mahale chimpanzees are absent in our autumn 1993 sample ($N = 367$ hours), and are on the whole seldom observed. This is

Table 6.3: Agonistic confrontations between males of the Taï chimpanzee community in 1993 ($N = 197$).

| | Agonistic confrontations against: | | | | | | | |
	Fitz	Kendo	Macho	Brutus	Darwin	Sartre	Marius	Gipsy	Total
Fitz	–	36	28	11	14	5	2	0	96
Kendo	3	–	24	12	13	3	5	2	62
Macho	4	4	–	2	4	0	2	0	16
Brutus	3	4	2	–	4	1	5	0	19
Darwin	0	2	0	0	–	2	0	0	4
Sartre	0	0	0	0	0	–	0	0	0
Marius	0	0	0	0	0	0	–	0	0
Gipsy	0	0	0	0	0	0	0	–	0

Agonistic confrontations include bobbing, hunch, supplant, directed displays, push away, slapping, short chase, chase, charge, and attacks.

Table 6.4: Separating interventions observed among Taï chimpanzees in 1993 (*N* = 28)

Intervener	Group	Number
Fitz	Kendo-Macho	10
Fitz	Kendo-Brutus-Darwin	4
Fitz	Kendo-Brutus	3
Fitz	Macho-Darwin	3
Fitz	Kendo-Darwin	1
Fitz	Kendo-Belle	1
Fitz	Fanny-Gipsy	1
Fitz	Brutus-Darwin-Ricci	1
Kendo	Brutus-Loukoum-Bijou	2
Kendo	Brutus-Bijou-Goma	1
Kendo	Brutus-Macho-Poupée	1

the case for mounting and embracing, which we observed very rarely in situations of extreme tension, but were seen 61 and 22 times respectively in Mahale chimpanzees (Nishida and Hosaka 1996; observations time = 390 hours). Similarly, kissing was seen regularly in Mahale chimpanzees (*N* = 21) but only twice at Taï. The Mahale and Taï studies were performed in a comparable way, with only naturalistic observations and for a similar duration. These intriguing differences suggest the possibility that different social behaviour elements are in use in the two populations.

Taï male chimpanzees possess a linear dominance rank order that, as in other chimpanzee populations, is revealed by pant-grunt directionality and aggressive behaviour. The most dominant individuals are the centre of most aggressive exchanges. The fission–fusion social structure and the regular use of coalitions (see below) do not prevent the males from ranking themselves in a linear order. Whether affiliative behaviour, such as grooming, is also concentrated around the two highest-ranking individuals is the next point we address.

Grooming interactions

Most animal species invest a certain proportion of their time in carefully cleaning their skin of ectoparasites and dirt, thereby reducing the parasite load that can have important negative consequences. Social grooming is regularly observed between social partners, and it often seems to last longer than would be needed simply for hygiene (Dunbar 1988). Grooming, which has been shown to reduce heartbeat rates both in horses and in some primates, seems to have acquired a secondary social-bonding function in primates. Time spent grooming increases with group size, grooming is more frequently observed after agonistic or sexual interactions, and animals invest more time in grooming when relationships are being established or threatened by rivals (Dunbar 1988).

Because grooming represents a certain investment of time and energy by the groomer, and benefits the groomed, it is a behaviour for which individuals could compete. In primates, it has been proposed that grooming represents a kind of payment that subdominants give to dominants for accepting their presence and possibly supporting them (Seyfarth 1977). We thus expect inferiors to prefer to groom dominants and females to prefer to groom males. Such expectations are supported by observations on monkey species (Silk *et al.* 1996). However similar studies in Gombe and Mahale chimpanzees do not support these expectations, in the sense that females groom rarely, and the older or the dominant males tend to be the ones grooming the most (Takahata 1990a,b; Bygott 1979, Simpson 1973). In Mahale chimpanzees, grooming is viewed as a service used by dominants to guarantee the fidelity of their lower ranking coalition partners (Nishida and Hosaka 1996, Takahata 1990b).

Social interactions in general, and grooming in particular, can be affected by changes in the composition and the size of a group of individuals (Goodall 1986; Takahata 1990b). With this in mind, we analysed and compared grooming interactions for two different periods, the first in 1988, the second in 1993. The Taï chimpanzees invest a lot of time in grooming (43% of the resting time, or 9% of the time budget). Because grooming reveals the affiliative interactions in which adults of both sexes are involved, it is a good measure of the intimacy of a society. It is common to see chains of up to six individuals grooming one another while others watch them.

In 1988, males were involved in 77% of the grooming interactions between adults (Table 6.5), as expected from the higher association between males. Still, females were part of 56% of the grooming interactions between adults. The time females groomed females represents 51% of the time males groomed males, and the time females groomed males represents 61% of the time they were being groomed by them. Females contributed significantly to the grooming interactions of the community, but when we correct for the adult sex ratio, as they outnumbered the males by about 4 to 1, it appears that females groomed one another less than males, that males were active groomers preferring to groom males, and that female–male grooming interactions were only as frequent as would be expected by chance (Table 6.6).

In comparing different chimpanzee populations, the special situation at Bossou helped us to realize an important factor influencing grooming interactions: the composition of the community. Grooming between adult females is strikingly higher at Bossou than at Taï (see Table 6.6). If directly compared with a chance distribution of grooming between group members, Bossou females groom themselves at chance level without showing any preference, contrary to what we might have concluded from a direct comparison between Taï and Bossou. Therefore, Table 6.6 compares, for all populations, the observed grooming distribution between the sexes with the expected random distribution based on the

Table 6.5: Grooming duration between members of the Taï chimpanzee community over a period of 3 months between September and December 1988 (the frequent grooming between mothers and offspring was excluded here)

| Groomer | Groomed | | | |
	Male	Female	Subadult	Total
Male	1'047'	504'	168'	1'719'
Female	308'	536'	34'	878'
Subadult	216'	29'	33'	278'
Total	1'571'	1'069'	235'	2'875'

Table 6.6: Grooming distribution between adults in different chimpanzee populations (⇔, ⇑, and ⇓ stands for observed value equal, higher, and lower than expected)

	Male/ male	Female/ female	Male/ female	Number of male:female
Chimpanzees				
Bossou	6% ⇔	62% ⇔	32% ⇔	2.5:7
Gombe	37% ⇑	13% ⇓	49% ⇔	6:18
Kibale	72% ⇑	0% ⇓	28% ⇓	8:12
Mahale	79% ⇑		21% ⇓	10:39
Taï	44% ⇑	22% ⇓	34% ⇔	7:27
Bonobos				
Wamba	16% ⇓	15% ⇓	68% ⇑	14:15
Lomako	19% ⇑	29% ⇓	52% ⇑	8:14

Bossou: Data from 1976–77 and 1982–83 (Sugiyama 1988)
Gombe: Data from 1978 (Goodall 1986)
Kibale: Data from 1989 (Wrangham *et al.* 1994)
Mahale: Data from 1981 (Takahata 1990a,b)
Wamba: Data from 1978 (Kano 1992)
Lomako: White (1988)

number of adult males and females existing at the time the grooming data were collected. Bossou chimpanzees are special as they are the only known chimpanzee population in which grooming distribution follows a chance distribution. In all other chimpanzee populations, males groom one another significantly more than expected, whereas females groom one another significantly less than expected. Some inconsistency is seen in the distribution of male–female grooming, that is as expected by chance in Gombe and Taï chimpanzees, but is less frequent in Kibale and Mahale chimpanzees. In all chimpanzee populations, grooming interactions between males are more intense than those between the females.

Interestingly, in bonobos, grooming interactions between females are also less frequent than expected by chance. The male bonobo data are not clear-cut, but in both populations under study there is a higher grooming frequency than expected between males and females. Therefore, the only consistent difference between the two species may be in the grooming relationship between the sexes, which is more frequent in bonobos. If we view grooming as a strategic investment in social partners, we could say that male chimpanzees invest in males, and male bonobos in females. This last point is, however, complicated by the possibility that some of these female bonobos are the mothers of the males and should thus be viewed as mother–son interactions (Kano 1992; Furuichi 1989; Ihobe 1992). The low frequency of grooming between females contrasts with the proposition that female bonobos have very strong affiliative interactions (White 1988).

Beside the sex ratio, another factor may affect grooming distribution in a social group, namely the association tendencies between individuals: individuals who often associate may groom each other more frequently. A comparison between grooming distribution (Table 6.6) and association within and between the sexes in different populations (see Table 5.9) indicates that, in Mahale and Taï chimpanzees, the more a class is associated the more they groom one another. At Taï, males more frequently groom males with whom they associate more frequently ($Kr = 24$, $p < 0.005$). A similar result was found at Bossou using proximity instead of association (Sugiyama 1988). Gombe chimpanzees stand out since male–female grooming is more frequent than expected from the association pattern between the sexes. In bonobos, association measures are not yet available.

In conclusion, the grooming distribution in chimpanzees and bonobos can be explained by some of the social characteristics of the groups in which the members live. Sex ratio affects the tendency of the sexes to groom with the opposite sex, whereas association between the sexes affects grooming frequency in that individuals most frequently groom their most frequent associates.

Reciprocity of grooming interactions

Our initial view was that grooming is a service provided to one individual. It could also be regarded, when reciprocal, as an exchange between individuals. At Taï, in 1988, grooming interactions between male chimpanzees were mutual, two individuals grooming synchronously one another, 73% of the time (Table 6.7). Some individual differences were observed in the amount given or received between the six males, but all of them showed a near reciprocity in the duration of the grooming they received and gave. This weakens the idea of grooming as a unidirectional service and emphasizes the reciprocity in grooming between the males at Taï during this period. The same was observed in the grooming interactions between females and between males and females, which were mostly a mutual exchange (Table 6.8). Five years later in 1993, the grooming interactions were similarly distributed: reciprocal exchange continued to dominate grooming exchanges between the adult community members (Table 6.7). Detailed analysis of the grooming interactions in Mahale male chimpanzees also suggest reciprocal exchange

Table 6.7: Grooming distribution among the males in autumn 1988 and 1993 in the Taï community: grooming can either be unidirectional (called 'received' when one passive individual is groomed by another, or 'given' when one individual grooms another, passive one) or 'mutual', when two individuals groom one another simultaneously.

Name of male	Received (mins) 1	Given (mins) 2	Mutual (mins) 3	Total (mins)	Reciprocity index 1/2	(1 + 3)/(2 + 3)
1988						
Macho	89	35	282	406	2.54	1.18
Kendo	7	69	156	232	0.10	0.72
Ulysse	60	68	408	536	0.88	0.98
Brutus	61	37	375	473	1.65	1.06
Rousseau	45	9	144	198	5.00	1.23
Darwin	21	65	162	248	0.32	0.81
Total	283	283	1'527	2'093	1.00	1.00
1993						
Fitz	189	206	280	675	0.92	0.96
Kendo	33	139	159	331	0.24	0.64
Macho	120	70	112	302	1.71	1.27
Brutus	61	43	95	199	1.42	1.13
Darwin	100	45	40	185	2.22	1.65
Total	503	503	686	1'692	1.00	1.00

Table 6.8: Proportion of mutual grooming in different chimpanzee populations according to the sexes involved

	Male/male (%)	Female/female (%)	Male/female (%)	All (%)
Bossou	25	55	48	–?
Taï 88	73	65	60	66
93	23	40	29	27
Gombe	16	0	19	17
71	21	–	–	–
Mahale	9	–	2	–

Bossou: Sugiyama (1988)
Gombe: Goodall (1986, Fig. 14.4, p. 403) and 1971: Bygott (1974)
Mahale: Takahata (1990*a,b*) and Kawanaka (1990, Table 8.3, 8.4)

(Nishida and Hosaka 1996): The ratio of grooming to groomed was 1.14 for the 9 males, ranging from 0.68 to 1.22.

Whereas mutual grooming remained more frequent in Taï than in other populations (Table 6.8), in the autumn of 1993 it was clearly less mutual than five years before. By then it was close to what was observed at Bossou but remained higher than in Gombe and Mahale chimpanzees. It is difficult to explain this difference. One possibility is that the more a community has to sustain conflicts with neighbouring communities the more the males have to cooperate, and the more they need to live in good terms with one another. The difference in mutual grooming would suggest that Taï males have to deal with stronger neighbours and need to be very cohesive, whereas Mahale males have weaker neighbours whom they do not fear. The M group at Mahale in 1983 and 1986 was about 100 individuals strong, with 13 to 15 males (adult and adolescent) and had possibly driven to extinction the neighbouring K group. At Gombe in 1978, the Kasakela community had successfully eliminated the small neighbouring Kahama community and controlled one of the largest territories during the observational period (Goodall 1986). In contrast, at Taï, the community had started to decline by 1987 with less than 70 individuals and 8 males (adult and adolescent) and was regularly encountering neighbouring communities (Chapter 7).

Grooming preferences between individuals

Grooming between males

The distribution of grooming between the individual males in autumn 1988 shows that unidirectional grooming tended to be reciprocal in all dyads ($Kr = -13$, $p = 0.09$) (Table 6.9). Mutual grooming was frequent in most dyads with three exceptions, all involving Kendo. The duration of grooming for all males correlates strongly with the amount of mutual grooming ($r_s = 1.00$, $p < 0.001$). Rank did not play a significant role in the distribution of grooming, but age was important, for old males received more grooming than younger males ($r_s = -0.99$, $p < 0.05$). Grooming in dyads was also strongly affected by the association between the individuals ($Kr = 24$, $p < 0.005$): the more two individuals were together, the more they groomed each other. However, once corrected for association, certain males showed strong preferences in grooming partners (Fig. 6.1).

Table 6.9: Dyadic grooming relations between the adult males in 1988. For the groomed male we give the time in minutes he was groomed and the total time mutual grooming was observed in that dyad.

Groomer	Groomed					
	Macho	Kendo	Ulysse	Brutus	Rousseau	Darwin
Macho	–	0–56	15–76	15–101	3–27	2–22
Kendo	10	–	29–89	19–9	11–2	0–15
Ulysse	38	1	–	14–140	10–20	5–83
Brutus	4	1	12	–	6–80	11–42
Rousseau	4	0	3	2	–	0–15
Darwin	33	15	1	14	15	–

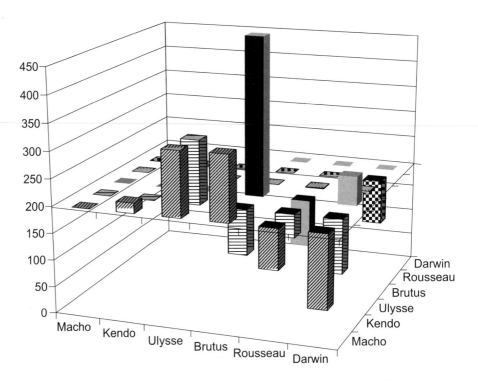

Fig. 6.1.   Grooming preference in Taï adult males: grooming duration corrected for the dyadic association tendency.

Four dyads groomed one another more than the other ones, and all involved Brutus and Ulysse. These two males were preferred grooming partners as well as eager groomers.

The distribution of grooming among the males in the autumn of 1993 differed strikingly from what we observed five years before. First, unidirectional grooming was no longer reciprocal within all dyads ($Kr = 12$, $p = 0.12$), whereas the amount of mutual grooming in dyads still correlated with the amount of unidirectional grooming ($Kr = 16$, $p = 0.04$) (Table 6.10). Second, mutual grooming represented only 40% of the total

Table 6.10: Grooming distribution among the males in autumn 1993

Groomer	Groomed				
	Fitz	Kendo	Macho	Brutus	Darwin
Fitz	–	23–110	82–92	39–57	62–21
Kendo	95	–	15–14	17–27	12–8
Macho	51	6	–	4–3	9–3
Brutus	20	3	3	–	17–8
Darwin	23	1	20	1	–

grooming time and was more important than unidirectional grooming only for one dyad. Third, although the amount of grooming time for each male still correlated with the amount of mutual grooming ($r_s = 0.90$, $p < 0.07$), rank was now the main determinant of grooming distribution: dominant males tended to have larger total grooming time ($r_s = -0.90$, $p < 0.07$) and were more involved in mutual grooming ($r_s = -1.00$, $p < 0.001$). Age was no longer important and tended to correlate negatively with rank, younger males being dominant.

Grooming between males and females

Males invested on average 27% of their total grooming time in the adult females. In 1988, Brutus, Ulysse, and Rousseau groomed adult females more than did the other males and the two former were the most active groomers among the males. This strong investment of Brutus and Ulysse in grooming females might relate to their common interest in hunting, and the help some females provided in assuring them regular access to meat (Chapter 8). Brutus, five years later, was still seen to invest much more time in grooming females than the other males (Table 6.11).

Grooming in Taï chimpanzees forms a complex social network of interactions. Its distribution among social partners can be affected by the period at which they are observed, and the individual personalities that are part of the community, as well as by some intrinsic properties of the community itself. This makes comparisons between populations quite uncertain. Nevertheless, we gained the impression that grooming distribution in

Table 6.11: Males' grooming interactions with adult females during the autumn of 1988 and 1993.

	1988				1993			
	1	2	3	4	1	2	3	4
Macho	50	19	69	25%	59	6	27	23%
Kendo	18	25	25	23%	76	8	20	24%
Ulysse	63	17	135	29%				
Brutus	35	26	159	32%	126	34	97	56%
Rousseau	15	11	70	33%				
Darwin	20	24	31	23%	18	33	14	26%
Fitz					45	27	19	12%

For each male, we give the time he groomed females (1), the time he was groomed by them (2), the mutual grooming time (3) and the proportion of his total grooming time devoted to females (4).

wild chimpanzees is partly opportunistic, stimulated simply by the simultaneous presence of individuals as well as the number of individuals of each sex present in the community. In addition, females seem to groom other females less than expected, as in bonobos. The frequency of grooming between males and females varies more than between other classes of groomers among populations of chimpanzees. In contrast to aggressive interactions, grooming is not centred around the highest-ranking individuals. At Taï, males possess clearly marked individual preferences among grooming partners independent of associations, and the two most active groomers in our analysis were active hunters not dominant males.

Coalitions among Taï male chimpanzees

Coalitions and alliances (long-term coalition with a same individual) have recently been a centre of interest for two reasons: First, how is it that an individual sacrifices its own interest while contributing to someone else's? Second, if coalitions allow access to more resources, how are the resources divided between the two partners? The answer to the first question is that either individuals get involved because they are closely related and also benefit from the other's gain (the kin selection argument), or because both fare better as a result of the cooperative action (the mutualistic argument) (Dugatkin 1997). The answer to the second question is easy as long as the benefits can be shared, but is more difficult for indivisible resources like mating partners. One conclusion of empirical studies is that coalitions are rarely observed between unrelated individuals in most animal species, except to gain access to sexual partners, as in lions (Packer *et al.* 1990) and baboons (Noé 1992; Bercovitch 1988).

In the Mahale chimpanzees, 29 coalition aggressions were observed in 390 hours of observations (Nishida and Hosaka 1996), illustrating how frequent they can be in this species. In Gombe chimpanzees, alliances between brother pairs have been described, and these have played a role in helping at least one individual (Figan) to reach the alpha position. The importance of coalitions in gaining social dominance has been assumed to be important, but data are deceptive, both because coalitions between unrelated individuals are not frequent and because the dominance rank order between the males remains linear in most studied populations despite the coalitions (Nishida and Hosaka 1996; Goodall 1986; this chapter).

During the autumn of 1993, the Taï chimpanzees relied on coalitions in 49 instances for 367 hours of observations (Tables 6.12 and 6.13). In Table 6.12, we specify whether the individual in the coalition who was attacked was dominant in relation to the attacker (dominant support) or if it was inferior (inferior support), if it was older or younger than the attacker and if the supporting individual was dominant or inferior over to attacker. Supporting coalitions, that follow an aggression against an individual, were observed in 30 instances. By definition, all included at least three individuals, and six of them involved four or five individuals simultaneously. Among Taï chimpanzees, coalitions are frequently observed but do not involve all males of the community equally: in all but two cases they involved the highest-ranking male, Fitz, and the oldest one, Brutus. Brutus took part in 20 of the supportive coalitions (66%) and was member of 14 of the joint attacks (58%). He was involved in five of the six coalitions including more than two individuals and was the only one regularly supported by the dominant females (in our sample

Table 6.12: Supporting coalitions in Taï chimpanzees in autumn 1993 (N = 30)

Supporter	Pro	Contra	D/I	O/Y	Dom	Result
Fitz	Belle	Macho	I	Y	D	S
Fitz	Macho	Kendo	I	O	D	S
Fitz	Marius	Loukoum	I	Y	D	S
Fitz	Brutus	Kendo	I	O	D	S
Fitz	Darwin	Kendo	I	Y	D	S
Kendo	Ricci	Brutus	I	Y	D	U
Kendo	Ricci	Macho	I	O	D	S
Kendo	Macho	Fitz	I	O	I	S
Kendo	Macho	Fitz	I	O	I	S
Kendo	Brutus	Fitz	I	O	I	S
Macho+Kendo	Brutus	Fitz+Darwin	I	O	I	S
Macho+Kendo	Brutus	Fitz	I	O	I	S
Macho+Bijou	Castor	Fitz	I	O	I	S
Brutus	Macho	Kendo	I	O	I	S
Brutus	Darwin	Bijou	D	O	D	S
Brutus	Loukoum	Kendo	I	Y	I	S
Brutus	Loukoum	Kendo	I	Y	I	U
Brutus	Loukoum	Kendo	I	Y	I	U
Brutus	Loukoum	Kendo	I	Y	I	S
Brutus	Loukoum	Kendo	I	Y	I	S
Brutus	Loukoum	Darwin	I	Y	D	S
Brutus	Loukoum	Goma	D	Y	D	S
Brutus	Bijou	Goma	I	Y	D	S
Brutus	Ricci	Darwin	I	O	D	S
Brutus+Darwin	Kendo	Fitz	I	O	I	S
Brutus+Darwin	Macho	Fitz	I	O	I	S
Brutus+Darwin	Perla	Kendo	I	Y	I	S
Loukoum	Brutus	Macho	I	O	I	L
Loukoum	Brutus	Kendo	I	O	I	S
Mystère	Dilly	Kendo	I	Y	I	U

Note: Pro: support for, Contra: support against, D/I: Pro being dominant or inferior over Contra, O/Y: Pro being older or younger than Contra, Dom: Supporter is inferior or dominant over Contra, Result indicates if the coalition succeeded (S), was undecided (U), or lost (L) against Contra.

Table 6.13: Joint attacks in Taï chimpanzees in autumn 1993 (N = 19).

Partners		Contra	Number or attacks	Dom.
A	B			
Fitz	Kendo	Group	1	D
Fitz	Kendo	Darwin	1	D
Fitz	Macho	Kendo	1	D
Fitz	Darwin	Goma	1	D
Fitz	Darwin	Brutus	1	D
Kendo	Macho	Fitz	2	I
Macho	Bijou	Fitz	1	I
Brutus	Fitz	3 females	1	D
Brutus	Kendo	Poupée+Goma	1	D
Brutus	Kendo	Fitz	1	I
Brutus	Kendo+Macho	Fitz+Darwin	1	I
Brutus	Macho	Fitz	1	I
Brutus	Darwin	Fitz	2	I
Brutus	Darwin+Perla	Fitz	1	I
Brutus	Loukoum	Kendo	1	I
Brutus	Loukoum	Darwin	1	D
Brutus	Ricci	Darwin	1	D

Contra: support against; Dom: supporter is inferior or dominant over Contra

Loukoum). Brutus was estimated to be 43 years old in autumn 1993 and was down to number four in the hierarchy. He must have had some special skill to acquire such a pivotal position in the coalition game of the community.

An alliance is a consistent support between two individuals in coalitions over a long period of time, and if present, alliances should be apparent here. Fitz was only seen twice to display in tandem with his older brother Kendo, indicating that the two brothers were then rivals rather than allies. This was confirmed by the high rate of pant-grunts and ago-nistic interactions observed between them. Fitz and Darwin displayed in tandem only four times, and this was the only trace of what had been an alliance when Fitz fought his way up the hierarchy to the alpha position in 1991. Kendo, as said before, had a very antagonis-tic relationship with Macho, whom he had defeated as an alpha in 1989, and they rarely coordinated their displays. We saw him support Macho twice, both times against his brother Fitz. Brutus seemed to have built two alliances: first with Kendo, for they dis-played in tandem three times and supported each other four times. A special case was the long-term support between Loukoum and Brutus, which may have originated in their common interest in securing meat, a situation in which they would invariably support each other, but which extended to other situations, as six of the nine coalitions they built were observed in non-meat-eating situations. This alliance was seen from 1988 and intensified with the death in 1992 of Ondine, who had been Brutus's main female support.

Most coalitions were unsolicited (22 out of 30 instances); supporters rushed sponta-neously to help screaming attacked individuals. If solicited, a favourable answer was obtained in seven out of twelve cases (only solicitations from Macho and Darwin were neglected). Almost all coalitions were successful, as the attacking individual stopped its attack ('U' in Table 6.12), or the coalition forced the attacked to retreat and scream ('S' in Table 6.12). Furthermore, coalitions of inferiors were about as successful in repelling dominant individuals as coalitions in which at least one individual dominated the oppo-nent (Fisher's exact test: $p = 0.32$). In all but one case the inferior coalitions were able to put the superior to flight or stop its attack. Most coalitions at Taï were carried out by inferior against dominant individuals, and they had a destabilizing effect on the dominance order.

Similarly, joint attacks were observed in nineteen instances when two chimpanzees ini-tiated an attack against one or more individuals. In ten cases, two inferiors attacked a higher ranking one (in nine cases the alpha and once the beta male). At Mahale, only five joint attacks were witnessed, and only one was against a dominant individual (Nishida and Hosaka 1996). Thus both coalition supports and attacks at Taï were frequently done by inferior against dominant individuals, as expected in an egalitarian type of dominance system (de Waal 1996).

Nine alliances have been described in Gombe chimpanzees as lasting up to nine years (Goodall 1986). Kinship seems to play the main role in the establishment of an alliance at Gombe: five of them were between brothers, one between an uncle and a nephew, and three between unrelated males. At Taï, the most stable alliances were observed between Brutus and one of the females: Brutus and Ondine for eight years until her death in 1992, and Brutus and Loukoum for the last five years. Similarly Kendo and his mother Ella were strong and efficient allies for many years until Kendo reached the alpha position (see Chapter 4). Alliances between males were of shorter duration: Darwin constantly supported Fitz for about two years in his rise to the alpha position, but Fitz rarely supported him. Kendo and Fitz were brothers, Fitz constantly supported his older brother, and they also displayed in parallel for fourteen months. Then Fitz started to challenge his older brother successfully, and they remained rivals until their death in 1994. In 1993, the coalition games were dominated by Brutus, and alliances apparently did not play an important role.

Coalitions as a social tool

Here we compare some details of coalitions to understand their role as a social tool in primate societies (Harcourt and de Waal 1992). Coalitions have been observed in many mammal species, but the classes of individuals at which they are directed varies among species (Table 6.14). Coalitions can confirm the dominance order or alter it, making it non-linear or unstable. This destabilizing role of coalitions has rarely been observed in carnivores or monkeys, where they are as a rule of the winner-support type, a supporter helping the dominant individual against a lower-ranking one (de Waal 1992). Such winner-support coalitions do not profit the supported individual and often seem opportunistic, in the sense that the supporter seeks an easy victory. Inferior-support coalitions directly help the attacked individual, who is by definition here inferior to the attacker (Table 6.14). Such coalitions tend to be rarer in monkeys, are usually observed when an individual supports his kin or friends, and confirm the established dominance system. In contrast, coalitions of low-ranking individuals against higher-ranking ones (called loser-support coalitions in Table 6.14) destabilize the hierarchy. Such coalitions are relatively rare in monkeys, reflecting a dominance style where dominant individuals can drive low-ranking individuals away in all circumstances.

Table 6.14: Comparison of coalition occurrences in different mammal species: in loser-support coalitions both members of the coalitions are inferior to the attacker

	Number	Frequency (per h)	Inferior support	Loser support
Carnivores				
Hyenas	500	2.6	~18%	–
Wolves	26	0.06	18%	–
Macaques				
Bonnet	849	1.54	25%	–
Rhesus	2053	–	49%	13%
Stumptail	535	–	67%	14%
Chimpanzees				
Arnhem Zoo	1504	–	28%	50%
Mahale	23	0.07	69%	17%
Taï	30	0.14	90%	60%

Hyenas: captive study, Zabel *et al.* (1992)
Wolves: captive study, Fentress *et al.* (1986)
Bonnet macaques: captive study, Silk (1992)
Rhesus, Stumptail macaques and Arnhem zoo chimpanzees: de Waal and Luttrell (1988), de Waal (1992).
Mahale chimpanzees: Nishida and Hosaka (1996).

In chimpanzees, both inferior- and loser-support coalitions seem to be more common than in monkeys (Table 6.14). In the Arnhem zoo chimpanzees made few inferior-support coalitions, and in half of the cases attacked higher-ranking individuals. This was presented as being in marked contrast to what is observed in monkeys (de Waal 1992, 1996). At Mahale, 69% of the coalitions were inferior-support. But in 19 of the 23 supportive coalitions seen at Mahale, one of the coalition members was dominant over the opponent. Thus, these coalitions confirmed the dominance ranking. At Taï, 90% of the coalitions were inferior-support as well, but out of 30 supportive coalitions, 18 included only inferiors attacking a dominant individual. The Mahale and Taï observations differ significantly (for inferior/winner support: Fisher exact test: $p < 0.02$; for dominant/looser coalition: $X^2 = 8.06$, $p < 0.003$). Thus, coalitions in chimpanzees appear to be population specific. Captive chimpanzees ignore dominance rank, whereas Mahale chimpanzee coalitions tend to support dominance rank, in contrast to Taï where coalitions disrupt dominance rank.

Two questions arise when considering coalitions in Taï chimpanzees. Why do the dominant males let an inferior destabilize the dominance order? What could individuals gain from such a destabilization? Concerning the first question, we note that Brutus attacked dominant individuals in 15 coalitions, but greeted them 17 times. He was not punished: dominant males did not attack him more often for his disrespect in coalitions, since they did so 24 times, which is less than the 52 for Macho and the 31 for Darwin. Similarly, Brutus was not groomed less by dominant individuals than expected for his rank, that is, 147 minutes, which is equivalent to the 140 minutes Darwin received. This indicates that coalitions may be used by inferiors for social purposes and that they may get away with it. What kept the dominant males from pushing Brutus harder to make him think twice before challenging them in coalitions? Was Brutus trading their tolerance for something else? One possibility is that dominant males tolerated his behaviour because they benefited from Brutus in another social context. The best candidate seems to be meat sharing (Table 6.15). Brutus shared more meat with other chimpanzees than all the other males together. Brutus was an important meat provider within the community, and most males gained more meat through him than they would in any other way. It appears he traded meat for tolerance from the dominant males when he threatened them with his coalitions.

Table 6.15: Amount of meat-sharing by the adult males (the amount of meat is estimated as the time an individual provides meat to other group members)

	Amount shared per hunt	
	Autumn 1993 (*N* = 14)	1987–91 (*N* = 88)
Fitz	10'	6'*
Kendo	9'	22'
Macho	5'	20'
Brutus	30'	50'
Darwin	2'	3'

* Fitz was not adult at this time and was allowed to have access to only a much more limited amount of meat (see Chapter 8).

Now we turn to the second question. What do inferiors gain by destabilizing the hierarchy? One possibility would be more sexual opportunities. Dominance has been proposed as allowing better access to sexual partners within a multi-male group (Dunbar and Colishaw 1992). We have already seen in Chapter 4 that males' reproductive success is difficult to evaluate, but the chimpanzees themselves can only use behavioural measures to assess their success. We analysed the matings observed during the two months from August to October 1993 when the males had access to seven adult females with regular oestrus and where we could judge *a posteriori* the real reproductive status of the females. Dominance and age were poor predictors of mating success. The frequency of a male's participation in coalitions was, however, a perfect predicator of mating success with fertile females ($r_s = 1.0$, $N = 5$, $p < 0.005$) (see Table 4.4). This suggests that the ability to enlist and assist other males is the key to getting access to fertile females. Thus, coalitions of inferiors against dominants in Taï chimpanzees is partly tolerated by the fact that dominants gain meat from those inferiors, and the benefit of coalitions is higher mating success.

Note the importance of individual personality for chimpanzee mating success. Brutus was for years the alpha male, and despite his old age remained a very gifted social manipulator. In most cases, he mated on the ground within a party, whereas the other males, including the dominant ones, mated mostly in trees or at the edge of a party. Brutus could rely on the social support of both females and males. Fitz and Kendo, much younger and less experienced, but in their prime, had less social skill and were unable to block Brutus. We know of a comparably strong personality from Mahale. There in the early 1990s, Ntologi, an experienced and socially gifted alpha male, was able to control the other males through aggression and coalition. During this period, in 1992, mating success at Mahale in males correlated with social rank (calculated from Table 17 of Nishida and Hosaka 1996; all mating: $r_s = -0.83$, $p = 0.017$, mating with females in stage 3 (near ovulation): $r_s = -0.70$, $p = 0.04$). Nishida and Hosaka (1996) suggest that the alpha male decides by his presence which other males can mate and thereby favours his coalition partners. A study done ten years earlier on the same community at Mahale showed that, for all females, mating rate correlated negatively with age, but did not correlate with the rank of the males (Hasegawa and Hiraiwa-Hasegawa 1990). However, males in their prime mated more frequently with females near ovulation than did younger or older males. Similarly, females in their prime copulated more frequently with higher-ranking males than did younger or older females (Hasegawa and Hiraiwa-Hasegawa 1990). Thus at Mahale, the personalities of the males present affect mating success. We suspect that some years earlier with Brutus still alpha, the mating rate at Taï would also have correlated with the dominance rank. This illustrates how flexible most aspects of the social interactions can be in chimpanzees and emphasizes the danger of generalizing from only one observation period.

Involvement of females in aggressive interactions by males

At Mahale, females are rarely involved in confrontations among males. Only three instances of females joining forces with males in an attack against another male were observed in a two-year period (Nishida and Hosaka 1996). Similarly Gombe females are described as almost never involved in male confrontations. In one year (1978) there were no observations of coalitions which included females, and only once was a female seen to attack a male (Goodall 1986). This is in striking contrast to the 17 attacks involving females against males in 30 coalitions, and the four joint attacks in a period of just two months at Taï.

Bonobo females are described as intervening in the males' interactions when their adult sons, who may gain dominance through the support of their mothers, are involved (Kano 1992; Ihobe 1992). This resembles the case of Ella helping her son Kendo to become alpha at Taï (see Chapter 4). Apart from this particular situation, female bonobos apparently do not get involved in the males' interactions, and we wonder why Taï females intervene more frequently. We have already seen that their social bonds are stronger as they are more frequently associated with both males and females than are females of other populations, but this does not explain why they should take a more active part in the males' interactions. We suggest that a clue to the puzzle might be found in the way the Taï chimpanzees solve the problem of hunting monkeys (see Chapter 8). The dense forest habitat forces them to cooperate to be successful, and this cooperation can only be stable if food-sharing rules assure hunters a fair share of meat. As these rules conflict with the existing hierarchy, hunters are guaranteed access to meat only with the help of the females, who are in turn granted a specially high position in meat access (Boesch 1994*b*; Boesch and Boesch 1989). The most active females in male coalitions are also the ones with priority access to meat (Loukoum and previously Ondine). Thus, shared interests between some adult females and male hunters at Taï may result in the females becoming active social partners.

Social interactions in wild chimpanzees can vary considerably between populations and within a population over time. This makes generalization difficult, as new observations can present new patterns. Two points, however, emerge clearly. First, chimpanzees have the potential for great social variability. The fission–fusion system has been observed in all wild populations, but the way social interactions are distributed between individuals varies greatly. It is difficult to be sure about the factors that determine differences between populations, but we can approach this question by looking at the variation in the grooming distribution at Taï. Between 1988 and 1993, grooming changed from being mainly reciprocal, mutual, and affected by the age of the males to becoming non-reciprocal, unilateral, and affected by the rank of the males. Between these two periods, the community lost two adult males, a young one reached adulthood, and Brutus invested much more time in grooming females. Thus, both the number and the personalities of the individuals present may affect social interactions.

Second, chimpanzees show capacities for cooperation, reciprocal interactions, and coalitional behaviour. The extent to which chimpanzees engage in coalitions varies both in frequency and, more importantly, in quality. At Taï, coalitions involving males as well as females are used as a social tool by low-ranking males to increase their access to sexual partners. Inferiors seem to buy tolerance of the dominants with meat. During the study period, this seemed to be mainly due to one individual, Brutus. Brutus was an old, experienced, former alpha male, with a low social position, who had to rely on alternative strategies to obtain large sexual gains. In the Taï chimpanzee society, alternative strategies were successful. From the comparison with the Mahale chimpanzees in the early 1990s, it is suggested that when experienced and skilful males are at the top of the social hierarchy, they control social interactions more precisely and favour their supporters (Nishida and Hosaka 1996). Chimpanzees use social strategies to gain certain social goals, and the strategies used can make dominance look despotic or egalitarian.

In Taï chimpanzees, females have a special position. They are regularly involved in coalitions with adult males, are directly involved in aggressive interactions with them (mainly outside the mother–son relationship) and build stable and long-term alliances with other females — in contrast to what has been observed in female bonobos. We suggest that shared interest in meat might account for the special social position of the females. Whatever the cause, the view of chimpanzees as a purely male-oriented society does not reflect the social life in Taï chimpanzees. This is also apparent in the stronger role of female choice in reproductive strategies (Chapter 4), in the importance of female friendship (Chapter 5), in their dominant position in gaining access to meat (Chapter 8), and in their greater involvement in territorial defence (Chapter 7). All this supports our view of the Taï chimpanzee society being bi-sexually bonded.

7 Inter-group aggression and territoriality in Taï chimpanzees

Forest scenes, questions of neighbourhood: In the early days of the project we had difficulty in estimating the size of the territory of the study community. We happened to be following for a certain distance a noisy group of chimpanzees that vocalized and drummed at regular intervals, in what we considered the northern range of their territory. After about two hours of moving northwards, we started to wonder how far they were going and tried to get into visual contact without scaring them by approaching too quickly. In striking contrast to the behaviour we had so far achieved in the habituation process, they immediately ran away at full speed before we could even get a faint glimpse of them. This was quite unlike what we expected. We expected them to be curious and, if they decided to leave, they would do so quietly and slowly. It became obvious that we had been following two communities, first the one under study and then a totally non-habituated neighbour community.

What is it that attracts two communities? Do they have violent interactions or do they peacefully intermingle?

27 June 1985: We were with a large party in the northern part of the territory. The chimpanzees quietly socialized after having fed for some time on the abundant fruit of *Chrysophyllum* trees. Macho had been trying for some time to lead the oestrus female Loukoum away from the other males, without much success. Suddenly, at 14.55, as he had just made another attempt to lead her away, he found himself, without any kind of warning, facing several stranger chimpanzees who rushed towards him. With a tremendous scream of fear, Macho ran towards the other males to be reassured by Brutus and Schubert, the alpha and beta males. They stared for a few seconds in the direction of the strangers, then the three of them followed by four other adult males charged with typical waa-barks against the strangers. Possibly surprised by the violence of this immediate reaction, the strangers fled and were pursued for many hundred metres before the two parties apparently settled down to exchange signals by drumming and calls.

How could the strangers so perfectly surprise our chimpanzees? What was the reason for this attack? In this chapter we analyse these and many related questions about the territorial behaviour of chimpanzees in the Taï forest .

Territoriality has been observed in many species of insects, fishes, reptiles, birds, and mammals. It is usually a behaviour that one individual, most often a male, performs to control a valuable resource that will give him an advantage in gaining a mate. The resource may be a rich food patch, a breeding site, or a conventional mating site (as indicated in the lekking behaviour of some birds and antelopes). Territorial possession can vary from several days to year-round ownership. Territories held by groups are rarer than those held by individuals, but groups can control large areas. Territory defence is generally strict, and aimed at excluding a certain class of individuals of the same species. For sexual territories, competing males are chased off, for social territories, most strangers are repelled, except for dispersers.

Animal and human territorial behaviour have been contrasted for two main reasons. In humans, territorial defence is primarily a communal activity. In all human societies, men are involved in the defence of territory in cooperative groups. This includes aggression aimed at weakening neighbouring groups and acquiring new mates. Aggression can include systematic attacks on weaker groups, which cause the extinction of these groups after most male members have been killed. Social anthropologists consider human territoriality to be special and call it 'war'. War is characterized by raiding, killing, kidnapping, and cooperative strategies to harass and defeat neighbours (Boehm 1992; Durham 1976; Dennen 1995; Chagnon 1988). Warlike activities, although highly emotional, include multiple rational actions, for example keeping a tally of acts of aggression or planning deliberate action to capture territory and make conquests. In addition, war is reciprocal in the sense that it involves an exchange of hostilities by two social groups that alternatively take the role of aggressor or defender (Boehm 1992). Tests of the purely human origin of war are limited by our ignorance of complex territorial behaviour in animals.

However, there have been some attempts to compare chimpanzee and human territoriality, and it has been suggested that they share features of inter-group aggression (Nishida *et al.* 1985; Goodall 1986). At Gombe, the destruction of a small community by a larger one, including systematic attacks and killing of individuals by a group of males from the larger community, has been observed, and the same process has been inferred at Mahale. Some descriptions of intentional controls of territory borders and the dominance relationship between communities are available (Kawanaka and Nishida 1974; Goodall 1986). On this basis, it has been proposed that human and chimpanzee territoriality have much in common (Goodall 1986; Manson and Wrangham 1991). All these observations come from two communities of chimpanzees living in Tanzania, while information from other chimpanzee populations with less detailed observations leave open the possibility that territorial behaviour may be less prevalent and especially that some females may be less rigidly attached to a single community (Reynolds and Reynolds 1965; Suzuki 1971; Sugiyama 1968; Wrangham *et al.* 1992). In some cases, these inferences are based on observations of unidentified individuals and cannot be reliable for judging individual attachment to a community. Detailed observations on more chimpanzee populations are needed to estimate the diversity of chimpanzee territorial behaviour.

Two functions have been identified for inter-group aggression. First, it increases access to fertile females (Chagnon 1988; Manson and Wrangham 1991). Second, it increases the amount of resources available by making land conquests and eliminating competitors for resources (Dennen 1995; Dunbar 1988; Durham 1976). The importance of each factor in explaining inter-group aggression is still unclear. Inter-group dynamics depend on the relations between two neighbouring communities. Unstable territorial relations are produced when there is a large discrepancy in the value of the resources defended, or in the power of the coalitions holding the territories (Manson and Wrangham 1991; Durham

1976). Where there are large discrepancies, aggressive interactions would be favoured in territorial species, including human warfare. The two examples of community extinction in chimpanzees (Goodall *et al.* 1979; Nishida *et al.* 1985) suggest that once the size of a community is reduced below a certain level, aggressive interactions can lead to fatalities.

In this chapter, we analyse the territorial behaviour of the Taï chimpanzees over a period of thirteen years, including 129 territorial interactions. Throughout the study, the dominance relationship between neighbours seemed to remain about even. This differs from the situation observed at Mahale and Gombe, where the power imbalance between communities was much more marked. We concentrate on the strategies of the macro-coalitions of males used to protect the territory and harass their neighbours. Because such observations are scarce, we describe some of them in detail.

Territory use in Taï chimpanzees

As at Gombe and Mahale, all males and adult females in Taï chimpanzees remained attached to one region for years. The fidelity to a territory is the rule for male chimpanzees. Females transfer between communities at adolescence, but once they have integrated a new community, they remain there until their death. Only at Mahale have some adult females been observed to transfer a second time (see Chapter 3).

Territory size

To study fluctuation in territory size and use, daily foraging routes of the target individuals were plotted on maps each month from 1982 onwards. A sample of 18 months was analysed. As can be seen from Table 7.1, territory size varies according to the method used to calculate it and can differ by 30%. The restrictive polygon (RP) method includes only the area in which we actually saw the chimpanzees; we use this method to compare territorial use between different months. The minimum convex polygon (MCP) method, which joins observations along the periphery, overestimates the area really used, but this compensates partly for the days we did not follow the chimpanzees. This method therefore yields estimates that are probably closer to the real territory size. For the figures, we used the MCP method, but remember that we rarely followed the chimpanzees for more than 20 days per month.

Thus, the estimated size of the territory depends on the method used. It also depends on the season and the year, independently of the method. Average figures for the autumn (September to December) covering the whole study period indicate that the territory varied from 16 km² to 27 km² (MCP method), the autumn was chosen for its high level of chimpanzee activity, which allowed us to follow them more consistently.

Territory use is irregular in that chimpanzees spend most of their time in the centre of the territory and use the periphery much less frequently. Table 7.2 suggests that on average the chimpanzees spent 5% of their time in the peripheral quarter of the territory and 75% of their time in only 35% of the territory (the core area).

Factors influencing territorial use

The use of the territory can depend on many factors. We analyse the influence of three of them, seasonality, annual variation, and demographic change.

Table 7.1: Territory size (in km²) of the Taï chimpanzee community, measured using two different methods. (RP = restrictive polygons, MCP = minimum convex polygons).

Period	RP	MCP
Autumn 82	18.1	19.5
Autumn 89	23.7	26.9
Autumn 95	15.5	16.5
Mean	19.1	21.6
Seasonal variations		
9–89	12.1	13.6
10–89	15.4	22.2
11–89	19.4	23.2
12–89	14.5	15.7
1–90	11.6	12.6
2–90	14.5	16.1
5–90	13.5	16.7
Mean	14.4 ± 2.5	17.2 ± 4.1
Yearly variations		
9–89	12.1	13.6
9–90	20.6	23.5
9–91	14.6	15.6
9–93	16.4	18.2
9–95	9.7	10.0
Mean	11.7 ± 4.1	16.2 ± 5.1

Table 7.2: Differential use of the territory by the Taï chimpanzees. Size of the area used by chimpanzees in km² (proportion of the territory in brackets)

	Size of area		
	100%	95%	75%
Autumn 1982	19.5	15.1 (77%)	7.4 (38%)
Autumn 1989	26.9	20.0 (74%)	8.7 (32%)
Autumn 1995	16.5	9.0 (55%)	5.7 (35%)

Seasonal variations

Food resources change with the season and their spatial distribution shifts enough to oblige the chimpanzees to use different areas of their territory at different periods of the year. This is particularly true when the chimpanzees feed on staple fruits that grow in different parts of the forest. For example, *Sacoglottis gabonensis* grows along rivulets in the low part of the forest and its fruits are eaten daily in September and October, whereas the fruits of *Coula edulis*, *Dacryodes gabonensis*, and *Trychoscypha arborea* grow mainly on the dry ridges and they are eaten daily in December and January. A comparison of the territory used from September 1989 to May 1990 shows that territory size changed very little (Table 7.1), and the centre of activity shifted only 500 meters in latitude and 1.5 km in longitude. The main shift occurred between September and October, while both the

Table 7.3: Distances travelled by the chimpanzees during different months of the year

Month	Hours followed	Distance per hour	Daily range	Territorial encounters (whole study)
December 88	35	850 m/h	10.2 km	0.169
January 89	50	940 m/h	11.3 km	0.337
February 89	32	290 m/h	3.6 km	0.102
September 89	65	620 m/h	7.4 km	0.191
October 89	79	720 m/h	8.7 km	0.125
November 89	66	560 m/h	6.7 km	0.123

fruiting season of a major staple fruit (*Sacoglottis gabonensis*) and the location of hunting activities remained similar.

This stability in the use of territory can be partly explained by the fact that the chimpanzees cover large distances each day all year round (Table 7.3). However, daily ranges vary extensively over the months in a way that is not easily explained. First, the months with most fruit production (December and January) are those during which the chimpanzees had the largest daily range, whereas they moved the least during February, when abundant *Coula* nuts were available. Thus, fruit abundance is not a good predicator of daily range. During the hunting season, they moved about 3 kilometres less per day than during the fruiting season (December–January), although they were searching for prey. In addition, monthly daily ranges did not correlate with territory size. All this suggests that neither the pattern of food abundance, nor the distribution of food patches completely explains territory use and daily range.

The second candidate is a social factor, that is interactions with neighbouring communities. In Table 7.3, we list the average frequency of territorial encounters for each month observed. Daily range length correlates positively with territorial encounter rate ($r_s = 0.83$, $N = 6$, $p < 0.05$) and territory size shows a tendency to correlate negatively with encounter rate ($r_s = 0.75$, $N = 6$, $p = 0.08$). Thus, territorial encounter risk, not food distribution or abundance, appears to explain the variation in territory size used by chimpanzees over short periods of time. Taï chimpanzees move over longer distances during periods when the probability of encounters with neighbours is high, but they tend to do so within a smaller area. Daily range during the fruit-scarce months of April to August 1988 (Doran 1997: 0.25 to 0.10 km/h) were comparable to February 1989 when the probability of encountering neighbours was low. Daily ranges at Taï are very high compared with those observed in other chimpanzee populations (Gombe: 3–4 km per day (Wrangham 1975); Kibale: 2–3 km per day, (Wrangham, personal communication)) and it might be explained by a lower frequency of territorial encounters in those populations.

Annual variations

Over 13 years, this community consistently used the same parts of the forest (Fig. 7.1). The size of the territory decreased in the 1990s, but the centre of activity remained very constant (Fig. 7.lb). From the autumn of 1982 through to the autumn of 1995, the centre of activity moved less than 300 metres which represents a shift of about 6% relative to the diameter of the territory. Although territory size has changed over the years, the community has always centred their foraging movements around a stable core. The overlap

(a)

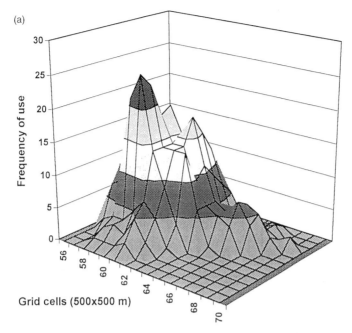

Fig. 7.1(a): Territory use by the Taï community: October to December 1982.

(b)

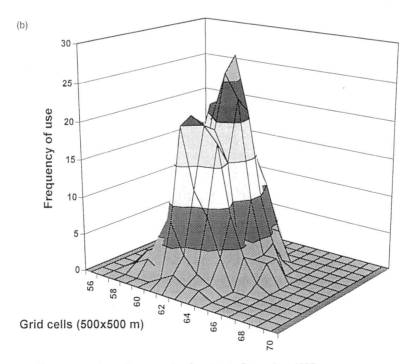

Fig. 7.1(b): Territory use by Taï community: October to December 1995.

between years was always over 90%. Such variations are within the range of the seasonal variations (Table 7.1) and show that neither season nor years are sufficient to explain the changes in territorial use over the 13-year period.

Influences of demographic changes on use of territory

A comparison between the same month over seven years showed a strong tendency for territory size to decrease with the number of adult and adolescent males present (Fig. 7.2). This tendency was less clear when we used only the number of adult males, and disappeared when we used the number of members within the community. Thus, it seems that the only steady shift we saw in territorial use in the study community over the 13-year period can be explained by the decreasing number of males, who represent the fighting power of the community (see below).

The fact that territory size did not vary in parallel with the size of the chimpanzee community resulted in a decrease in the density of the chimpanzees living in this area (Table 7.4). This must have led to a decrease in feeding competition within the community, and reduce the need to forage in the periphery.

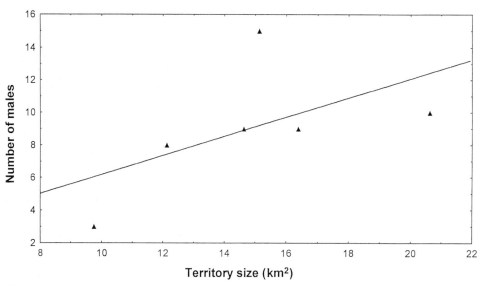

Fig. 7.2: Territory size in relation to the number of males in the community ($r_s = 0.78$, $N = 6$, $p < 0.06$).

Table 7.4: Territory size and demographic changes in the Taï chimpanzees over a period of 13 years

Period	Size (km²) a	Number of males	Community size b	Chimpanzee density b/a
Autumn 1982	18.1	9	74	4.1
Autumn 1989	23.7	6	66	2.8
Autumn 1995	15.5	2	29	1.9

In conclusion, the variations in the size of the territory used by the Taï chimpanzees, both within a year and between years, seem to be explained principally by the relationships with their neighbours. Both the risk of territorial encounters and the intrinsic power of the community affect the territory size and use.

Overlapping zones and territory borders

In some territorial species, neighbouring territories overlap extensively, making it difficult to draw clear boundaries. The parts of a territory in which members of other social groups can be encountered are called the 'overlapping zone'. This zone can be considered as part of the territory of both communities, whereas the exclusive zone is used only by members of one community. In 1989, 53%, and, in 1995, 56% of the territory formed part of an overlapping zone. In Taï chimpanzees, overlapping zones include important parts of the territory. The core area is defined as the area within the territory where the owners spend 75% of their time. In the autumn of 1989, the core area included 32% of the territory, of which 89% was never entered by neighbouring chimpanzees. Similarly, in autumn 1995, the core area entailed 44% of the territory, of which 82% was never entered by neighbouring chimpanzees. Thus, it seems that the territorial use is greatly affected by the risk of encounters with neighbours and that the area in which the chimpanzees spend most of the time is the one that is safe from encounters with neighbours.

Territorial relationship between neighbouring communities

Taï males often invest time and energy in defending their territory or locating their neighbours. In 45 months between 1984 to 1991, we observed 74 territorial activities, or about 1.6 per month. By correcting for those we missed because we did not find the males every day, we estimate that at least every two weeks the males engaged in territorial activities. Table 7.5 shows how territorial activities towards the neighbouring communities were distributed.

The distinction between north, east, south, and west communities is arbitrary, as we have not yet identified individuals in them and have based the distinction on the geographical location of the interactions. Only in the south could we test our assumptions, since the southern community is in the process of habituation with the help of students

Table 7.5: Distribution of the territorial activities by the Taï chimpanzees from 1982 to 1995, when at least auditory contact was established with one of the neighbour communities: 1 stands for the study community, and 2 for the neighbouring communities

Neighbour community	Auditory contact	Visual contact	Attack initiative		Incursion	
			1	2	1→2	2→1
North	30	16	8	7	31	7
East	36	13	9.5	7.5	35	7
South	20	7	5	11	8	13
West	5	3	1	2	3	2
Total	91	39	23.5	27.5	77	29

and field assistants. Our assumption was supported by their presence in the area at the time of the territorial encounters with the 'southern' community. Our study community lives on the western side of Taï National Park, and we believe that the western neighbour community has suffered badly from poaching by farmers, which would explain their low frequency of interactions and generally much weaker response in territorial activities. The conflicts with the northern and the eastern communities were more frequent in all respects: more frequent incursions into their territory were made, and they were heard and seen more frequently, which is also a sign that the neighbours were in the vicinity more frequently. The study community attacked these two communities as frequently as they themselves were attacked, giving us the impression that while the tension between these two neighbouring communities varied, their forces were about equal, and the situation remained basically stable.

Without individual identification of the members of a neighbouring community, it is impossible to estimate precisely its size and the number of males. Therefore we used visual and auditory cues to draw limited conclusions. Most of the time, direct attacks by our community were conducted by five to six males, while other community members held back. This roughly corresponds to the number of stranger males we could distinguish during such attacks. Our impression that the western community is the weakest of the five is supported by the very weak responses we could hear when the study community entered their area, and also by the few visual contacts we had, once with one male and once with two males. It is intriguing that the study community did not expand westward. However, pressure from poaching and illegal agricultural clearings within the territory of the western community may have inhibited the invasion of this region. We suspect the study community lost four adult males in a row there in 1984, probably to poaching, emphasizing the devastating impact of humans at the western border of the park.

Territorial strategies of Taï chimpanzees

Taï chimpanzees were followed during 129 territorial activities, of which 91 included at least auditory contact between the two communities. Because we only followed one community, our observations are biased. We also judged the actions and the reactions of the other communities mainly by sound. We could recognize different strategies used by chimpanzees to control their territory. They occurred at two stages: before or after strangers were heard (Table 7.6), and after they had decided to make visual or physical contact with them (Table 7.7). We discuss them in some detail so as to understand the conditions under which a strategy is used.

Patrols

In 29% of the territorial activities, the males of the community actively searched for signs of the presence of the neighbours by patrolling along or within the territory of the strangers. Patrols are initiated by the males prior to any contact with strangers (Table 7.6). In 87% of the patrols, the chimpanzees made deep incursions into the territory of their neighbours, sometimes more than one kilometre. Typically, when a group is foraging near the edge of their territory, the males may leave the females without any special calls

Table 7.6: Frequency of different strategies used by the Taï chimpanzees during 129 territorial activities

Strategy	Auditory contact	
	No	Yes
Patrol	36	2
Check	–	32
Avoidance	–	14
Attack	–	45
Total	36	93

Table 7.7: Strategies used by Taï chimpanzees when attacking neighbouring communities

Strategy	Numbers observed
Frontal	16
Back/forth	9
Rearguard	9
Lateral	6
Commando	5

obvious to us and start a patrol. The adult males are the main patrollers; their disappearance is often sudden and silent. They may then wait at a distance for more males to join them, for as a rule, fewer than four males do not go on patrol (96% of the patrols had four or more males, see below).

A patrol is typified by the resolute and silent progression of the chimpanzees: they advance rapidly in a line one behind the other and stop regularly to listen and search for signs of chimpanzees. The farther they intrude into the neighbours' territory, the more carefully they progress and the more regularly they stop to listen. In this forest, sight does not help to locate strangers, but listening is very important. They remain silent during the whole patrol, very occasionally eating a leaf or two. At some crucial point, such as a ridge, they may stop and listen some time before going on. They cover large areas within the stranger territory, and normally come back into their territory at a different point. They regularly smell at tree trunks and leaves, even more so if they come upon fresh traces of chimpanzees, such as a wadge of fruit, a nut-cracking atelier, or a nest. Some may climb to a fresh nest but leave it undisturbed. The end of a patrol, when re-entering their territory, is usually signalled by very loud drumming. The only feeding activity that we have regularly seen during a patrol is hunting for colobus monkeys (12 hunts on 43 patrols). In strong contrast to hunting within their own territory, they remain silent throughout the hunting episode and if a squabble happens between them, the screams that would normally be so loud are totally suppressed.

Patrols were usually not very successful, in the sense that they found the neighbours in only 26% of them. Patrols are risky, for there is no additional help to count on if outnumbered by the strangers. Because party size in Taï chimpanzee is on average ten individuals, whereas 66% of patrolling parties include less than five, mostly only males, being outnumbered is quite likely. It seems that the goal of patrols is not to win a battle against neighbours, but rather to gain information on their location, and, if they find them, to try to unsettle them.

Response to auditory contact with neighbours

The study community advanced towards the strangers they heard in all but 14 cases of 98 auditory contacts. In two of the avoidance reactions, the chimpanzees we followed were in a small party of two to three individuals, and in two other cases they had just captured a monkey and were eating meat. In these cases they silently moved away, the strangers most probably not being aware of their presence.

In one third of the auditory contacts, no visual contact was made between the two communities, but they exchanged aggressive drumming for up to 30 minutes. When they heard strangers, the males usually gathered to reassure each other, sometimes uttering little screams, and then moved silently towards the strangers. This distance to the strangers varied from 80 m to 2 km, depending both on the distance between the two communities when they were heard, and on the motivation of the chimpanzees to approach them. They then usually started to drum very loudly and maintained this auditory contact for up to about 30 minutes (Table 7.8). Then they progressively moved apart, with the males joining the females that had stayed behind. Drumming during territorial encounters is especially loud with markedly more hits on buttresses than usual and is accompanied by typical climax screams following the pant-hoots that contain many loud high-pitched waa-barks that are absent in the pant-hoots performed in ordinary social contexts.

We classify such drumming exchanges between members of different communities without visual contact as 'check', since two types of information are transferred. First, they make it clear that they consider this area belongs to their territory by repeatedly drumming at the same spot for 20 to 30 minutes. Second, by drumming and screaming all together, they convey information about their number. In this context, their calls are especially loud, and they support each other vocally much more systematically then they do in other social contexts. In addition, females regularly join in drumming on these occasions. Female drumming exaggerates the force present, for it is not possible to differentiate it from the males' drumming, neither are voices easy to distinguish between males and females when barking and hoohing. This loud display is reciprocal between the commun-

Table 7.8: Duration of contacts, party size, and frequency of social support in different types of territorial encounters in Taï chimpanzees.

Activity	Number	Auditory contact (minutes)	Visual contact (minutes)	Party size	Support*
Patrol	38	2	0	8.2	11%
Check	32	29	0	7.0	38%
Avoidance	14	37	0	6.1	0%
Attacking	23	64	8.5	9.4	61%
Being attacked	20	47	6.7	9.1	65%

* Under 'support' we give the percentage of cases in which the leading party received support from other group members.

ities. By listening carefully, the chimpanzees can probably judge the strength of the neighbour and decide whether to move forward or not. And one might expect the neighbours to use the same tactics with regard to exaggerating the size of their fighting force by including the females in the drumming.

Attack strategies

In 48% of the auditory contacts, the communities came into visual contact with one another. The decision to enter into visual contact with the strangers was taken very quickly, possibly within minutes of noticing them. The number of males present plays a decisive role (see below and Table 7.9). The result of these 'attacks' was visual contact that normally lasted just a few minutes (Table 7.8). In all instances, the strangers that first saw the approaching chimpanzees uttered long loud screams and ran away. These screams have a special intensity. Within seconds the other community members know what is going on, and they immediately rush in to support. In our sample, we never observed an attack without a counter-attack, and support was regularly provided by nearby community members. When the study community was surprised by drumming from strangers, the males reassured each other, uttering soft screams, and within minutes they silently attacked the strangers. In these cases, support was frequently provided by other community members (Table 7.8). During the fights, additional support may be provided by females and males, depending on how the battle is developing.

When the study community attacked, we distinguished five strategies (Table 7.7).

The frontal attack

The most straightforward strategy is the frontal attack. The chimpanzees simply head directly towards the strangers, and, when they see them, start to charge. These frontal attacks were mainly used when the study community contained more than eight adult males, which suggests that this is a strategy for strong communities. Except for the first emotional screams, the march towards an attack is always totally silent. The males walk closely one behind another, regularly turning towards the following male with a fearful open-grin face and seeking reassurance. The progression after the first hundred metres is done with a typical 'Indian walk', similar to that seen sometimes during the last metres before a hunt starts. The males avoid stepping on branches or leaves that could produce a cracking noise and place their feet and hands cautiously, typically shifting their body weight gradually from one side to the other as they move on. The same male regularly

Table 7.9: Territorial strategies used depending upon the number of adult males present

	1–3 males	4–6 males	7–9 males
Check/avoidance	67%	35%	17%
Patrol	17%	37%	20%
Attack	17%	28%	63%
Total	18	76	30

Statistics:
  1–3 males versus 4–6 males: $X^2 = 5.88$, $p = 0.05$
  4–6 males versus 7–9 males: $X^2 = 11.7$, $p < 0.005$

leads the progression throughout a whole attack. When more calls are heard from the strangers, they carefully listen and adapt their progression, always remaining totally silent. When approaching into almost visibility distance of the strangers, the attackers slow down even further and fan out so as to charge in a line. Of course, following behind, we cannot judge exactly what the chimpanzees see. They seem to make a first judgement about the number of chimpanzees. If they estimate them as numerous, they start the charge from farther away and make aggressive waa-barks earlier than if they have esti- mated them as being small in number, in which case they then go in much closer, start the attack silently, and try to catch them.

During a frontal attack, all males rush towards the strangers, giving loud attack calls, powerful high waa-barks made with the mouth wide open, but the teeth fully covered by the lips. We heard these 'attack calls' only in attack and counter-attack situations. The surprise effect is impressive and attacked individuals always retreat at first. These initial attacks are also the ones in which physical contacts are most frequent. Bad bites can be suffered within a minute. The ideal situation is to hide and surprise the strangers that are feeding in a tree when they come down. Once, Macho, alone in a tree, was caught like that unexpectedly by two males from the southern community. He was almost immediately rescued by an adult female and three males that arrived to support him within the next minute. However, Macho had already been bitten nineteen times and he was bleeding from all the wounds. None looked really bad, but one bite had missed his eye by about a centimetre.

As they retreat, the attacked individuals probably judge how close and how numerous the aggressors are. Sometimes they stop their flight after about 50 metres, at other times they flee for up to 400 metres. If the retreat is long, the situation is clear and counter- attacks are brief. The communities rapidly part. If, however, the strangers stop, the pur- suers do so too and wait for a reaction. The strangers might be more powerful and may chase the attackers over a long distance. In this case too, the situation is clear and the communities part rapidly. If, however, the powers present seem equal, probably judged by the mutual abundance of screams and the number of individuals that are seen (it is impossible in this forest to see all of them), then the frontal attack changes into a back- and-forth attack (see below).

Females with and without infants may join a frontal attack or counter-attack, but they tend to avoid direct physical contact with members of the other community. We have seen mothers of the study community taking part in an attack as well as stranger mothers with some males attacking the study community. Owing to poor visibility in the forest, the chimpanzees have not always the time to check on the number of adult male pursuers and may run away although they may be wrong in their estimate of potential attackers.

Case study 1, a frontal attack: 1 July 1985: At 12.45, a party of 25 chimpanzees including all adult males moves as a compact group north-west in the northern part of the territory (I suspect they heard something I did not). Ten minutes later, I hear the strangers drum not too far to the north. Without a sound Brutus heads directly towards them. Five minutes later, one of the chim- panzees makes some fear shrieks. Brutus and others immediately quieten him. At 13.10, the strangers call from the north-east. They all turn silently in this direction, walking so as to avoid any twigs that might crack under their feet. At 13.35, we hear calls and screams of normal social activity among the strangers. Brutus reassures Schubert, the beta male, and they move on. The adult females regularly seek reassurance from the adult males. All scrutinize the vicinity carefully for any stranger. Now we can hear them feed on leaves in a tree. Four of the ten adult females turn back and head southwards. The males, from a distance of about 40 metres of the strangers, fan out in a line, with Brutus to the extreme left and Schubert to the right. More females now head backwards. Schubert seems to hesitate, while Brutus, with very loud aggressive calls, starts the attack, followed by the males beside him. Schubert follows, but some metres behind. The last females now leave. The males attack on their own. The strangers that are now climbing down from the trees are very numerous, and within a minute the attackers are running away screaming with four stranger males on their heels. A second wave of strangers follows, but they see me and make a detour before running to support their front males. In the second wave of counter attack, I see two stranger mothers with their infants on their backs. The attackers face the strangers some 100 metres away, they counter-attack briefly, but the strangers pursue them again over some 200 metres, and the attackers retreat rapidly.

Back-and-forth attack

In this case, both parties present are about equal in power and motivation and they will both attack one another (Table 7.7). In an extreme situation, one can observe two lines with all the adult males and some females facing one another, the attacks alternating from one side to the other. In other situations, they are more spread out in the forest, and we have seen parties of two to three males attacking the other side. These attacks seem to be coordinated vocally through the attack calls. In two of these back-and-forth attacks, lasting over twenty minutes, the opponents calmed down, just facing and threatening each other. Five young oestrous females quietly crossed the lines to join the males on the other side, mated with one or two of them, and returned calmly back to their community. These were the only cases in which we witnessed sexual activity during inter-community aggression. None of the females transferred for good, nor did these visits result in conception in the females of the study community.

Rearguard-supported attack

During encounters in which many females are present, the males fighting at the front may receive the support of the females, who start moving towards the battle line, making very loud waa-barks and drumming a lot (Table 7.7). They listen to the calls of the males, and each time their males attack, they amplify their calls and rush forwards. Often one of the older males or adolescent males joins in with the females and they produce a very impressive rearguard that progresses towards the front. If battle screams increase, they accelerate, and the young males may even rush to support the fighting males at the front. This appears to generate significant support for the males, and there is no way for the strangers to know for sure how many of the drummers in the rearguard are actually males that might provide physical support for the fighting males.

This strategy may be intentionally deceptive when some males remain at the back, drumming and repeatedly calling loudly, while other males move silently towards the opponents. There, the front males may wait silently for the strangers advancing to surprise the noisy rearguard, unaware of the close presence of the silent males who may then attack unexpectedly.

Lateral attack

Once the study community had less than nine adult males, Brutus, at the time alpha male and clear leader in most inter-community encounters, started to lead lateral attacks instead of frontal ones (Table 7.7). A lateral attack occurs when the advancing and strictly silent males aim their progress not straight towards the audible opponents but laterally. In this way they avoid the noisiest and possibly also largest party and look for individuals in smaller parties that they might defeat. In three such lateral attacks, they found a small party of strangers and chased them for many hundreds of metres. The main party, containing probably most of the males, knew by the screams that some of their community members were being attacked and had to head backwards to support them.

Case study 2, a lateral attack, 3 October 1987: At 14.25, a large party, including six males, is resting under a group of red colobus to the far eastern side of the territory when they hear strangers drum further far off to the east. Responding with some screaming, they all move towards the drumming. Rapidly a seventh male joins them, while eight females also follow. After five minutes of silent progression, Brutus leads them slightly more to the north-east, and rapidly all the females with Kendo and Ulysse leave them. The five males, together with one juvenile male and an adolescent female, now progress more carefully, listening for any signs of the strangers. They also look toward the south-east where a few calls of the strangers can be heard, apparently unaware of their neighbours' approach.

Fifteen minutes after the first signs of the strangers, Brutus starts a quick and silent run followed by the six others. I follow, trying not to make too much noise. Some 100 metres further, they appear to have found some strangers as I hear the violent barking and anxious screams of two chimpanzees. The pursuit lasts for a minute, eventually all the barking concentrates on one point. I join them as Kendo overtakes me, running and barking aggressively.

A stranger female with a three to four-year-old infant has been trapped and is surrounded now by six males, while the adolescent female is drumming nearby. They prevent the stranger from making any move. Her infant is clinging to her belly. She screams and barks at the males. The males hit and bite her on the head, shoulders, and legs repeatedly. I don't see blood on her or on the ground. They never torment her for more than a minute, let her rest for a moment, but pull at her leg when she tries to move away. Whenever the males hit her, she lies down face to the ground, and covers the infant with her body. The hands and feet of the infant remain visible during the whole attack, but I never see any of the males grab or bite the infant. Five minutes after her capture, the stranger males arrive to rescue her, aggressively barking and screaming. Immediately, Brutus and Rousseau face them in a line with Kendo and Snoopy. In no time, the female rushes towards her males. For some seconds, the two parties face one another, the strangers being three or four in number, all males. Then, Brutus, followed by all the others, attacks and the strangers disappear without a sound. The chase unfolds over about 200 metres. While Brutus and Rousseau return quickly towards the west, the younger males with the adolescent female remain in the east drumming and barking for another 25 minutes before joining the group, which then heads back north-west.

The lateral attacks appear to be intentional; while going sidewise, the males constantly look towards the direction of the stranger calls. This strategy allows a community in decline to win fights that might have been more difficult to win in a frontal attack. This strategy was initiated by Brutus, the best hunter of the community. He is the one who made most of the complete anticipations during the hunts and the only double anticipations we saw (Chapter 8), demonstrating an understanding of the effect of other hunters on the reactions of the escaping colobus monkey. Brutus might have understood that there were not enough adult males in his community any more for frontal attacks and then favoured the lateral strategy.

Commando attack

The last strategy, and the most dangerous one, is the commando attack (Table 7.7). In a commando attack, a group of adult males makes a deep incursion into a neighbouring territory looking for strangers. Such incursions last up to six hours. When they find strangers, the goal seems to scare them as much as possible, rather than attempting to defeat them physically. Therefore, they attack by surprise and pursue strangers if they run away, but when they themselves are counter-attacked, they simply flee until they stop being pursued and return to their own territory. Such a commando attack is very silent and invests much time in searching for strangers. On two occasions, they found isolated mothers and kept them prisoners, preventing them from moving away, biting and hitting them. The bites were concentrated on the head, shoulders, feet, and oestrus, and did not appear to be serious.

Case study 3, a commando attack, 23 September 1993: At 9.45, I hear drumming far to the east. At first, the chimpanzees do not react. At 10.00, the four males I was following, Macho, Kendo, Fitz, and Darwin, move silently towards the east (at this time only five adult males were alive in the community). They move decisively, crossing a valley and following the ridge on the other side. They listen regularly, sometimes sitting for about three minutes. Then they enter a valley two kilometres inside the strangers' territory. Forty-five minutes later, they show some interest in black-and-white colobus monkeys and make two hunting attempts. They then turn north. At 11.45, they all rest silently in a natural windbreak. Suddenly at 12.10, the strangers scream, not too far north in the valley. Immediately, the four get up with bared teeth, reassuring one another. Then they all move towards the strangers, but after only 100 metres stop to listen, to make sure that they are going to surprise them and not the other way round.

All morning, Darwin has led the party and they seek regular reassurance with one another. After 10 minutes of a very careful approach, they start to watch where they place their feet and hands, avoiding making any noise. I do my best, but they regularly look at me after a dead twig cracked under my weight. After 18 minutes, they spot the strangers in a tree. They still progress and at 12.32, I see that the strangers are eating in a fig tree, some infants playing in the branches, all totally unaware of the presence of other chimpanzees. Fitz sits and looks at them, partly hidden by the foliage. Darwin and Kendo place themselves on each side of Fitz, while Macho remains a bit behind them. Obviously they are waiting for the chimpanzees to come down, but they are resting by now and we hear their lip-smacking as they groom one another.

Suddenly, at 12.55, Darwin gets up and silently threatens the strangers by arm waving. Immediately, Fitz, Macho, and Kendo bark and rush to drum on the fig tree. Kendo climbs the neighbouring tree to threaten them at closer range. The strangers are totally silent. At 12.56, all the strangers bark and scream aggressively. Kendo rushes down, but the strangers don't move and the four males join up again under the fig tree and drum. Once more the strangers call aggressively without moving. At 12.59, an adult male rushes to the ground, running away. Kendo and Macho immediately chase him towards the north and by the intensity of the calls seem to catch up with him. Hearing that, the strangers call and rush down the tree in a close pack. I count five mothers, two adult males, and three more chimpanzees, all probably going to rescue the first male, pursued by Darwin. Fitz remains under the tree looking up and threatening those that remain in it. The fight in the north has changed into a pursuit. Now an infant is screaming in the fig tree, blocked by Fitz. Three stranger males counter-attack against the four which allows the infant to rush down with another adult male that had remained silently with him in the tree. Now for three minutes, the two groups of four males chase one another, the two brothers, Kendo and Fitz, leading the attacks. Soon, the four males of the study community head back westwards, drumming repeatedly and loudly. The strangers move east and after 10 minutes are not heard any more, while the four still drum for five more minutes. They join the group of the morning one hour later.

Many patrols were probably aimed at finding and attacking strangers, but were not classified as commando attacks as they did not find any. The commandos we followed were impressive by the intensity with which the males searched for strangers and the importance of the incursion within the stranger territory. Sometimes they waited and listened silently for hours before an attack. We twice saw the study community being victim of a commando attack, in one of which Macho escaped with nineteen wounds.

Assessing the imbalance of power between communities

We should expect violent encounters between groups when either the resources or power between two neighbouring communities are out of balance. Imbalance of resources would motivate a poor group to improve its conditions. Imbalance of power would signify low costs for the stronger group to attack the weaker one. Lions hearing roars of strangers within their territory will approach them at a rate that depends upon the number of intruders and the size of their own group (Grinnell *et al.* 1995). Males approach more slowly when they are outnumbered. Animals tend to forage in groups larger than the expected optimal size if they face the possibility of an inter-group confrontation (Zemel and Lubin 1995).

Table 7.9 shows the use of territorial strategies according to the number of males taking part in the action. Support from other members of the community depends upon the action that is going to ensue and is most frequent in the case of attacks (Table 7.8). We also see in Table 7.8 that small groups of males concentrate on checking for the presence of strangers by drumming and listening to the response. Parties of between four and six males more commonly go on patrol inside the strangers' territory, which is the dominant strategy of parties of that size. Large parties of six or more males mainly carry out attacks. When considering adult males and females present in the party, no significant difference is observed. This suggests that Taï chimpanzees take into account the number of adult males and not the total number of chimpanzees that are present when deciding a strategy. At various times we saw parties of four males waiting, we presumed, for more males to join them, but as none came, they limited themselves to check or patrol the area.

In the course of our study, the community declined in size and in the number of adult males (Chapter 2). As the males became less numerous, they changed their strategy (Table 7.10), which is what one could expect if intrinsic power plays a role. When in 1988, the number of adult males fell below eight, they started to be more careful, investing more time in going on patrol and less in direct confrontation. Strikingly, from 1992

Table 7.10: Chronological variation in the territorial strategies used by the Taï chimpanzees

Period	1982 1985	1986 1987	1988 1989	1990 1991	1992 1995
Community size	78	79	70	53	41
Number of adult males	9	9	7	6	4
Check/avoidance	30%	40%	18%	17%	55%
Patrol	17%	28%	52%	50%	10%
Attack by 1*	19%	22%	26%	7%	25%
Attack by 2	35%	10%	4%	27%	10%
Total	26	25	27	30	20

Statistics:
1982–87 versus 1988–91: X^2 = 9.61, p <0.01
1990–91 versus 1992–95: X^2 = 13.20, p <0.002

1 and 2 as in Table 7.5.

onwards, for the first time since it was possible to make observations, avoidance behaviour, in which the chimpanzees retreated in silence from the strangers, became the dominant strategy and avoidance replaced the checks used before. This suggests that they were aware of the imbalance of power and avoided taking risks.

There are many indications that chimpanzees seem aware that the power of the coalitions facing one another determines who will win a confrontation. The macro-coalition of males in a community adapts their behaviour to play the best game with the cards they hold. When a confrontation is expected, supporting individuals join the front party. The front party then adapts its strategy to its own power, attacking only when large enough. They seem to do this without having a precise knowledge of the power of their opponents. When the overall number of males at Taï decreased after 1990, they became more careful when facing strangers, but the frequency of encounters with neighbours remained the same, and the territory size decreased only moderately (Table 7.4). It seems that by forming patrols to surprise strangers and by replacing direct confrontations with lateral attacks, the males developed a confrontation strategy for holding their territory despite the reduction in their intrinsic power. This was only possible to a certain extent. When the group declined further to four or even two adult males, they started to avoid confrontation (Table 7.10). The higher frequency of avoidance tactics by small groups illustrates how the territory size reduces, and how neighbours progressively extend their territory without encountering resistance.

Males seem able to evaluate the power they represent and choose the strategy in relation to the risk they face, taking into account both the real power of the party (number of males present; Table 7.9) and the potential power of the community (number of males within the community; Table 7.10). This provides a mechanism that explains how under certain circumstances chimpanzees territorial confrontations can become so violent and systematic. The question remains: What do they gain from such territorial violence? Why do males invest so much in territorial defence?

According to the defence of resources hypothesis, we should expect a community to invade neighbouring territories when they face a shortage in one of their two important resources, food and mating partners. Food might be limited due either to an increase in the density of chimpanzees within a territory or due to an unfavourable distribution of highly clumped food types. In the Taï forest, food is only weakly clumped, in the sense that some trees are more often found in some parts of the forest, such as ridges or swampy areas, but not strongly clumped in the sense of being found in only a few places in the forest. This is apparent in the chimpanzees' use of the territory. In addition, the decline of the study community has had no clear effect on the number of attacks they make against neighbouring communities or on the number of attacks received from neighbours (Table 7.10). Territorial encounters did not become rarer, nor did the attack rate decrease with density (we would expect it to be at its highest in 1982 and decrease until 1995 [Table 7.4], whereas it was at its highest in 1988–89 [Table 7.10]). In addition, we never had the impression that any of these confrontations were specifically over food. When fights started under food trees it was always because the attackers had been able to fully surprise their enemies in an ideal position for taking prisoners (see below). The spatial distribution of the encounters did not specifically aggregate near food sources, which would indicate feeding competition.

Sex could be the second resource that might affect territoriality. The absolute number of females has strongly decreased in the study community (see Table 2.2), which could be a strong incentive for the males to search for sexual partners in other communities. Contrary to expectation, they predominantly avoided neighbours at that time (Table 7.10). However, the search for sexual partners might be more subtle and complex. We suggest that both sexes constantly compete for sexual partners and that this affects their territorial behaviour. Males can gain new sexual partners in two distinct ways: first, males can force females to join them either by kidnapping them, as observed once at Gombe (Goodall 1986), or they can force them to leave their group by killing their infants, as seen in Mountain gorillas (Fossey 1983). This last way might be the reason for the inter-community infanticide observed in chimpanzees (see Table 7.14). Second, males can impress neighbouring females during encounters by being specially active and fearless. Females might then select them. That this might be the case is suggested by the fact that over 50% of all infants born in the last five years in the Täi community were fathered by males not belonging to that community (Gagneux *et al.* 1997). Territorial aggression might not only benefit males who can convince neighbouring females to copulate, but also the females who can select 'better' males (see Chapter 4). Preliminary analysis of the interactions of three communities at Taï show that males tend to make more use of the periphery of their territory when oestrous females are present in the neighbouring community, and that generally females make more use of the periphery of their territory when they are in oestrus (Herbinger *et al.*, in press; Egger and Boesch, in preparation.). Thus, both sexes may have an interest in inter-group encounters, and because reproduction is not seasonal in Taï chimpanzees, we expect inter-group aggression to occur throughout the year. One additional aspect points to the importance of sex in inter-community relationships. Female immigration patterns seem to be influenced by the power of the community, so that fewer females enter communities that contain fewer males (Chapter 2). Sex would also explain why Taï males do not harm stranger females seriously when they capture them.

Thus, several observations suggest that the imbalance of power between communities regulates the patterns of inter-community relationships, but that the ultimate reason for these interactions are that they increase the likelihood for both sexes of finding more or better sexual partners.

Territoriality in various chimpanzee populations

All long-term studies on chimpanzees have shown aggressive inter-community inter-actions and active defence of territory. The only exception is Bossou, where the study community lives in an isolated forest and with little or no connection to other wild chim-panzees. At Gombe, the history of the Kasakela community (the main study community) shows that aggression is a very important factor with regard to territory size, and that it fluctuated greatly according to the size of this community. The smaller the community became, the more the neighbouring community pressed it and the smaller the territory became (Goodall 1986). At Mahale, the larger M unit-group dominated the smaller K unit-group and eventually the K unit-group became extinct, with the M unit-group absorbing the former K range into its territory (Kawanaka and Nishida 1974; Nishida *et al.* 1985). Thus, imbalance of power varying through community size is an important aspect of territoriality.

Territorial behaviour between Gombe and Taï can be compared with some quantitative data (Table 7.11). Such comparisons are uncertain because the definitions of terms by various observers can differ, and even more so for observers working at different sites with different visibility conditions. Despite this, some intriguing similarities and differ-ences emerge from Table 7.11. First, the frequency of male patrols is quite similar in the two populations. Patrols reflect the interest of the males in checking their territory, and this tendency appears to be similar for male chimpanzees living in low and high visibility environments (Gombe is a high visibility environment, because on the savannahs and open woodlands near and on the ridges one can have a detailed view of the opposite slopes of the valley).

Second, strangers are heard or seen more frequently at Taï than at Gombe. This prob-ably has a simple ecological cause. The large buttress trees that chimpanzees use to drum are comparatively rare in Gombe woodlands, where sounds carry well within a given valley but poorly between valleys. In contrast, in Taï forest buttress trees are common, often very large, and the flat topography allows sound to carry much farther. Therefore,

Table 7.11: Some aspects of territorial behaviour in Gombe and Taï chimpanzees

	Gombe	Taï
Days observed (>6 hrs)	758	846
Visit to boundaries	17.6%	–
Patrol	5%	5%
Stranger male heard/seen	5.5%	11.6%
Stranger female encountered	4%	0.2%

Gombe: Data from 1977 to 1982 (Goodall 1986). Percentages are overestimated because they include days females were followed, but these were not included in the observation days (Goodall 1986, Table 17.1).
Taï: Data from 1984 to 1995. Visits to boundaries were not analysed.

the chance of hearing drumming of stranger males is higher at Taï than at Gombe and most probably compensates for the lower visibility of the forest habitat.

Third, stranger females are encountered more rarely at Taï than at Gombe. In fact, in thirteen years at Taï there was only one instance when the target chimpanzees met two lone stranger females, whereas at Gombe stranger females were encountered thirty one times in six years. Two differences in the sociality of the females in the two populations might explain this: Gombe females are much more solitary than Taï females (Chapter 5) and they use a more limited range than Taï females. Therefore, the chances of encountering lone females are higher at Gombe, and when they are encountered the chance that they are far away from the males of their community is greater. When Gombe males encountered stranger females, they were very aggressive, attacking them violently. In 10% of the cases the females' infants were killed. Support by group members of these females are not mentioned, suggesting that they were alone. In the one instance at Taï, community members appeared in support and the attack was interrupted after eight minutes (see case study 4).

Thus, although males in the two populations are equally keen to patrol their territory, differences in the environment as well as in social organization seem to affect the probability of encounters with strangers, making encounters with males less frequent at Gombe, while encounters with silent females were more frequent.

Death due to territoriality

One much publicized feature of territoriality in Tanzanian chimpanzees is that both at Gombe and Mahale, one community was driven to extinction through the repetitive and aggressive attacks of a larger community, which also included infanticides. Detailed descriptions are available only for Gombe chimpanzees, and they show that attacks can become extremely dramatic and violent (Goodall *et al.* 1979). At Mahale, the extinction of a small community by a larger one was suspected, but there were no direct observations of the aggressions. In both situations, the imbalance of power was pronounced, as the small community was less than half the size of the larger one (Gombe: 18 individuals for the Kahama community versus 50 for the Kasakela community, Mahale: 20 individuals for the K-community versus over 100 for the M-community). Once a community falls below 30 group members, it seems unable to resist neighbours or to assist community members in danger, and fatal aggressions are observed (Table 7.12).

At Taï, lateral attacks and commando attacks are strategies that were used by small communities to bluff larger ones effectively. Why were they not observed in Tanzania? Comparison are difficult and not straightforward because detailed descriptions of territorial strategies are missing from other chimpanzee populations and no clear conclusions can be drawn. From Gombe we know that males may sit high up in a tree, when on the ridge between two valleys, and thus gain a full view of the valley in front of them and monitor it for presence of monkeys, as well as chimpanzees (Goodall 1986, personal observation). In this way, Gombe males can check for neighbours and after some time gain a fairly accurate knowledge of the composition of a neighbouring community. Under such circumstances, the surprise effect is more about the timing of an attack than about the number of adversaries. This is very different from the situation in the level terrain of

Table 7.12: Number of casualties due to inter-community aggression in different chimpanzees populations. Figures are given for observed cases, inferred cases are given in brackets

	Death toll		
	Adult male	**Adult female**	**Infant**
Gombe	4 (1)	1	2
Mahale	0 (3)	0	2 (3)
Kibale	0 (2)	0	0
Taï	0	0	0

Gombe: Goodall (1986)
Mahale: Hamai *et al.* (1992), Nishida *et al.* (1985).
Kibale: Wrangham (in press).

Taï forest, where chimpanzees can never find a position in the continuous canopy cover to monitor animals moving on the ground. Neither the power present nor the strategies to be used can be accurately foreseen in the Taï habitat.

In Gombe chimpanzees, the physical attacks were much more violent than at Taï. The Gombe chimpanzees gave the impression that they wanted to kill or badly harm the prisoners, for example each victim was immobilized against the ground, arms and legs were dislocated by twisting them round and round, they tore off pieces of flesh, and they drank their victim's blood. The attack could last for up to 30 minutes, and those attacked were barely able to move afterwards. They were all assumed to have died of their wounds (Goodall 1986). When mothers with infants were trapped, the infants were killed and often eaten. This dramatic situation might be the result of the combined effect of a high imbalance of power between the opponents and the fact it happened within an rather open environment in which it is risky to provide support. The sex of the adult victim may also explain this difference between Taï and Gombe; a male is a competitor while a female is a potential mate. At Taï, all prisoners were females, while at Gombe most of them were males. Similarly, in Mahale chimpanzees, only males were thought to have been victims of inter-group aggression (Nishida *et al.* 1985). In addition, the frequency of extra-group paternity within a population may affect how often a female is considered a mate rather than a food competitor. If extra-group paternity was rarer at Gombe than at Taï, this would explain the higher frequency of infanticide at Gombe and the strange fact that Madame Bee was killed (as a competitor), while her daughter was forced to integrate into her mother's killers' community (mate).

The difference in visibility is also a decisive factor in the support that can be expected from others. With good visibility, noisy supporters can make as much row as they want, their number can be evaluated from a distance without risk. In a low-visibility environment, the supporters will be seen only once they are close, and it is risky to remain close if one is unsure about the overall strength of the opponents. At Gombe, this might explain the relative rarity of support given to outnumbered individuals. At Taï, support was the rule in all inter-community attacks.

This is especially important when prisoners have been taken. From Gombe the taking of prisoners has been reported (Goodall *et al.* 1979), but there is no mention of rescue by other community members. The fact that support was provided at Taï even by clearly less powerful individuals than the attackers suggests that imbalance of power alone does not explain why support is provided in certain cases and not in others. The environment might play a role, for it is very difficult to be sure of the number of individuals providing support in a low-visibility environment like the tropical rainforest. However, in some instances, we were aware of the difference in power (and thus the chimpanzees must have known it as well), and support was nevertheless provided from the weaker side (see case study 4). It may be that at Taï, the chimpanzees never really know when and how many more will come in support. When Macho was wounded by two stranger males, a single female, Goma, went to rescue him. The stranger males did not run away, but let him go while they searched for who else might be following Goma. They were quite right to be cautious, for within a minute a party of three adult males came quickly and completely silently to attack them. This illustrates how deceptive a silent reaction of chimpanzees can be in a dense forest. Vocalization is an unreliable indication of the power of the attacking party. Only direct observations, which in the forest mean close proximity, are reliable, but they are at the same time very risky. Thus, with better visibility, support may pay only if you are really more powerful, and this could explain why it was so rarely provided in the instances described from Gombe.

Thus, visibility conditions and imbalance of power might explain some of the differences observed between populations in the death toll due to such inter-group aggressions.

Female participation in group aggression

Female involvement in territorial encounters has been proposed to be rare in humans and in chimpanzees (Alexander 1979; Goodall *et al.* 1979; Nishida 1979; Manson and Wrangham 1991). Taï chimpanzees are an exception to this statement, for females are part of the attack parties in more than two-thirds of the cases, and 35% of the members of such parties are female. They are significantly more often in parties going to make a frontal or a rearguard attack than for a commando or a lateral attack ($X^2 = 4.24$, $df = 1$, $p < 0.05$) (Table 7.13). Thus, their contribution seems to be specially important for certain strategies, and males attack more readily when females are present. Females may, however, lag behind during the last minute of the attack, when it comes to physical contact. In other situations, such as case study 4, when prisoners were taken with relatively low risk, we have seen mothers taking directly part in the attacks. In a low visibility environment, vocal display is important, and

Table 7.13: Female contribution to the territorial activities in Taï chimpanzees according to the strategy used by the community

Strategy	Proportion of parties with females	Average number of	
		Females	Males
Attack	72%	3.4	6.1
Frontal	80%	3.2	5.7
Rearguard	83%	3.8	6.7
Lateral	50%	3.3	6.5
Commando	40%	1.5	4.4
Patrol	57%	2.2	5.2
Check	57%	2.5	4.8

females always contribute by aggressively barking and frequently drumming, making it quite impossible for strangers to estimate the real power of the opponent community. Since females may benefit from territorial activities by getting to know potential neighbouring sires, we should not be surprised to see them so active.

In one case, Ondine, the alpha female, a mother with a young baby, was with a party of four young males progressing towards the strangers far in the north of the territory. At one point, when she started to lag behind whimpering softly, the males repeatedly waited for her to follow. Eventually, the males headed back following Ondine and they walked back until they met two additional males, including the alpha. They then returned all together and successfully surprised and chased the strangers away. Not only did this mother come along with the males, but she seemed able to judge the danger of attacking the strangers with so few.

Infanticide in territorial conflicts

Infanticide can be important in some species, and it has been proposed to be a major factor affecting sociability (van Schaik 1996). In lions, take-overs are bloody, and stranger males constantly threaten to kill infants (Schaller 1972; Packer *et al.* 1990). Langurs and other one-male group primates also show a similar pattern with new immigrant males killing newborns (Hrdy 1977; Hiraiwa-Hasegawa 1988). This has been proposed as a reproduction strategy by the males to achieve higher reproductive success, as the females whose infants were killed come into oestrus much more quickly than if they waited for the infant to be weaned. This time gain is decisive as a male's tenure of a group of females is always limited (Packer *et al.* 1990; Hrdy 1977). A similar interpretation has been suggested for the infanticides that were observed in chimpanzees (Hamai *et al.* 1992; Goodall 1986) and in gorillas (Fossey 1983). A further complexity exists in chimpanzees where infanticide is perpetrated by males both from the same and from stranger communities.

At Gombe, infanticide has been observed on some occasions when males encountered lone stranger females, succeeded in stealing the baby, and partly ate it (Goodall 1986). Beside this type of cannibalism, one female at Gombe, Passion, ate many baby chimpanzees from mothers of her own community. She attacked them with the help of her adolescent daughter, Pom, in what was considered abnormal behaviour (Goodall *et al.* 1979). As this stopped when Passion had her own baby, it could indeed have been an exceptional pathological behaviour (but see Pusey *et al.* 1997). At Mahale, males have recently attacked females from their own community, killing and partly eating their sons ($N = 7$). This was suggested as being a reaction of dominant males towards females that copulated frequently with low-ranking males of the same community, and that might have fathered them (Nishida and Kawanaka 1985; Hamai *et al.* 1992). Infanticide has also been seen in Budongo chimpanzees (Suzuki 1971) and once in Taï chimpanzees (see Chapter 2).

> *Case study 4; two female prisoners, 24 August 1994*: At 6.48 drumming from the eastern neighbours is heard immediately after Kendo and Fitz had drummed. Following this, Brutus, Kendo, and Fitz with the oestrus female Perla and her son Papot start a long patrol first to the

Table 7.14: Comparison of infanticide occurrences between different chimpanzee populations

	Infanticide within community	Infanticide between community	Stranger females encountered
Gombe	8[1]	3	25
Mahale	7[2]	2	?
Taï	1[3]	0	7

[1] Passion and Pom were responsible.
[2] Dominant males were responsible.
[3] We have no clear proof that it was really a within community case.
Gombe: Goodall (1986).
Mahale: Hamai *et al.* (1992).

east and than to the north. At 12.26 they eat some fruit in the territory of the northern community. Suddenly, they must have seen or heard something. With their fur bristling they move silently westwards. At 12.28, they charge with aggressive barks, and I hear the frightened screams of a chimpanzee. I hurry to see that they have captured a possibly isolated stranger female with a three-year-old infant on her back. The three males hold her on the ground, hit her with big arm movements, and bite her. None of the males tries to touch the infant still on her back. During an interruption of the attack, she tries to move away, but they immediately bite her, holding her arms and legs.

At 12.29, in one of the most dramatic examples of support, a second previously unnoticed stranger mother with a four-year-old male infant on her back runs barking wildly to the rescue of the prisoner. This was pure sacrifice, as Kendo and Fitz immediately turn against her and bite her. However, she escapes and climbs a small tree, while her son watches from a neighbouring tree. Fitz simply hangs himself from her foot and she slides down and is then attacked by all the males and Perla. The female is mainly slapped all over her body and bitten around the neck. Her son has come down and sits near Papot, Perla's son, looking at his mother. Three times she succeeds in moving away for some metres, but is immediately caught up by the males and bitten. Her son follows the movements staying close to her and the males. Perla regularly hits and bites the female.

After six minutes, the victim succeeds in joining the first female, who has been waiting for her. Perla and the males chase after them with noisy barking and aggressive screams. The second female is caught once more and bitten again, but the attack is short as Brutus, Fitz, and Papot chase after the first female. Kendo remains near the second one who pant-hoots rapidly to him as he smells her, possibly licking her wounds but I cannot see precisely. When she sees me, she runs away but has to come back as her son is seeking reassurance from Kendo, stretching his hand towards him. She has to take him on her back to run away. Kendo did not show any aggressive movement towards the youngster. Brutus and Fitz try to catch up with one of the females while Kendo waits for Perla. At 12.38, we hear for the first time drumming from the strangers of the northern community, probably approaching to rescue the females. But the two are able to escape and without a call the four intruders and young Papot join up and head back south to their territory.

The one common infanticide pattern observed at Gombe, Mahale, and Budongo is the killing and partial eating of infants stolen by males from stranger females (Suzuki 1971;

Goodall *et al.* 1979; Hamai *et al.* 1992). In 18 years of observations, we have never seen such a case at Taï. On the contrary, we saw that Taï males interacted in a neutral or friendly way with the seven infants of stranger females that were attacked or captured (see case studies 2 and 4). One possible explanation for this difference is that at Taï extra-group paternity is common (Chapter 6), and males might risk killing their own infants. In addition, this might explain the lesser violence of the attacks. The males probably try to impress the females with their power, but do not want to seriously harm them, for the females might decide to visit them later when searching for extra-community fathers.

We should expect females that face high infanticide risk to limit this risk as best they can (van Schaik 1996). However, contrary to this expectation, Gombe females that suffer high risk are also the ones that are more solitary, which in turn increases the risk of encountering infanticidal males. Why do females not react as expected? The most straightforward explanation is that infanticidal costs are lower than other costs chim-panzees face in their daily lives. We proposed in Chapter 5 that intracommunity competi-tion and low predation risk may explain the low sociality in Gombe chimpanzees. The cost of infanticide may not be high enough at Gombe to compensate for the higher feeding competition costs if females were to become more social. At Taï, females are more social, and this cannot be explained by infanticide cost, as it is almost non-existent.

Territoriality or warfare in chimpanzees?

A minimal definition of war in human includes the following elements: (1) it is a group enterprise; (2) it is directed against a second community; (3) it is directed against any members of the opposing community; and (4) it is aimed at serious injury or killing members of the other community (Prosterman 1972; Chagnon 1988; Dennen 1995; Durham 1976). However, other authors have reserved the term 'warfare' for hostilities fought in the pursuit of national policies by organized forces (Malinowski 1941), exclud-ing from it the collective and organized fighting between groups of similar size observed in pre-agricultural human societies. Our point here is not to provide an extensive discus-sion of the impressively large spectrum of different opinions that exist about war in humans, but to note that warfare can be conducted at different levels of complexity and sophistication. It is, then, logical to search for the ancestry of human warfare and to try to understand how such a behaviour evolved.

Warfare at any level is primarily the collaborative aggressive action of groups against other groups, which shows some planning (Chagnon 1988). Other groups are often per-ceived as totally alien, so that social rules applying within groups do not apply absolutely to them (Dennen 1995). Descriptions of warfare in primitive human groups show a constant use of the following properties that could be said to characterize the most simple level of human warfare (Dennen 1995; Boehm 1992):

1. Reciprocity of the aggression: each human group may attack the other, although in fact we might see more uni-directional interactions, depending upon the circumstances.

2. Score-keeping: the distribution of the attacks is not random, but follows some rules, for example concentrating on a particular group over a certain period of time or acting out of revenge, by returning an assault received from one particular group (vendetta).

3. Planning in order to reduce risk: war is not completely unrestrained with unlimited casualties, instead the number is kept within bounds by such actions as vendettas or short raids. Pitched battles with few or no casualties, but which provide a very vociferous and emotional spectacle, are common, though there are also ambush attacks with more fatalities.

4. Strategic decision-making: different strategies are used which follow intentional decisions made by the fighting force.

It is very difficult to say when the first signs of warfare are seen in our own history. Neanderthal fossils as well as modern *Homo sapiens* fossils show signs of violence, with ample evidence of killing, body mutilation, and cannibalism (Bailey 1987). That this was the result of warfare or the mutilation of corpses after death is not likely to be proven with certainty from fossils that are between 15 000 and 150 000 years old.

Mobbing is often observed in animals, and is distinct from warfare in that different individuals join forces against a threat, most commonly a predator. This has been observed in birds, ungulates, carnivores, and primates and is mainly an inter-specific phenomenon. Warfare is, however, intra-specific. Intra-specific inter-group agonistic behaviour has been observed in many insects, in one species of birds (the Arabian babbler), two species of aquatic mammals (dolphins and sealions), six species of carnivores (dwarf mongoose, lion, hyena, wolf, cheetah, and wild dog), and forty-nine species of primates (Dennen 1995). In primates, it can take highly variable forms, ranging from being rare in seven species, through very relaxed and relatively 'peaceful' encounters, to lethal raiding. In most cases, mutual avoidance is observed with the help of long-distance calls, and contacts are of a ritualized type with injuries and fatalities virtually unknown. Such inter-group aggressions seem to be aimed at securing either food or reproductive chances.

Do inter-group aggressions in chimpanzees have anything in common with human warfare as has been suggested (Goodall 1986; Manson and Wrangham 1991)? What are the special features of inter-group aggression in chimpanzees? We discuss several points that seem relevant.

Aggressive macro-coalitions

Descriptions of chimpanzee strategies when engaged in inter-community aggression are still fragmentary, but all descriptions available from Taï and Gombe point to the action always being performed by a group, mainly composed of adult males acting together to search for and attack individuals from stranger groups (see above; Goodall 1986; Goodall *et al.* 1979). This is true for the patrols as well as for the attacking parties. So, group action is the rule in such acts of aggression. In addition, the dramatic descriptions from Gombe show that these aggressions can be very violent and that they are performed with the object of killing members of the stranger groups (Goodall 1986; Goodall *et al.* 1979). At Gombe and Mahale the result of such group violence was the destruction of a stranger group (Goodall 1986; Nishida *et al.* 1985). The situation at Mahale might be more complicated, if Nishida's suggestion is true that within-group infanticide is due to conflicts between males over paternity. Males within a community then become competitors as well as cooperators. It is difficult to understand how such males would be able to cooperate systematically in inter-group aggression.

Strategic planning of the attacks

The destruction of a stranger group at Gombe was the result of a series of attacks concentrated against it (Goodall 1986). Detailed descriptions of the inter-group aggressions are mostly missing. From Gombe, some descriptions of patrols and commando attacks have been provided, suggesting that the strategies we described at Taï may be in general use. At Taï, chimpanzees used four strategies to react to the presence of strangers and used five different strategies when attacking strangers. The use of these strategies was not random, but context-dependent, demonstrating a dynamic decision-making process among the males of a given coalition. The strategic planning of attacks in chimpanzees includes a precise evaluation of the forces present, and a precise collaboration between the males when applying these strategies. Patrols and commando raids are similar to specific actions described in human warfare (like raids in humans). The strategies used by Taï chimpanzees show strong collaborative actions between males in very risky enterprises.

Planning is also evident in other aspects. Territorial activities to improve border control should be spaced-out evenly so that incursions are detected and neighbours encountered on a regular basis. If, however, patrols also have the aim of frightening or harassing neighbours, we should expect that the encounters will be unevenly distributed. Whoever has an advantage against a community, would be expected to exploit it and make the most of it. Table 7.15 analyses the distribution of encounters by the study community with their neighbours. It appears that chimpanzees follow two strategies at the same time, first checking all sides of the territory regularly to prevent incursion (continuous analysis), while, at the same time, concentrating the attacks on a given neighbouring community, checking or attacking it up to nine times within a same month (month-interval analysis).

Strategies used by the Taï chimpanzees in inter-group acts of aggression are elaborate and reflect a dynamic evaluation of the forces present and an anticipation of the consequence of their actions. Some of these faculties are very similar to those shown in the hunting context (Chapter 8), and it might not be coincidence that Brutus, the most expert hunter in the community, regularly played a decisive role in inter-group attacks. Brutus,

Table 7.15: Distribution of encounters with neighbours by the study community

	Successive encounters with the same community								
	1	2	3	4	5	6	7	8	9
Continuous	35	19	5	2	3	–	2	–	–
Month-interval	13	10	7	5	7	–	–	1	1

Statistics:
  Continuous versus Month-interval: $X^2 = 11.6$, $df = 3$, $p < 0.01$

The continuous analysis considers two encounters to be successive against a same neighbour only if no other community was encountered in the meantime. In the month-interval analysis they are considered as successive as long as they occur within the same month, independent of the fact that they might have encountered others in the meantime (e.g. if the northern, the southern, and the northern once again are encountered within a month, it would count as 3 independent encounters under the first analysis, and as one with one encounter and one with two successive ones in the second analysis).

over 35 years old at the time, would follow behind younger males very keen to attack strangers. He was the one who regulated the speed and the direction of the attacks for the younger males based their moves on his. And later, when the community size was reduced, he initiated all the lateral attacks, being followed by the other males. When he was over 40 years old, he no longer participated in the final onslaught on the strangers, but he was still making all the decisions about the speed and the strategy to be used. In a sense, Brutus had the status of a 'chief warrior' and younger males still adapted their moves to his actions.

Prisoners in wild chimpanzees

Chimpanzees trap stranger individuals and prevent them from running away, as humans do with a prisoner. At Taï, all the attackers built a circle around the stranger mothers (all seven prisoners were mothers). Prisoners were regularly beaten with hands and bitten. The wounds inflicted never looked very serious or extensive, although this was difficult to ascertain precisely as we never approached closely. In most cases there was blood on the ground. What happens to the prisoners? At Taï, in all cases, the males of their community came to their rescue and freed them within minutes. Once, three females of the western community were trapped and beaten by both males and females of the study community, while the remaining males chased the other strangers. Nevertheless, the males of that community, which we supposed was quite small, kept counter-attacking until the last female was freed, then disappeared without a single call.

Prisoners have also been observed at Gombe (Goodall *et al.* 1979), but rescuing by other community members was not reported. The observation that little Bee was forced to follow her male aggressors and was eventually integrated in their community illustrates how taking prisoners can be part of a reproductive strategy for the males, as has been observed in human populations (Dennen 1995). Male prisoners that were all killed at Gombe, were probably considered as competitors, and their elimination might also be part of a long-term reproductive strategy.

In conclusion, in chimpanzees inter-group aggression possesses several features considered typical for human warfare, such as a group enterprise of large coalitions concentrated against neighbour groups. They also possess various strategies for attack that are adopted after a flexible decision is taken, depending upon the forces present, and they also include an anticipation of the possible outcome.

Territories in Taï chimpanzees are controlled by the owners all year round on a weekly basis for decades. Fluctuations in the size of the territory mainly reflect variations in the fighting power of the community (number of adult and adolescent males). The chimpanzees use various strategies to protect their territory and to repel the neighbour groups. Macro-coalitions of males choose these strategies according to a precise evaluation of their own force and of the outcome that they expect. Clearly the tropical rainforest with its low visibility increases the benefit of surprise attacks, and the chimpanzees seem to use them in several ways. Strategies used in these interactions have no direct benefit and are performed apparently with an anticipation of future rewards, such as more access to fertile females. Such long-term planning of actions is typical in chimpanzee territoriality and is evident in the distribution of the interactions between different neighbouring groups.

At Taï, cooperation between males is standard in such situations just as in hunting, and all males of the community were keen to take part in attacks. The coordination of the actions was carried out by the older males, especially Brutus, whereas the lead in the attacks was always taken by young fully grown adults. The coordination between the males is easier in inter-group interactions than in hunting as they remain in visual contact, and often even in very close proximity, until the actual attack on the strangers. In hunting, the coordination is mainly done between males dispersed in different trees where they rarely see one another. However, the cost associated with cooperation in inter-group aggression could be much higher than in hunting, for unlike colobus monkeys, chimpanzees are redoutable opponents. Cooperation in territoriality is complicated by the evaluation of forces present before deciding which strategy to adopt, whereas in hunting, it is the anticipation of the reaction of the prey that makes cooperation difficult. In both situations, cooperation between adult males is the rule in Taï chimpanzees and brings great benefits. In both situations, the need for cooperation may also appear unexpectedly, without much possibility of anticipation and both contribute to the Taï chimpanzees' high degree of sociality.

8 *Hunting behaviour in wild chimpanzees*

Forest scene: 4 January 1994: Attracted by the deep and loud calls of diana monkeys that are usually associated with other monkey species, Fitz, Kendo, Darwin, Macho, Ella, Bijou, and Gala move silently to the north. After three minutes, they stop and listen. Further on we hear the typical sound of branches bent by monkeys moving in the trees. With a glance towards Fitz and Darwin, Kendo leads them in this direction. After 150 metres, we locate the agitating foliage. Soundlessly, the chimpanzees place themselves underneath and sit and scrutinize the foliage. Directly above diana monkeys utter little contact calls, unaware of the chimpanzees. About 60 metres ahead, we see branches being moved, probably by heavier monkeys, possibly red colobus. Darwin and Macho rush there silently, climb immediately, and run after the fleeing monkeys. As so often happens, no monkey is caught during this first move. The diana monkeys immediately give alarm calls and most monkeys climb higher in the trees and flee in all directions.

The hunt has started. Darwin pushes before him a few red colobus that he has singled out. Kendo and Macho follow their progress from the ground. Fitz has anticipated the direction they will take by about 80 m and has already climbed into the higher canopy to be ready to move towards them on their arrival. Kendo and Macho see Fitz and accelerate on the ground, trying to anticipate the reaction of the monkeys to Fitz's actions. When Fitz reaches the top of the trees above the canopy, everything accelerates and I have to run in order to keep up with Fitz. Darwin, still in the trees, follows Fitz in a distance. About 200 metres ahead of us, Kendo and Macho climb a tree. The monkeys seem trapped, but both chimpanzees are too slow and most of the colobus escape. Just one unfortunate colobus mother and her juvenile has fallen into the trap and moves towards Fitz and manages to avoid him, only to find herself facing Darwin who has followed Fitz. Kendo and Fitz immediately see the benefit of this move and both press hard on her. To escape, she almost runs into Darwin. He grabs her while she is probably petrified with fear and does not even try to bite him. The three males kill her and start to eat straight away. Macho and the three females join in rapidly, uttering noisy food grunts.

We analyse in this chapter whether the main characteristics of this hunt, that is prey selection, acting in group, and the use of different roles by different hunters, are commonly observed in chimpanzees and how young chimpanzees learn such techniques.

Hunting affects several aspects of the evolution of social behaviour. First, predation promotes group living in prey, because individuals living in groups suffer less from predation (see Chapter 5). Second, hunters have to overcome prey defences to remain successful, which may lead to an arms race between prey and predator (Endler 1992). One outcome can be predators that hunt in groups. The outcome of such an arms race is affected by the environment, and we expect prey–predator relations to be variable. In this chapter, we describe in detail the hunting behaviour of one population of chimpanzees and show the complexity of the social adaptations that make hunting profitable and stable. Then, by comparing the hunting behaviour of different chimpanzee populations with other social carnivores, we assess the diversity of hunting strategies and the factors that affect it.

Hunting for meat is the rule for carnivores and is observed much less frequently in omnivores with a varied diet. Most primates are omnivores; beside fruit, leaves, and other plant food, most of them eat some animal material, usually insects. It is not only chimpanzees that eat meat. Several other primate species have occasionally been seen to eat vertebrates (reviewed in Boesch and Boesch 1989). Thus meat-eating may be widespread in primates, and the meat-eating behaviour found in all chimpanzee populations may not be exceptional.

We have previously published some data on the hunting behaviour of the Taï chimpanzees (Boesch and Boesch 1989; Boesch 1994*a*, *b*). Here we present new data extending the observational period from 1984 to 1995 and a new analysis of hunting strategies, the stability of cooperation, and how males learn to hunt. For the comparisons with other sites, all data have come from published results, except partly for Gombe, where Christophe Boesch made observations during two three-month stays in 1990 and 1992 to document differences in hunting tactics.

Hunting behaviour in Taï chimpanzees

In the early part of the study we thought hunting was rare, because we found only one faeces out of 381 containing vertebrate bones, in that case those of a francolin partridge. We have never seen the chimpanzees capture a francolin or any other bird, but these particular faeces suggest that birds must be captured and eaten occasionally. Because we did not regularly analyse faeces, we probably missed captures of small prey, such as birds, mice, or frogs by lone individuals. Once the chimpanzees were well enough habituated and we could follow them without much disturbance, we noticed that Taï chimpanzees hunt frequently. In a twelve-year period, we saw them hunt vertebrates 413 times, leading to 267 captures (Table 8.1). Here we use only hunting data from 1984 onwards, when the identification of individuals had been achieved and we were able to follow the hunts without disturbing the hunters. Only data collected by the authors are presented here, because data on hunting behaviour collected by observers who are not concentrating specifically on hunting strongly bias the observations towards successful hunts, thereby affecting all data on hunting frequency, failures, successes, and hunting strategies. The data in Table 8.1 describe the hunts we saw and do not reliably estimate hunting frequency. They illustrate the chimpanzees' strong specialization in colobus monkeys, which made up 93% of all prey: 80% were red colobus and 13% black-and-white colobus. The primate community in the Taï forest includes ten species, of which we saw seven preyed upon (the Bosman's potto, not listed, was eaten four times, once before full habituation in 1982 and three times later seen by Ivorian field assistants). Medium-sized mammals other than primates are regularly seen throughout the year and include six species of duikers, of which we encountered the most common ones more than once per

Table 8.1: List of the primate species preyed upon by the Taï chimpanzees in the 1984–1995 period. We give the number of hunts and, in brackets, the number of captures for each species. *N* = 413 hunts and 267 prey caught

Year	*Colobus badius*	*Colobus polykomos*	*Colobus verus*	*diana*	*Cercopithecus petaurista*	*mona*	*Cercocebus atys*
1984	38 (18)	5 (1)	–	1 (2)	–	–	1 (1)
1985	19 (19)	5 (6)	–	–	–	–	–
1986	27 (22)	2 (1)	2 (2)	1 (1)	–	–	–
1987	40 (27)	11 (7)	–	–	–	1*	–
1988	30 (15)	13 (8)	–	–	1 (1)	–	–
1989	30 (21)	6 (5)	–	–	–	–	–
1990	50 (26)	3 (2)	2 (2)	2 (1)	1 (2)	–	1
1991	41 (24)	1 (0)	1 (0)	1 (0)	1 (0)	1 (2)	–
1992	10 (8)	–	–	1 (1)	–	–	–
1993	21 (18)	7 (2)	2 (1)	–	–	–	–
1994	15 (8)	2 (2)	1 (1)	–	1 (1)	–	–
1995	15 (9)	–	–	–	–	–	–
Total	336 (215)	55 (34)	8 (6)	6 (5)	4 (4)	2 (2)	2 (1)

* Poupée, an adult female, released a juvenile *Cercopithecus mona* unhurt while being mobbed by adult members of the mona group.

day (see Tables 8.8 and A.2). Other potential prey include the abundant squirrels (at least three species), mongooses (three species), civets, genets (four species) and birds.

Taï chimpanzees not only specialize in colobus monkeys, they also hunt much more frequently in September–October, these months constituting a real hunting season (Fig. 8.1). It starts around mid-August and ends towards mid-November, lasting roughly three months. Monthly hunting rate and capture rate differ significantly (respectively $X^2 = 40.43$ and $X^2 = 25.58$, df = 10, $p < 0.01$), due to the higher rate for the months of September and October. The more frequently they hunt, the higher the success rate

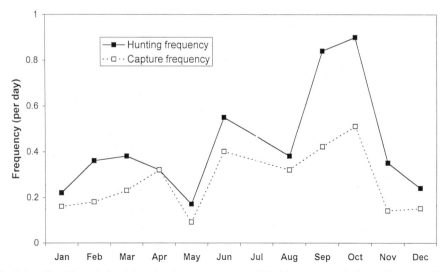

Fig. 8.1: Monthly variation in hunting frequency among Taï chimpanzees during a 12-year period.

($r_2 = 0.87$, $N = 11$, $p < 0.001$). Thus, the hunting season is characterized by a higher frequency and greater success of hunting during the months of September and October. During these months, the chimpanzees hunt every day and in some years did so twice or more per day. During the rest of the year, they hunt on average about once per week.

Monkeys live in groups all the year round, are territorial, and thus are available to the chimpanzees all the time. What, then, causes the pronounced seasonality of hunting? Several factors play a role. First, September and October are the two months of the year with the highest rainfall (see Fig. 1.3), and during the hunt the monkeys seemed less stable than the chimpanzees on wet slippery branches. The chimpanzees regularly and vigorously shook the branches on which the monkeys were sitting until they jumped off to escape or fell to the ground. The wetter the branch, the more easily they fell. In one extreme case, the chimpanzees discovered a group of black-and-white colobus during a heavy shower and gave their typical loud capture calls before even starting to hunt. They appeared to be completely confident of making a capture. When the hunt started, the colobus remained almost motionless in the pouring rain, and an adult was captured by a juvenile male chimpanzee.

However, when the rainy season arrives late (as it did in 1985), the chimpanzees still start to hunt as usual during the second half of August. The very low availability of food in June and July may have an influence. During this period, the chimpanzees forage in small parties of one to three individuals, are unusually silent, and very difficult to find (Doran 1997). After this long period of low social activity, the males gather and start to hunt; this attracts some females so that the size of the party progressively increases. The availability of food has not yet improved and some more weeks of rain are needed before this happens. Another factor may also favour this season for hunting: September and October tend to be the birth season for the red colobus monkeys (Bshary 1995), with most adult colobus females being at this time either at the end of their pregnancy or

Fig. 8.2.   Kendo eating the upper part of an adolescent red colobus after it was divided in two. In contrast to the Gombe chimpanzees, the head with the brain is not a preferred piece of the prey and Kendo will give it to a female later.

accompanied by a new baby. The mothers are heavier and less agile, making them easier prey, as are their young.

Thus, three factors, rainy season, reduced availability of other food sources, and birth season in colobus combine in shaping the seasonality of hunting behaviour in Taï chimpanzees.

The end of the hunting season in November can be more easily explained by the beginning of the nut-cracking season. Nuts provide all group members with a very large daily return in protein and calories (Chapter 9), and one that rivals the benefits derived from hunting. It is not uncommon in late November for a party of males looking for monkeys to arrive at an early producing *Coula* nut tree and start pounding nuts, their mood for hunting having suddenly disappeared.

Hunting in other chimpanzee populations

All populations of chimpanzees that have been studied are known to hunt for meat, as documented either by direct observation or analysis of faeces (McGrew 1992; Kuroda *et al.* 1996). However, the size of prey differs between populations. At Taï, half of the prey are adults (Table 8.2). Of the 24 adult prey captured between 1991 and 1995, 6 were males, 15 females, and 3 could not be sexed. Even more adults were captured when chimpanzees hunted black-and-white colobus (60%), despite the fact that males of this species are bigger than red colobus and regularly attack the chimpanzee hunters. At Gombe, chimpanzees are afraid of adult colobus (Goodall 1986; Boesch 1994*a*; Stanford *et al.* 1994b), and they more frequently captured infants, regularly snatching the baby from its mother's belly without harming her, a behaviour we saw only twice at Taï. Mahale chimpanzees also appear to prefer to hunt infants (Table 8.2).

In Gombe chimpanzees, an increase in hunting in September and October has been observed recently, perhaps because an increase in the size of the party and a larger number of oestrus females occurred at the same time, and males might be trading meat for sex (Stanford *et al.* 1994a). Co-occurrence and causality are not the same, and it remains to be shown that within a season, males hunt more frequently when more oestrous females are in the party and that they trade meat for sex. At Gombe and Mahale, higher availability of fruit has been shown to correspond with a higher hunting rate and a larger

Table 8.2: Comparison of the age classes of the captured prey in three chimpanzee populations (A = adults, I+J = infants and juveniles)

	Taï		Gombe		Mahale	
	A	**I+J**	**A**	**I+J**	**A**	**I+J**
Colobus badius	92	121*	29	101	12	29
Colobus polykomos	19	13	–	–	–	–
Other primates	3	12	0	50	2	5
Ungulates	–	–	0	90	4	22
All prey	114	146	29	241	18	56

* 24 dependent infants were captured with their mother.
Gombe: Goodall (1986)
Mahale: Uehara *et al.* (1992), Takahata *et al.* (1984)

party size (Stanford *et al*. 1994a; Takahata *et al*. 1984), and it may well be that the availability of fruits (Chapter 5) and the number of oestrous females is the driving force behind
the increase in party size (see Wallis [1995] for a similar argument about synchrony in
the oestrus cycle of Gombe female chimpanzees being due to nutritional conditions). At
Taï, the number of oestrus females is high for most months of the year, except near the
low fruit productivity period between May and July (see Fig. 5.10). The peak in hunting
frequency does not correspond to the peak in the number of oestrus females at Taï. In
both Taï and Gombe, however, higher fruit productivity coincides with a birth season
in red colobus, which might also be important for the Gombe chimpanzees that have
specialized in preying upon infant monkeys (Table 8.2).

The question of hunting frequency

It appears that all chimpanzee populations hunt, but it is difficult to determine whether
they all hunt with the same frequency. At first it seems that some populations hunt rarely,
others regularly. For example, in four years, Budongo chimpanzees were observed to
capture only two prey (Reynolds, in press) and Bossou chimpanzees have only been seen
to capture vertebrate prey five times in eight years (Sugiyama and Koman 1987), whereas
at Gombe and Taï hunts were seen every month. Faecal analysis may reveal the presence
of hunting behaviour in a population (McGrew 1992), but it is an unreliable measure of
hunting frequency, since in two years we found only one faeces out of 381 which contained animal remains, while we now know that Taï chimpanzees hunt regularly. In none
of the field projects was it possible to follow all potential hunters at the same time, all day
long, throughout the year. Therefore, we have to rely on estimates based on different
assumptions for hunting frequency. Estimates are made by correcting the observed
hunting frequency for the time chimpanzees could have hunted but were not followed by
observers. The basic assumption is that observers do not influence the hunting frequency,
and that chimpanzees hunt as frequently alone as when they are followed by observers.
Whereas fully habituated chimpanzees may only be minimally influenced by human presence, an observer may have a strong influence on unhabituated prey. Colobus monkeys
are naturally wary of human beings, running away whenever they spot one. The only
way to measure such an effect is to observe precisely the reaction of the colobus monkeys
and try to determine whether their reaction is elicited by the chimpanzees or by ourselves. We tried this at Taï and found that in 3% of the encounters between chimpanzees
and colobus monkey our presence made some of the monkeys move away. This is probably an underestimate. The trees reaching above the canopy in the Taï forest are very
high, so that if we approach carefully, it is possible to move under the colobus monkeys
without being spotted. When they do spot us, they usually move higher in the trees,
becoming there a less easy prey for the chimpanzees. In contrast, at Gombe, the forest is
much more open, the trees are smaller, and the undergrowth is very dense. When the red
colobus see the human observers, they are at a much closer range than in Taï forest. At
Gombe, the presence of observers influences the chimpanzee–colobus interactions in
30% of the cases, and humans regularly put a whole group of colobus to flight by merely
approaching them (Boesch 1994*a*; Goodall 1986). A comparison between the two sites
shows that human presence tends to improve the hunting performance of the Gombe

Table 8.3: Variation in the hunting frequency in the Taï chimpanzees with diminishing numbers of adult males

Number of adult males	7	6	5	4	2
Frequency of detours	–	2.01	1.41	0.87	0.81
Frequency of hunts	1.02	1.14	0.74	0.45	0.42
Success of the hunts	0.53	0.50	0.58	0.63	0.54
Number of hunters	3.78	2.98	2.61	2.89	2.00
Adults captured (in %)	57	34	65	25	33

chimpanzees and that they regularly use the impact of observers on the prey to their own benefit (Boesch 1994*a*). At Gombe, chimpanzees hunt in 65% of the encounters with red colobus when the presence of humans leads to a double predator effect, whereas when the human presence is not noticed, they hunt in only 13% of the encounters (Boesch 1994*a*).

Thus, the assumption that human observers do not affect the hunting frequency of chimpanzees does not always hold, and we should correct for this effect. After correction, we estimated that Taï chimpanzees kill 72 colobus during the hunting season and 53 colobus in the rest of the year (Boesch 1994*a*). This gives for Taï chimpanzees an annual killing rate of 125 colobus in 250 hunts. For Gombe chimpanzees, after correcting for the effect of human presence, we estimated that they kill 66 colobus monkeys per year. In addition, they are known to have very irregular and sudden 'crazes' during which they hunt daily for a week or more (Goodall, personal communication; Stanford *et al.* 1994*a*). To allow for this we added 14 colobus monkeys to their annual kills, yielding 80 colobus monkeys killed per year for 160 hunts (Boesch 1994*a*). In contrast, if we assume that human observers do not interfere with the hunting behaviour of the chimpanzees at Gombe, the estimates yield much higher figures, about 115 colobus monkeys killed per year (Wrangham and Bergmann-Riss 1990; Stanford *et al.* 1994).

Hunting frequency is also affected by the number of males present in a community, an effect that became clear at Taï in recent years. We followed the hunting behaviour as the number of males dropped from seven to two (Table 8.3); the hunting frequency diminished dramatically. Surprisingly, the decrease was not associated with reduced success, but with fewer detours towards noisy colobus monkeys before starting to hunt. This suggests that the hunting decision is taken very early, mostly when monkeys are out of sight.

Despite the impression that hunting frequency might be quite different in chimpanzee populations, we have few detailed data to answer this question. In some regularly observed populations hunting is probably much rarer than at Gombe or Taï, especially in Bossou or Budongo chimpanzees. This would imply that hunting is not important for the survival of chimpanzees.

The importance of meat in the diet

Meat consumption is not distributed evenly between group members, and there are always individuals that have no access to meat at all, while others eat a lot. At Taï, we recorded for each individual the time we saw them eat meat. Then, using an estimate of the time chimpanzees need to eat one kilogram of meat, we evaluated the quantity of

Table 8.4: Meat consumption by Taï chimpanzee during the 1987–1991 period (N = 88 successful hunts)

	Number	Individuals' success in receiving some meat for all captures	Amount of meat eaten per successful hunt (g)			Average daily intake of meat (g)
			Mean	Median	Range	
Male	8	48%	477	547	139–741	186
Female	30	15%	132	74	0–516	25
Adolescent	14	–	62	6	0–265	–

meat eaten by each individual (see Boesch 1994*b*). Each male ate on average about half a kilogram of meat per successful hunt and had access to meat in 48% of all captures (Table 8.4), demonstrating the importance of meat in the diet of some of the chimpanzees. Females, on the other hand, had access to meat in 15% of the captures and ate less than a third of the males' share, with most of them having less (median = 74 g). Adolescents ate much less meat than the adults, mostly recovering pieces the adults had discarded. We could not quantify the quantity of meat eaten by infants, as we saw them mostly suckle or nibble on small pieces they obtained from their mother. They appear to eat less meat than most adolescents. At Taï, chimpanzees have been estimated to capture 125 colobus per year, suggesting that males eat on average 186 grams and females 25 grams of meat per day throughout the year (taking the median).

From Gombe, there is no detailed quantification on the amount of meat eaten. Both sexes have a similar access to meat as at Taï: Gombe males eat some meat in 36% of the successful hunts and females in 12% (Goodall 1986, Table 11.11 and 11.15). A rough estimate for Gombe by using the Taï figures but correcting for the fact that Gombe chimpanzees are estimated to eat 80 colobus per year (Boesch 1994*a*) and that prey are about 2.2 times smaller (Boesch 1994*b*, Tables 1 and 5) results in Gombe males eating 55 grams and females 7 grams of meat per day, which is close to the average figure of 22 grams of meat consumption for Gombe chimpanzees (Wrangham 1975).

Thus, the individuals that one would expect to need more of the rich content of the meat, the subadults and the nursing females, have access to much less meat than the adult males. This is consistent with our impression that meat is necessary neither for survival nor for normal growth. This is further supported by the fact that bonobos do not rely much on meat (Badrian and Badrian 1984; Sabater-Pi *et al.* 1993), nor do gorillas or orang-utans appear to eat any meat (so far only one anecdotal report exists of an orang-utan eating a bird [Sugardjito and Nuhuda 1981]). Thus, within the hominoids, only chimpanzees and humans eat meat regularly, but in both, some populations or individuals hardly eat any. Nevertheless, meat intake can be an important part of their diet for some individuals. Adult males at Taï, the hunters, eat an average 180 grams of meat per day, which represents an important source of high energy food and directly explains why these individuals invest so much energy in this behaviour. Meat is not only a source of energy, but is a very nutritious food, and in it male chimpanzees have access to an important source of energy, vitamins, minerals and other essentials. Even if 25 grams per day for the females does not seem a lot, it is highly nutritious and it may play a role in balancing the females' diet.

Chimpanzees as predators

Predator–prey interactions help to shape the structure of the community and determine the abundance of prey species (Begon *et al.* 1990, 1996). The stronger the pressure from predation, the more benefit the potential prey derive from living in groups, since each individual will see its risk of being caught decrease in proportion to the number of individuals present at the time of the attack (Pulliam and Caraco 1984; Caraco and Wolf 1975). For this reason, even different species may associate in large groups to reduce the risk of predation. Taï forest is characterized by multi-species groups of diurnal monkeys, and it has been suggested that the main reason they associate is to detect and defend themselves from predators (Struhsaker 1975; Holenweg *et al.* 1996). However, because larger groups of prey may be noisier, they may also be easier to detect by predators in an environment with low visibility (Terborgh 1983; Boesch 1991*c*), and they may suffer more frequent attacks (Frause and Godin 1995).

We first quantify predation pressure on the monkeys by the chimpanzees and then ask whether multi-species groups are an effective anti-predator strategy.

1. Chimpanzee hunting pressure on monkeys

In Taï forest, primates are abundant (Table 8.5) and they are attacked by three predators: the chimpanzee, the eagle, and the leopard. Our observations of chimpanzee hunting behaviour allows us to evaluate precisely the hunting pressure exerted by chimpanzees on each of the monkey species (Table 8.5). We have used two estimates of monkey abundance, which can vary with location. The first estimate was made within the territory of the chimpanzee study community (Galat and Galat-Luong 1985), the other south of it (Bshary 1995). The strong specialization of the chimpanzees on monkeys is reflected in predation pressures of 3 to 7% for the two most hunted colobus monkey species, but less than 1% for other monkey species.

2. Polyspecific association as an anti-predator strategy

Polyspecific associations in forest monkeys have been observed in many African forests (Struhsaker 1975; Gauthier-Hion *et al.* 1983; Holenweg *et al.* 1996; Cords 1987), and the

Table 8.5: Abundance of diurnal primates in the Taï forest (Galat and Galat-Luong 1985) and evaluation of the hunting pressure they suffer from chimpanzees. (The territory size of the chimpanzee community was 25 km²)

Species	Density (ind/km²)	Abundance in chimp territory (Nb ind.) A	Annual predation toll (Nb ind.) B	Predation pressure (B/A)*100
Cercopithecus diana	17.5	437–1880	2	0.5–0.1%
Cercopithecus mona	15	375	1	0.3%
Cercopithecus petaurista	29.3	733	2	0.3%
Cercopithecus nictitans	–	187*	–	0%
Colobus badius	66	1650–3877	125	7.6–3.2%
Colobus polykomos	23.5	587	16	2.7%
Procolobus verus	21	525	3	0.6%
Cercocebus atys	10	250	1	0.4%
Total primates		4744	150	3.2%

* Assuming for the *C. nictitans* a density similar to the *C. mona* and knowing that their distribution includes only the northern half of the chimpanzee community (the monkey study was performed in the south of this community).

evolution of these associations has been explained as resulting from either increased foraging efficiency or better predator avoidance. Some confirmation of the improved foraging efficiency hypothesis could be demonstrated in some species under some situations
(Gauthier-Hion *et al.* 1983), but no advantage could be found in the Taï forest (Holenweg
et al. 1996). This has been used, by default, to support the predator-avoidance hypothesis
(Holenweg *et al.* 1996; Noé and Bshary 1997). Monkeys are quieter when real predators
are around (Boesch 1994*a*) and also when predator calls are played on loudspeakers (Noé
and Bshary 1997). However, chimpanzees have a varied repertoire of vocalizations, and
in those playbacks there were no hunting calls. Thus, these tests were of the monkeys'
reactions to the presence of chimpanzees, not specifically to chimpanzees that were
hunting. In addition, the chimpanzee is a pursuit-hunter, using ambush only as part of the
strategy at the beginning of the hunt, whereas the eagle and the leopard are pure ambush
hunters. Surprise is much more important to the success of ambush hunters than it is for a
pursuit hunters. Therefore, the benefit resulting from polyspecific association might differ
depending upon the type of predator. A direct test of the efficiency of polyspecific associations in counteracting chimpanzee predation should measure whether the association
decreases predation success against individual monkeys or decreases the predator attack
rate.

Chimpanzees use the fact that red colobus monkeys associate with diana monkeys to
find them more easily (Boesch 1994*a*). But do they move towards colobus monkeys more
frequently when those are alone or when associated with other monkey species? Once
heard, red colobus monkeys are hunted more frequently when they are members of a
polyspecific association ($X^2 = 3.44$, $df = 1$, $p = 0.06$) (Table 8.6). The data show that Taï
chimpanzees move significantly more frequently towards polyspecific associations than
towards pure red colobus groups ($X^2 = 18.96$, $df = 1$, $p < 0.001$). The tendency to capture
a red colobus monkey once a group is seen is similar for the two types of groups
($p < 0.1$). It is important for the chimpanzees to hunt quickly once the decision to hunt
has been made (Boesch and Boesch 1989), and therefore it may not be decisive for them
whether the prey is alone or associated with another species. What counts is knowing
where the prey is, and the noisy diana monkeys with whom red colobus mainly associate
help the chimpanzees to find their prey.

Does this tendency to hunt associations of monkeys increase predation pressure on the
individual red colobus? When facing a chimpanzee hunter, red colobus monkeys do not
profit from polyspecific associations, because monkeys rarely defend themselves against

**Table 8.6: Chimpanzee hunting strategies in the Taï National Park after
hearing mono- or polyspecific monkey groups (1993 to 1995)**

Chimpanzee behaviour	Red colobus monkey	
	Polyspecific association	Alone
Acoustic encounters	196	143
Detour towards the monkeys	64	18
Hunting	33	14
Hunting success	58%	64%

We considered here only situations in which red colobus monkey were present.

chimpanzees, and only colobus monkeys were seen to do so. Early detection by the attentive dianas could benefit the colobus, yet chimpanzees manage to surprise them in most hunts and were never seen to interrupt their hunting attempt because they had been spotted by monkeys. Monkeys may try to escape from the chimpanzees by running away, but their escape route can be followed quite easily on the ground for as long as needed. Red colobus monkeys do not profit from a dilution effect when in a polyspecific association, for they remain the chimpanzees' only target, the black and white colobus being rarely found in associations including red colobus and vice versa. Finally, since the group size for red colobus monkeys is the same when they are alone or associated with other monkeys (Holenweg *et al.* 1996), the risk of predation by chimpanzees for individual red colobus is larger when associated with other monkeys species than when alone. Thus chimpanzees appear to profit from the polyspecific associations because it makes it easier for them to detect their prey. There is one rare situation in which red colobus profit from polyspecific association: when chimpanzees discover black-and-white colobus, they systematically prefer them, which increases the cost of association for black-and-white colobus. This species has been seen to avoid polyspecific association when red colobus are present (Bshary 1995), which might be a reaction to higher predation risk.

Chimpanzees as predators have an important effect on the colobus community. They are probably their most important predator, killing a similar amount of the colobus to that observed at Gombe, where the chimpanzees capture 8 to 13% of the red colobus population annually (Goodall 1986; Stanford *et al.* 1994*b*). We conclude that the polyspecific association observed in the monkey species of the Taï forest did not evolve to improve protection against chimpanzees, since chimpanzees actually profit from these associations. They may have evolved to decrease predation by leopards and eagles, as both species prey on all species of monkeys (Hoppe-Dominik 1984; Dind 1995, Jenny 1996).

Specialization in prey species in chimpanzees

A marked specialization in hunting red colobus, a propensity to capture adult prey, and seasonality in hunting are distinctive attributes of the Taï chimpanzees' hunting behaviour. These attributes raise questions about the chimpanzees' hunting strategies and how such characteristics evolved. In all habitats, there are middle-sized mammals that are not hunted by chimpanzees (Table 8.7); and many more potential prey are ignored by the chimpanzees at Taï than at other sites, suggesting that they show a particularly strong selectivity towards potential prey.

Prey detectability depends both on the abundance of the prey and on how easily they are spotted by the predator. We compared the detectability of diurnal primates for Gombe and Taï chimpanzees (Table 8.8). For a hunting chimpanzee, the profitability of all these primate prey species is very similar: a similar amount of energy is required to catch any of them in the trees and they are of a similar size. Hence, there is no obvious reason why chimpanzees should not hunt all of them. A precise comparison between Table 8.8 and effective hunting frequency (Table 8.7) suggests that neither of the two chimpanzee populations hunt prey according to their detectability (Taï: $r_s = 0.18$, $N = 7$, $p > 0.05$, Gombe: $N < 5$ too small for a statistical test). Taï chimpanzees totally neglect diana monkeys,

Table 8.7: Comparison between potential and actual middle-sized prey species within three chimpanzee populations.

	Taï	Gombe	Mahale
Colobus badius	81%	55%	53%
Colobus polykomos	13%	–	–
Papio anubis	–	6%	*
Cercop. ascanius	–	2%	4%
Pan troglodytes	0.3%	2%	3%
Other primates	5.7%	*	3%
Bushpigs	*	16%	7%
Bushbucks	–	12%	7%
Blue duikers	*	–	20%
Squirrels	*	0.5%	2%
Rodents	*	0.5%	1%
Birds	0.3%	6%	*

If hunted, we indicate the proportion of prey they constitute within the population. A dash means the species is absent, while an asterisk means the species is present but was not seen to be hunted

Table 8.8: Detectability, in number of calls heard per hour, of the different monkey species within the Gombe and Taï chimpanzee territory, while following a party of chimpanzees

Monkey species	Taï	Gombe
Colobus badius	0.39	0.14
Colobus polykomos	0.13	–
Colobus verus	0.01	–
Papio anubis	–	0.73
Cercopithecus mitis	–	0.10
Cercopithecus ascanius	–	0.10
Cercopithecus diana	0.59	–
Cercopithecus petaurista	0.05	–
Cercopithecus mona	0.06	–
Cercopithecus nictitans	0.01	–
Cercocebus atys	0.16	–

A dash indicates the monkey species is absent at the study site.

which are amongst the most abundant and the noisiest monkeys in the forest, and Gombe chimpanzees hunt baboons much less frequently than they encounter them.

The chimpanzees' choice might be affected by the defence potential of the target primates, as has been shown for the baboons at Gombe, which were less frequently hunted once they started to defend themselves more vigorously (Goodall 1986). The same argument seems to apply to bushpigs, which can defend themselves much better than primates, and were never hunted at Taï. At Gombe, infant bushpigs were caught only if a panic was provoked among the defending adults. Prey defence can explain why larger prey are not caught by chimpanzees, but not why some smaller ones are neglected. Taï chimpanzees may have a sort of 'search image' that excludes some potential prey. For example, blue duikers are hunted by the Mahale chimpanzees, and they are also the most abundant duiker species in Taï forest where the chimpanzees encounter them at least three or four times daily. Adult Taï chimpanzees were never seen to make any intentional movement to capture one, even when a duiker happened to be running towards them and they had to step aside to avoid it ($N = 20$). At most they made a soft bark and it fled. In five cases, we were with the chimpanzees when they encountered duikers that were resting between the buttresses of tree trunks, easy victims indeed, but they just looked at them. Youngsters generally try to chase away any animal in the vicinity. On three occasions we saw them discover a young duiker (two blue duikers and one zebra duiker). They played with it, pulled and carried it around, threw it to the ground and stamped on it, watching carefully the reaction of the duiker to this rough treatment. But in all three cases, they finally released it without trying to kill it, for example with a neck-bite, nor did they try to eat part of it. In one case, the young duiker died, possibly of stress (Boesch and Boesch 1989). In the two other cases, they were left unhurt. Thus, despite numerous encounters, Taï chimpanzees never seemed to consider duikers as prey.

Taï chimpanzees also seem to have a taste for certain types of meat. Red and black-and-white colobus monkeys are always eaten completely, from skin to bones, when enough chimpanzees are around. We actually gained the impression that, if given a choice, they would always prefer black-and-white colobus to the red ones. On the other hand, they seem to have mixed feelings about the meat of *Cercopithecus* monkeys. All infants captured were eaten to the last bit, but they ate only parts of the two adults they caught: parts of an arm and of the tail, and none of the viscera they like so much in colobus monkeys. Both captures occurred in a similar sudden surprise rush up a tree in dense vine tangles against a mixed group of monkeys, including red colobus and we had the impression these captures of *Cercopithecus* were 'mistakes'.

Taï chimpanzees are only rarely scavengers. They found the dead but fresh bodies of several species (three mangabeys, five duikers, one genet, and one flying squirrel) and although they did show some interest, they did not feed on them. In the case of the flying squirrel, two youngsters played with it, holding it by the tail and swinging it around. The mothers looked at the scene for two minutes with some interest, then went on, and the youngsters left the squirrel behind and followed. They have been seen four times to eat primates killed by the crowned hawk-eagle. In three more instances, they robbed an eagle of a captured red colobus, before the eagle had killed it. It happened just a few times that they returned to a place where they had eaten a colobus monkey a day or two before, and very occasionally they ate some of the remains ($N = 3$). Gombe chimpanzees have been seen to rob baboons of fresh kills and eat them (Morris and Goodall 1977), and Mahale chimpanzees accepted and ate some dead animals presented to them by humans (Nishida *et al.* 1992).

Search image as well as a taste for a given kind of meat could explain the choice of prey, but both may have developed after the preference became established. There are proximal mechanisms for the maintenance of a behaviour, but its evolution is ultimately explained by the difference in benefit. That Taï chimpanzees concentrate on colobus monkeys is apparently an adaptive choice: once a group of chimpanzees has decided to hunt, they need to find a prey within a reasonable period of time, and they will search for it intentionally in half of the cases (Boesch and Boesch 1989). Colobus monkeys are noisy monkeys and among the most abundant prey in the forest (Tables 8.5, 8.8), they are the largest of the abundant prey (the weight is about 15 kg compared to the 4 kg for

Cercopithecus monkeys), and they are less agile than the *Cercopithecus* monkeys (Boesch 1994*a*).

Evolution of cooperation in hunting

Hunting by social animals sometimes involves the combined action of many individuals. Thanks to group actions, much larger prey can be subdued and hunts can happen more frequently than when animals are acting alone (Schaller 1972; Kruuk 1972). Why is group hunting not the rule in social animals, and what do participants gain from group hunts? Observations of the Gilgil baboons have nicely illustrated the dilemma (Strum 1981). In this study, as long as a particular alpha male was the major hunter, others participated and could recover some scraps of prey (baboons do not share meat intentionally). Once he lost his alpha position, all the captures he made were stolen by the new alpha male. As a consequence, he stopped hunting, and so did the others. Hunting disappeared rapidly within this population.

Cooperation is expected to evolve by one of three mechanisms: mutualism (Maynard-Smith 1982), kin selection (Hamilton 1964), or reciprocity (Trivers 1971; Axelrod and Hamilton 1981). In mutualism, the participants in the interaction have a higher gain than when acting alone, and understanding its evolution seems natural and simple. In kin selection, the gain from a joint action for the two participants might be unequal but is compensated by the close relatedness of the two individuals. In reciprocity, only one of the participants gains at a time, and it is through repetitive interactions between them, and alternation of the gains, that the benefit can be distributed equally between the participants. Few tests of these models exist, as the main problem lies in quantifying precisely the different gains of all possible strategies for each participant (Dugatkin 1997; Clements and Stephens 1995).

A constant threat to the evolution and maintenance of cooperation is cheating (Maynard-Smith 1982). Individuals may try to profit from the action of others without themselves paying the cost. This has proved to be a major problem, for in theory it is almost always better to cheat than to cooperate. In hunting, meat is often shared, and non-hunting individuals regularly try to gain access to it, thereby diminishing the benefit of the hunters. An understanding of the mechanisms and the benefit of cooperation requires, therefore, that benefits and costs are calculated for all participants for each hunt to sort out the possible pay-offs of the different strategies. This is possible only with identified individuals and precise observation during the hunts and the meat-eating episodes.

Cooperative hunting in social carnivores has been widely observed (*Lions*: Schaller 1972; Packer *et al.* 1990; Stander 1992, Orsdol 1994; Cooper 1991; *Wild dogs*: Estes and Goddard 1967; Fanshawe and Fitzgibbon 1993; Creel and Creel 1995; *Hyenas*: Kruuk 1972; Mills 1990; *Wolves*: Mech 1970). Detailed analysis of the individual food intake across hunt group size in some carnivore populations shows a tendency for U-shaped distribution, with one maximum for single hunters and sometimes a second maximum for the largest groups (Packer *et al.* 1990; Caro 1994). In some studies, the pack size of carnivores lies between these maxima near the minimum of the curve, and this has been used to suggest that cooperative hunting does not explain sociality in lions or other predators (Caro 1994; Packer and Ruttan 1988).

However, this evidence is not sufficient to support the conclusion for two reasons. First, such an analysis should be based on the net benefit per individual, including both the cost and the benefit of hunting, which is not available from most of the studies. Second, giving too much weight to results from only one population might be misleading, as the environmental conditions prevailing for different populations may require different hunting strategies. To settle this issue, data on benefit and costs need to be collected for individuals, allowing to analyse the net benefit for all available strategies. This has proved to be quite arduous, either because the hunts happen at night (for example in some lion studies) or because the animals followed are not all individually known (for example studies on other lions, wild dogs, and hyenas). In addition, measuring how much meat is eaten by each individual is also very difficult. Recent, more detailed, studies have shown that individual lions do not all invest the same amount of energy in the hunt and may perform different roles (Stander 1992; Scheel and Packer 1991). Pack size, which has been used in most lion studies as a measure for the hunting-group size, does not differentiate between hunters and non-hunters, and this is misleading. In Etosha, all females within a pack hunt (Stander 1992), whereas in the Serengeti the number of female 'cheaters' increases with the ease of a hunt (Scheel and Packer 1991). Including the costs of hunting into the analysis can notably alter the pay-off (Boesch 1994*b*; Creel and Creel 1995). We discuss this point below.

Description of hunting behaviour in Taï chimpanzees

Taï chimpanzees have specialized in colobus monkeys, abundant, middle-sized mammals, which live in groups in the highest strata of the forest canopy. Adult male red and black-and-white colobus weigh 13 and 20 kilograms respectively, and they can sit on branches in trees that would not carry an adult male chimpanzee of 40 to 50 kg. Thus, finding the prey and then capturing it presents a double challenge.

First, find the prey. In the low-visibility forest, chimpanzees search uniquely by sound for their prey. Red colobus monkeys are noisy (Table 8.8). But whenever the chimpanzees fail to hear any, they were found to take advantage of the frequent high-pitched calls of diana monkeys that are regularly associated with the colobus monkeys (Boesch 1994*a*). When they heard the diana monkeys, they often waited quietly until they heard some calls of red colobus, or looked under the diana monkey group for red colobus. If unsuccessful, they moved further for a while and listened again attentively, often hearing another monkey association within 30 minutes. In half of the hunts, chimpanzees searched intentionally for their prey, and such searches lasted about 16 minutes (Boesch and Boesch 1989). The chimpanzees searching for prey remain silent. They are not always successful in finding their favourite prey, and the mood for hunting may then disappear, despite the presence of other primate species. If they do find a colobus group, they face their second challenge.

Second, capture a red colobus. In almost all hunts, the chimpanzees surprise the prey by approaching soundlessly, remaining on the ground beneath the prey, scrutinizing the vegetation for the colobus, and concentrating their attention on those that are most numerous and lowest in the trees before starting to climb. When the colobus are really low in the trees, some of the chimpanzees may rush up in an attempt to catch one by surprise, which is rarely successful. Otherwise, one of them starts slowly to climb about

5 meters high, usually unnoticed by the colobus (a second chimpanzee may climb another tree in coordination with the first one, but this is rare). The others move on the ground in anticipation of the possible escape routes of the colobus and ready to join the pursuit. Once the climber is seen by the monkeys, he makes a rush upwards, which makes them move. His contribution then consists mainly in keeping them moving in a given direction, while the others on the ground follow and undertake different blocking moves, checking regularly where the climber moves. This one, acting as the driver, usually follows the monkeys in the branches without trying to capture one on his own.

At this stage the colobus usually still form a large group. The chimpanzees try to keep them moving in one direction. If the colobus try to escape in two or more directions, they may find that a blocker may have climbed up to block an escape route with his mere presence. As the hunt progresses, some take turns in performing the driver movement as they climb up under the escape movement of the colobus, while others assume a chaser role attempting to catch a monkey by a rapid pursuit. The chimpanzees usually select and try to isolate an individual, often a mother with her baby, or a small group of individuals. Once they have separated them from the main group, the hunt accelerates with chasers coming up from different directions. But the most difficult task remains to be done, requiring that a chimpanzee anticipates the location of the tree to which the colobus are going to flee and to be there and block before the first monkey arrives. The chimpanzees that have chased the monkeys up to now have little chance of achieving a capture themselves by straight pursuit in the trees above the canopy. The ambusher (or encircler) is the hunter who anticipates the escape route of the quarry long enough in advance to be able either to force it to turn backwards towards its pursuers or to move downwards into the lower canopy, where chimpanzees have a very good chance of catching it, because in the continuous tree cover at this level chimpanzees can run faster than colobus monkeys.

This describes an 'ideal' hunt that reaches its conclusion through the complete involvement of the hunters, but a capture may occur at any moment during such a hunt. We call such 'ideal' hunts 'collaborative hunts', that is the hunters perform different complementary roles all directed towards the same prey. However, group hunting does not have to be as collaborative as this, and we classified a group hunt as a 'similarity hunt' if all hunters concentrated similar actions on the same prey, with no spatial or temporal relation between them. However, at least two of them always act simultaneously. We called it a 'synchrony hunt', when they at least try to react to each others actions in time and a 'coordination hunt', when they relate their actions both in time and in space (Boesch and Boesch 1989). Table 8.9 compares the level of cooperation observed when chimpanzees hunted in groups. At Taï, three quarters of the group hunts were collaborative.

Table 8.10 organizes all the hunts we observed according to the number of hunters, their hunting success, and average hunting time. Taï chimpanzees have low success when hunting alone, and the increase in success when hunting in groups is a clear incentive for them to collaborate (Boesch and Boesch 1989; Boesch 1994*b*). Similarly, the more hunters are present, the longer they hunt. The increase in their hunting success with more hunters indicates that they profit directly from group hunting. The increase in collaboration with group size suggests that the more hunt together, the better they organize themselves and the higher their success becomes (Table 8.10). Thus, the above description of the 'ideal' hunt is the rule rather than the exception in Taï chimpanzees.

**Table 8.9: Group-hunting tendencies in Taï, Gombe, and Mahale chimpanzees.
For each hunt, we present only the highest level of organization observed**

	Solitary hunts	Group hunts	Collaboration
Taï	52 (16%)	274 (84%)	211 (77%)
Gombe	55 (64%)	31 (36%)	6 (19%)
Mahale	14 (28%)	37 (72%)	0 (0%)

Mahale: Takahata *et al.* (1984), Uehara *et al.* (1992).
Gombe: Includes two years of red colobus hunts (Busse 1978) and one year of baboon hunts
(Teleki 1973).

**Table 8.10: Effect of hunting group size on hunting success and hunting time in
Taï chimpanzees between 1984 to 1994**

Number of hunters	Number of hunts No.	(%)	Hunting success (%)	Hunt duration (minutes)	Collaboration (%)
1	52	16	17	4.6	0
2	70	21	26	8.9	47
3	70	21	53	10.6	74
4	58	18	69	13.9	93
5	40	13	63	18.1	90
6	18	5	61	30.7	100
+6	18	5	89	38.5	100

Comparison with other chimpanzee populations

Do all chimpanzee populations hunt similarly or do populations also differ in hunting as in social contexts? Comparing data from different populations is possible only with standard definitions. We use the term 'hunter' only for individuals that actively take part in a hunt by placing themselves in positions where they could perform a capture — in the case of colobus hunts mostly by climbing in trees at the height where prey are or are going to be (Boesch and Boesch 1989). In contrast, some authors classify party members that are with hunters as hunters on the basis that they are looking at what is happening (Teleki 1973), others merely on their passive presence (Stanford *et al.* 1994*a*). The assumption that the number of hunters directly correlates with the number of individuals in the party (Stanford *et al.* 1994*a*) was not supported in Taï chimpanzees (Boesch 1994*b*). We consider that individuals present during a hunt, but not taking actively part in it, are 'bystanders' and not hunters. Our comparisons are limited to the Gombe, Mahale, and Taï populations for which detailed and comparable observations on the hunting behaviour are available. The tendency to hunt in groups and to collaborate when hunting in groups varies greatly between the populations, both being most prevalent at Taï (Table 8.9). Greater specialization in prey goes in hand with more collaboration between groups of hunters. Is this only a coincidence?

Gombe chimpanzees are very efficient hunters, and lone hunters at Gombe capture red colobus monkeys more than five times more quickly than lone hunters at Taï (Table 8.11). Taï chimpanzees capture heavier prey, but this does not fully compensate for the longer time needed to achieve a capture, and the chimpanzees at Gombe gain more per minute

Table 8.11: Comparison of the hunting performance for red colobus of lone hunters in Gombe and Taï chimpanzees (after Boesch 1994*b*)

	Gombe	Taï
Time to capture	7.2 min	39 min
Weight of the prey	1.23 kg	9.5 kg
Benefit	1'179 kJ/min	794 kJ/min

than those at Taï. This performance results from the high hunting success of lone hunters that effect a capture within less than four minutes for every second hunting attempt. Obviously it is difficult to improve on such success, and there is no pressure for the Gombe chimpanzees to do more hunting in groups. The differences in success are associated with the conditions in which the chimpanzees hunt: at Taï, red colobus are mostly in the trees reaching above the canopy, which are over 40–50 metres high, and present a good chance of avoiding chimpanzees. This increases the costs for a capture in terms of energy expenditure. In contrast, at Gombe, a woodland-savanna, the canopy is lower, and when the colobus are in the highest trees this generally means that they are about 15 metres high, with limited possibilities of escape, so that the costs of pursuing them are much smaller (Boesch 1994*a*).

Cross-population comparisons of the hunting strategies of chimpanzees reveal a lot of variation that may be related to the different conditions under which they try to capture their prey. The more difficult the conditions, the more time is needed to achieve the capture, and the more the chimpanzees are forced to organize themselves to make hunting worthwhile. The more individuals hunt together, the more meat is necessary to satisfy all hunters, and the larger the preferred prey has to be. This prediction needs to be tested with data from other chimpanzee populations.

Evolution and stability of cooperation

When a group of individuals is hunting, non-hunters may try to profit from the gain. For cooperation to be stable and profitable, the hunters not only have to benefit from their investment, but this benefit must exceed that of cheaters. Cooperators are expected to react against cheaters in two ways (Maynard Smith 1982; Axelrod and Hamilton 1981). Either they become cheaters themselves, which causes cooperation to disappear, or they retaliate selectively against cheaters. This is only possible if individual recognition exists and individuals are judged on the basis of their hunting behaviour. This may work in chimpanzees, as we know that they recognize each other and that they probably have the cognitive abilities to judge one others' involvement.

Cheaters are a real problem in Taï chimpanzees, where they constitute 47% of the meat eaters (Boesch 1994*b*). How can cooperation at Taï remain stable? Taï hunters rely on individual recognition and identification of individual contributions to limit the success of cheaters, and hunters obtain significantly more meat than bystanders and latecomers for most hunting group sizes (Fig. 8.3*a*). Each individual may change strategies during a hunt, and group members adjust the amount of meat an individual receives according to its contribution to the hunt. In addition, as the quantity of meat available increases from

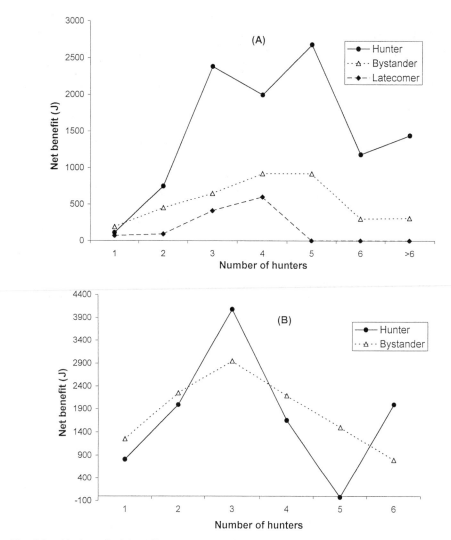

Fig. 8.3:   Net benefit of the different strategies for (a) Taï and (b) Gombe chimpanzees when hunting colobus monkeys.

lone hunts to group hunts with three to four participants, there is a strong incentive in favour of group hunts. In addition to hunting involvement, both age and dominance play a role in meat access. Dominance is more important than age: old dominant hunters gain more meat than young dominant ones, who in turn have more meat than old subdominant hunters (Boesch 1994*b*). Below we shall see that age correlates with the type of hunting tactics, and so age *per se* may play no role in meat access.

Thus, Taï chimpanzees have a social mechanism that favours hunters at the cost of the dominant or the older ones. This is a potential source of conflict in the group, as a dominant individual that arrives at a kill site will always try to have a share and will try to obtain it by force if necessary. Despite this, dominants receive less than hunters, and the

Fig. 8.4:   Brutus, in the centre, shares meat of an adult colobus with Macho on his right, while Rousseau on his left, who did not hunt, is looking at them without obtaining a share.

group has developed some flexibility that allows such conflicting rules to coexist without escalation. The stability of cooperation relies on this flexibility of the adult males, who are the hunters in 85% of the cases (Boesch and Boesch 1989). Females are not constrained by these rules, although they directly contribute to enforce them by supporting hunters at the meat-eating site when conflicts occur with dominant individuals. Some females obtain a large amount of meat independently of their contribution during the hunt (Boesch 1994*b*).

Cheating and group hunting in other chimpanzee populations

At other study sites the chimpanzees have a tendency to hunt less in groups than those at Taï, and we propose that this is because the hunting conditions are such that lone hunters are very successful. For these populations, however, group hunts still occur (Table 8.9). How do they solve their problems with the cheaters? No clear mechanism for punishing cheaters emerges from the Gombe data. It appears that cheaters gain more meat than hunters, and this difference is significant for groups of five hunters (Fig. 8.3*b*). Compared with Taï, age and dominance play a more important role in meat access at Gombe (Boesch 1994*b*; Goodall 1986; Busse 1977). Hunting participation does not guarantee more meat to the hunters, and cooperation is not stable. Hence, we should expect group hunting to be rare. Why are Gombe chimpanzees still seen to hunt regularly in groups? Descriptions from Gombe indicate that, when hunting in groups, the chimpanzees start to hunt on the same group of prey, but as a rule, each one follows a different target and they do not coordinate their movements. Thus, group hunting at Gombe is better described as a set of simultaneous solitary hunts than a true group hunt.

In Mahale chimpanzees, the alpha male took part in 8% of the hunts, but he got hold of the prey in 31% of them, and he was seen to share meat in order to favour coalition partners, that is close, middle-ranking old males (Nishida *et al.* 1992). Thus, the social mechanism of meat sharing required for cooperation to be stable is absent in Mahale chimpanzees as well, and it is not surprising that cooperation was rarely observed (Table 8.9).

In comparing chimpanzee populations, we have to remember that Gombe chimpanzees can hunt with the same most collaborative strategy used by Taï chimpanzees. They have been observed to do so on six occasions. It is thus not a question of a difference in ability, it is a question of a difference in the strategy needed to achieve a capture. Systematically hunting by collaboration, when it does not bring higher success than lone hunting, would be a wasteful high-energy-demanding solution. In contrast, it pays Taï chimpanzees to cooperate, with the consequence that they developed elaborate meat-sharing rules to make cooperative hunting stable. Wild chimpanzees are thus able to develop demanding social rules to guarantee the stability of cooperation if needed, or to use meat-sharing rules as a tactic to favour social allies, stabilizing the alpha male's social status.

Cheating and counter-strategies in Taï chimpanzees

To counter efficient meat-sharing regulations, some individuals may cheat by using more subtle ways to limit their investment of energy, such as hunting for only short periods of time or by performing less demanding hunting movements than others. To understand which contribution during the hunt guarantees the best meat access, we tried to observe the hunts as precisely as possible, recording their development and the contribution of each hunter once a first chimpanzee has initiated a hunt.

When the colobus monkeys notice the hunters, their major concern is to avoid them. We, the observers, then start to move within the group of chimpanzees on the ground to be able to monitor constantly the action of all the hunters and all the males present. We divided the hunts into several hunting moves during which the hunters aim at a given prey or group of prey. If they do not succeed, the hunters pause and reorganize themselves. Those in the trees may choose another quarry and start a new hunting attempt, or, if no prey is at close range, all the chimpanzees may come to the ground and reposition themselves under a larger party of colobus to start again. A hunt may last up to one hour and include five to seven such moves. For each of them, we recorded the tactic of each hunter in the 4 roles outlined in the ideal hunt description above and whether the tactics used were done by anticipating the movement of the monkeys or not (see below the section on 'learning of the hunting behaviour' for a description of the anticipation).

We found two strong correlations (Table 8.12). First, the more moves a chimpanzee participates in, the longer he hunts. Second, the more a chimpanzee tries to ambush the prey, the more he anticipates its escape movements. The other factors did not correlate. There seems to be no easy way for individual hunters to assure themselves of capture. Hunting for longer periods does not guarantee an individual more captures, nor do the more demanding tactics, such as anticipating prey movements and ambushing, increase

Table 8.12: Correlations between five characteristics of tactics observed in hunters among Taï chimpanzees 1986–1995: hunting time of a hunter, number of his moves, whether he made a capture or not, number of his ambushes or encirclements, level of his anticipations, and number of hunters taking part in the hunt.

	Time	Move	Capture	Ambush	Anticipate
Move	0.695**				
Capture	−0.107	−0.087			
Ambush	0.097	0.316	−0.024		
Anticipate	0.064	0.252	−0.004	0.734**	
Number	0.258	0.105	−0.230	0.098	0.054

To account for multiple comparisons, we considered only ** = $p < 0.01$)

the likelihood of achieving a capture. Hunting is a collective undertaking, and an ambush can increase the likelihood of a capture only if the other roles are simultaneously fulfilled. If that is not the case, however well the ambushing is done, it will not increase the chance of a capture for this hunt.

Meat access per successful hunt is significantly affected by several factors (Table 8.13). The number of hunters reduces the amount of meat eaten per successful hunt, as the number of meat eaters increases with the number of hunters. This does not contradict the results of Fig. 8.3, where individual meat access reaches a maximum at three to five hunters, as we include all hunts in that figure, the successful and the unsuccessful ones, and the latter are more frequent for smaller hunting groups. The three factors that improve meat access are: being the captor, ambushing the prey, and anticipating prey movements. Of these, being the captor is most important, but the other two roles are also significantly rewarded. Both ambushing and anticipating the prey movements allow the hunters to prevent escape; in most cases they bring hunters close to the prey, a basic condition for a

Table 8.13: Meat intake of hunters for successful hunts according to the role performed

Tactics	Number	Amount of meat eaten (time in min)	Amount of meat secured (time eating and sharing)
Bystander	314	27.8	32.2
Captor*	139	58.6	85.4
Driver/chaser	181	24.2	31.5
Ambush			
Half anticipation	86	27.3	35.5
Single anticipation	70	44.8	61.8
Double anticipation	17	54.3	84.8

Factor	df	F-value	P-value
Captor	1	17.57	0.0000
Ambush	1	14.04	0.0002
Number of hunters	1	4.60	0.01

Results of an analysis of variance considering factors affecting meat intake is shown taking interactions into account (Procedure GLM, SAS 1985). Only significant results ($p < 0.05$) are shown.
* Drivers, chasers and ambushes that succeed in capturing a prey are presented only in the captor category.

capture. Thus, it seems that certain tactics during the hunt are valued more highly by group members than others, for those who perform these tactics are allowed access to more meat. This indicates that group members monitor precisely what other individuals do during a hunt and can assess the contribution of each hunter. This is intriguing because, as we will see later, some of the anticipation tactics are very demanding and most group members seem unable to perform them, but nevertheless value them highly.

These meat-sharing rules strongly restrict the possibility of cheating. Individuals that pretend to hunt or that hunt by performing less demanding roles, or move in trees but without aiming at a prey, will receive less meat than other hunters. This penalizes un-talented adult hunters. It also penalizes young chimpanzees that may be keen to hunt but still have to learn how to do the right thing at the right time — hunting time is not corre-lated with meat access, and youngsters often hunt for longer periods of time than prime males. Cheaters that pretend to hunt cannot gain access to much meat, as ambushing or achieving a capture are the two keys to meat access. An individual that wants meat needs to perform the tactics that directly increase hunting success. The precision of this anti-cheating strategy suggests that cheating is a real problem in chimpanzees, but its import-ance is hidden because the meat-sharing rules ensure that cheating is not profitable.

For youngsters, this leads to the contradictory situation that although they regularly hunt for long periods of time performing the cognitively less demanding roles (driving or chasing prey), they have limited access to meat. In contrast, an adult that may hunt for a short time by ambushing will gain much more meat. Simply driving prey is rarely suc-cessful, but efficient encirclement often leads to a capture. The Taï meat-sharing rules acknowledge the role of the ambushers and stabilizes cooperation by guaranteeing both more meat to hunters than to the others, and more meat to hunters making an important contribution to the hunt than to those performing less decisive roles.

Individual hunting tactics

Some males were never good or keen hunters, some individuals were especially gifted, while others changed their hunting behaviour over time. Here we describe individuals that hunted in a special way.

Brutus was the oldest male of the community and regularly performed the most cogni-tively demanding hunting tactics. However, when no other individual was hunting, he also readily assumed the simplest role of initiating a hunt, climbing towards the prey to put them in motion. Once the hunt was well engaged, he would then come back to the ground to prepare himself for an ambush. Brutus showed the most elaborate anticipatory moves (see below for some descriptions of them), and we sometimes had to wait to see the result of one of his moves to understand them. Throughout our study of hunting, Brutus was the best meat provider of the community. His self confidence let him remain at ease even in the centre of the wildest meat sharing clusters, and it was not unusual to see Brutus with a prey surrounded by more than ten males and females all eating from his share, while he remained in control of the situation. Eating episodes of adult colobus monkeys with Brutus as the owner were especially noisy, as if the other chimpanzees knew their chance of receiving a share were especially good. When his decline in the hierarchy was complete, at the age of about 43, Brutus still dominated the meat-eating episodes, and, when he was owner of the prey, was still ready to share extensively with many other chimpanzees.

Ulysse, in contrast, showed little interest in social dominance and was not often seen trying to improve his position. He had quite an unusual personality: in 1982 he was a poor hunter, always present during a hunt and interested in meat but rarely participating. In March 1987, Schubert disappeared, leaving the beta rank to Macho, a keen hunter. Ulysse immediately started to hunt much more frequently and revealed himself to be a specially gifted hunter. This was striking in 1989, when for some reason the males were generally less keen to hunt. Ulysse demonstrated his intelligence by developing a personal hunting tactic: he provoked red colobus males to mob him. He would climb very slowly towards them to give the impression he was hesitating, then, if they showed no reaction, would even slow down in the most critical passages on thin branches between trees, until they would mob him. There, the chimpanzees are normally in a weak position and cannot protect themselves easily, and the red colobus often start to mob them successfully at this moment. Once we saw Ulysse move back and forth in such a passage until he was closely attacked by the colobus. But Ulysse was absolutely fearless and would invariably capture one right then. Black-and-white colobus males are powerful, weighing up to 20 kg, and they usually attacked hunters without much hesitation, but Ulysse invariably took advantage of their approach to capture one by turning it on its back or by making it fall to the ground by pulling wildly on one of its legs. Red colobus were more difficult to manoeuvre, and he often used the passive presence of another chimpanzee in a tree to help him corner one of them. Ulysse stopped using these special tactics when the other males started to hunt with him again.

Some years later, Fitz, then about 17 years old, developed a new way of capturing colobus monkeys that proved very successful. He took advantage of the fact that once a red colobus is pursued long enough, it stops running, both from exhaustion and stress, sits motionless, and simply lets the hunter approach and seize it (this reaction is very similar to the one described for exhausted wildebeest chased by wild dogs). Fitz was a young powerful male who could pursue an individual colobus fast enough and long enough until it was exhausted. He could then pick it like a ripe fruit. In 1992, Fitz started this efficient technique and managed to capture mostly adult monkeys. He still needed the help of the others in order to isolate the right target, but once this was done, he would accelerate, the others just following on the ground during these particularly long rushes. He used this technique until his death in 1994. The bystanders would start to make the capture calls when they saw the quarry was not running away any more, sometimes even when Fitz was not yet in the same tree.

Beside these gifted hunters, there was Rousseau, who hunted only when it seemed that without his contribution there would be no capture at all; otherwise he just followed the hunts from the ground, giving some hunting barks when it looked as if a capture was about to happen. Darwin did even less and for a whole year (in 1993) we did not see him hunt even once. Both of them had, nevertheless, a keen appetite for meat and always tried very hard to obtain some somehow. Both usually succeeded, but they received less than when they hunted. We think their lack of motivation had a physical basis. Darwin was badly handicapped on both feet, which hindered his grip on tree trunks. This made him slow in climbing trees, made him a poor ambusher, and restricted him usually to the driving role. Rousseau had had a dislocated left shoulder for some weeks, and although it must have been repositioned somehow and he was able to walk and run normally again, we noticed an obvious weakness in his left arm and shoulder when climbing tree trunks.

The learning of hunting behaviour

Hunting is a challenge for the chimpanzees; it is both energy demanding and risky to catch a colobus monkey, and the prey does all it can to elude the chimpanzees' attacks. To be successful, the chimpanzees have to be aware of the particular physical properties of the quarry, which are of different size and weight, and move in different ways, as well as reacting differently to the presence and actions of hunters. The hunting tactics we described above vary in how much is required in terms of understanding prey movement: a driver only needs to follow the prey from a distance, not a cognitively demanding action. A chaser also only follows the prey but has to adapt his speed to catch up with it, which requires judgement of speed and distance. A blocker or an ambusher has to position himself in places where the prey may come, and this demands anticipation of the prey's reactions.

Anticipation is further complicated by the fact that the hunter not only has to anticipate the direction in which the prey will flee (recorded as a half anticipation), but also the speed of the prey so as to synchronize his movements to reach the correct height in the tree before the prey enters it (recorded as a full anticipation). This is more complex than it sounds as the hunt happens in a three-dimensional space, and the hunter has to convert the speed of the colobus into the distance he has to run ahead of the colobus, mostly out of sight on the ground, plus the time he will need to climb a tree high enough to be able to lay an ambush, unnoticed by the monkey. To distinguish between half and full anticipation is quite easy, for a chimpanzee that makes a half anticipation typically waits for the colobus to confirm his prediction by waiting at the base of the tree for the monkey to reach it and then rushes up as quickly as possible. Those making a full anticipation climb into the tree before any colobus enters it and do so slowly and carefully so as not to alert the monkeys to their presence by moving branches. We also recorded a double anticipation when a hunter not only anticipates the actions of the prey, but also the effect the action of other chimpanzees will have on the future movements of the colobus, that is he does not anticipate what he sees (the escaping colobus), but how a future chimpanzee tactic will further influence the escaping monkeys. We saw only eight moves of double anticipation, and five of them were made by Brutus. However, we probably underestimated the frequency of double anticipations, for the observation conditions prevailing during the hunts usually made it difficult to determine that the conditions for double anticipation were fulfilled.

We shall try to understand how hunting tactics are learned by analysing the propensity of hunters to use either complex movements (ambush and block) or anticipation of the prey movements. It is typical of chimpanzees' hunting behaviour that it develops late. Youngsters are attracted to hunting, and we regularly saw very young chimpanzees climb towards groups of colobus monkeys while their mothers rested. However, it is not until they are six to eight years old that they progressively approach the colobus. However, when they get close up they are invariably chased by the adult colobus, and they run

Fig. 8.5:   Brutus, on the right, shares meat from the rump of an adult red colobus with the dominant female, Ondine. Ondine's daughter looks at Tarzan, a six-year-old male, adopted by Brutus, who allows him to take some meat from the carcass.

away screaming with fear. Some reckless ones persist and approach again and again, and this mock chase between young chimpanzees and adult colobus can last for up to half an hour. We did not count such interactions as hunts, and they are not analysed here. When eight years old, some young males progressively master their fear, stop screaming when threatened, and start to chase small colobus, but still run away quickly when attacked by adult males. In the real hunting context, when the adult chimpanzee males are active, these youngsters may contribute progressively to the hunt by eliciting flight movements among the colobus. This sort of participation by the four youngest males in a hunt were included in our analysis. Ten-year-old chimpanzees start to hunt more efficiently. They are less afraid of the colobus and can become effective drivers, for example Marius drove prey in 70% of his contributions and Ali did so in 60% of his.

Learning hunting behaviour is a long process, as most aspects of the hunt are progressively acquired over twenty years (Fig. 8.6 and 8.7). The young males are keen to hunt, and there is even a tendency for young males to perform more movements per hunt than the older ones ($r_s = 0.51$, $N = 13$, $p = 0.07$). The proportion of ambushes and blocks performed by males correlates strongly with age (Fig. 8.6). The older males performed these demanding roles three to four times more frequently than the young ones.

The correlation with age is equally strong for anticipation. Half of the anticipations were used quite early, and their frequency increased in old adolescents and then rose quickly, eighteen-year-old males performing half anticipations most frequently. The two

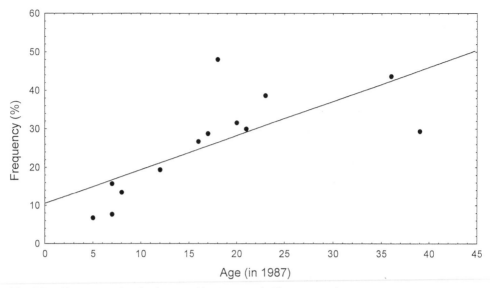

Fig. 8.6:   Frequency of ambushes used by hunters of different ages from 1987 to 1995 in Taï chimpanzees ($r_s = -0.85$, $N = 13$, $p < 0.001$).

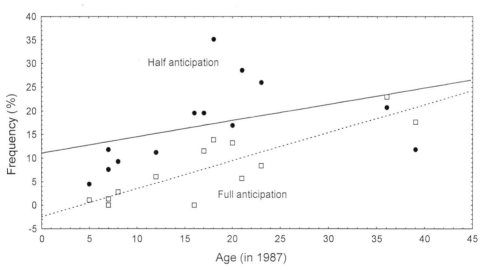

Fig. 8.7:   Frequency of anticipations used by hunters of different ages from 1987 to 1995 in Taï chimpanzees (half: $r_s = 0.76$, $p < 0.02$; full: $r_s = 0.82$, $p < 0.001$).

oldest males used fewer of them, because when they anticipated, they did so fully. The frequency of full anticipations increased steadily with age and only the oldest males (over thirty years) used them frequently (Fig. 8.7). It is striking that full anticipations were routinely performed only by males over thirty years old. Twenty-year-old males did them occasionally but all males made errors occasionally in these situations, either by selecting trees to which no colobus were coming or by changing their mind and climbing more than one tree during the same move. Long-distance anticipations, in which the hunter chooses a tree far ahead of the colobus, were seen only in older males so confident of their choice that, unlike younger ones, they did not need to look constantly at the colobus. Similarly, double anticipations were only performed by the two oldest hunters.

Thus, the learning of hunting behaviour is an exceptionally slow process, for it starts only when the young males are nine to ten years old and then lasts for about twenty years. Three aspects of hunting might contribute to make this learning process so long. First, the mother is not the model for hunting, and second an individual can survive without meat. These peculiarities apply, however, to other social behaviour patterns and do not explain a twenty-year-long apprenticeship. Third, collaborative hunting requires of the hunter a complex understanding of another species as well as the coordination of his own actions with those of other group members. This might well explain a twenty-year learning process. Not only have the young hunters to overcome their fear of the colobus' reactions, which may not be obvious, if we consider the general fear of adult red colobus seen in other chimpanzee populations, but they have to realize that colobus monkeys have different physical possibilities and reactions from their own. These aspects of the hunts are apparently understood by fifteen to twenty-year-old chimpanzees, as seen in all studied chimpanzee populations. The most demanding aspect of collaborative hunting is to coordinate actions both in time and in space with those of other hunters. As we shall see in Chapter 10, predicting the reaction of another chimpanzee and its influence on the reaction of another species is a very demanding task, and it is not surprising that only a few individuals are able to do it. Note that twenty- to twenty-five-year old chimpanzees are still uncertain about their predictions (performing mainly half anticipations), and that males still learn these elaborate tactics later.

Is the learning of hunting similar in other chimpanzee populations? At Gombe, in 1992, Frodo was seventeen years old and by far the best hunter of the community. He was involved in all three collaborative hunts I saw, in each of which he performed the driver role. In two of those hunts, his oldest brother Freud, twenty-two years old, was apparently trying to predict the colobus' reactions and made half anticipations. In the last hunt witnessed, Evered, forty years old, made a clear full anticipation of the colobus' reaction. This suggests that learning of the hunting behaviour at Gombe may follow the same time course as at Taï. Our estimates of the learning abilities of chimpanzees are extended by these observations on hunting. Nut-cracking behaviour, which involves difficult tool use (Chapter 9), takes five to seven years to be learned. Hunting, with its late start at about ten years and long apprenticeship of about twenty years, takes us into another dimension. These observations on hunting question the classical idea that learning in adult animals is impossible. Not only are twenty-five-year-old chimpanzees still learning more about hunting, but what they learn is remarkably complex.

The learning of hunting behaviour in humans has been followed in detail in two hunter–gatherer societies, the Ache from Paraguay and the Hiwi from Venezuela (Kaplan *et al.* in press), and it shows surprising similarities with the Taï observations. Young men start to hunt at about fifteen, and meat production by men peaks at around thirty-five years. This suggests a comparable learning process of twenty years as in chimpanzees.

Cooperative hunting among animals

Group hunting has often been observed in various species of birds, fishes, and in all social carnivores (Packer and Ruttan 1988; Dugatkin 1997). The published material is of uneven quality, ranging from purely qualitative descriptions to quantified reports. Nevertheless, the descriptions of hunting techniques are sufficient to establish that different strategies are used in different species. For example, when hunting for hares, two to three Harris' hawks chase the prey and attack it from different directions so as to encircle it and close off all escape routes (Bednarz 1988). Those that hunt in such a way achieve greater success than lone hunters.

Several questions arise about cooperation. Does cooperation really exist within a species? Does cooperation benefit the individual performing it. Does cooperative hunting help to explain sociality? Table 8.14 shows that group hunts are regularly observed in all social carnivores and chimpanzee populations, but the tendency to hunt in groups varies between populations of the same species. Carnivores hunt larger prey in larger groups (Kruuk 1972; Caro 1994; Fanshawe and Fitzgibbon 1993; Mills 1990), and thus the observed variation might partly reflect the tendency of some populations to hunt rela-

Table 8.14: Cooperation among animals: hunting strategies of social predators classified according to four levels of organization by groups of hunters

	Number of group hunts	Similarity	Synchrony	Coordination	Collaboration
Primates					
Cebus	26	58%	←– – – –	42%	– – – –→
Baboon	14	←– – – – 100% – – –→		–	–
Chimpanzee					
Taï	273	4%	8%	11%	77%
Gombe	31	←– – – –	81%	– – – –→	19%
Mahale	37	←– – – –	100%	– – – –→	0%
Social carnivores					
Lion					
Serengeti	523	←– – – –	91%	– – – –→	9%
Etosha	795	–	29%	30%	40%
Hyena					
Serengeti	46–164	24%	91–11%	–	–
Wild dog					
Ngorongoro	54	–	←– – – 100% – – –→		–
Serengeti	131	77%	23%	–	–
Wolf	103	←– – – 94% – – –→		2%	4%
Cheetah	238	60%	26%	14%	0%[1]

[1] The very few hunts that looked as being collaborative were explicitly said to be such by chance due to the spatial disposition of the individuals (Caro 1994: p. 274).
Cebus: Rose (1997).
Baboon: Strum (1981).
Lion: Serengeti: Schaller (1972) including all hunts he observed.
       Etosha: Stander (1992).
Spotted hyena: Kruuk (1972) presenting hunts on two different prey species to illustrate the variability of their hunting behaviour; zebra hunts are the most elaborate ones in this predator species.
Wild dog: Ngorongoro: Estes and Goddard (1967).
          Serengeti: Fanshawe and Fitzgibbon (1993).
Wolf: Mech (1970).
Cheetah: Caro (1994).

tively larger prey than others. This factor does not, however, explain the variations observed among chimpanzee and lion populations. Table 8.14 describes how well the individual hunters coordinate their actions according to our operational definition of levels of group hunting (Boesch and Boesch 1989). Detailed observations of hunting strategies at the individual level remain rare. In chimpanzees, as well as in social carnivores, collaboration has been regularly observed, but high levels of collaboration occur in only a few of the populations observed. Our suggestion that collaborative hunting is peculiar to chimpanzees (Boesch and Boesch 1989) no longer holds, since, for example, lions have been seen to collaborate regularly. Is there a common explanation for the distribution of collaboration in those populations? Based on our observations of chimpanzees, we suggest that collaboration arises as a reaction to the challenges faced: the harder it becomes to achieve a capture, the more hunters will organize themselves.

In Table 8.15, we have tried to test this hypothesis by estimating the ease of capture for several species in various environments. Ease of capture is a relative notion, since it

Table 8.15: Group hunting and individual success

	Group hunt	Group size		Ease of capture	Individual benefit as a function of hunt group size		Best strategy	Group size favoured
		Hunt	Foray		Total gain	Net gain		
Primates								
Chimpanzees								
Taï	68%	4.2	10.0	Low	Increase	∩-shape	Coop.	4–5
Gombe	34%	2.3	5.6	High	NS	∩-shape	Cheat	1
Mahale	23%	1.8	6.1	High	–	–	Cheat	1
Social carnivores								
Lions								
Serengeti	29%	2.5	>3	High	U-shape	–	Cheat	1;5–6
Etosha	94%	3.5	–	Low	Increase	–	Coop.	2–3
Uganda	64%	2.5	–	Low	Increase	–	Coop.	3–4
Wild dogs								
Serengeti	79%	4	–	High	NS/∩-shape	–	No/Coop	4
Selous	100%	10	7.7	Low	U-shape	∩-shape	Coop.	12–14
Spotted hyenas								
Serengeti	44%	3.2	>4	High	NS/increase	–	No/Coop	2–3
Kalahari	85%	3.6	3.0	Med.	NS	–	None	No
Chobe	76%	4.6	–	High	–	–	?Cheat	3–8
Cheetahs								
Serengeti	57%	2.8	1.7	High	Increase	–	Cheat/Coop	3

Total gain is based on different measures of either hunting success or the amount of meat eaten per individual per day or per hunt. Net gain subtracts the costs associated with the hunts from the total gain. Whenever possible we mention which strategy was favoured and what hunting group size had the best reward.
NS = individual benefit not affected by hunt group size

Lions: Serengeti: Packer *et al.* (1990).
  Etosha: Stander (1992), Stander and Albon (1993).
  Uganda: Queen Elisabeth N.P., van Orsdol (1984).
Wild dogs: Serengeti: Fanshawe and Fitzgibbon (1993).
  Selous Game Reserve: Creel and Creel (1995), Creel (1997).
Hyenas: Serengeti: Kruuk (1972).
  Kalahari: Mills (1990).
  Chobe N.P.: Cooper (1990).
Cheetahs: Serengeti: Caro (1994).

compares the conditions between different populations of a same species and should not be used to make comparisons between species. The Serengeti National Park is recognized as a prey-rich region compared to Etosha in Namibia or to the Queen Elisabeth National Park in Uganda (Stander 1992; van Orsdol 1984). As in Taï chimpanzees, the fact that Etosha lions hunt habitually in groups and collaborate frequently seems to be related to the habitat where hunting is more difficult. A similar tendency is observed in wild dogs, in which a population living in the Selous Game Reserve, a habitat that is possibly less rich than the Serengeti, was observed to hunt more frequently in groups. Thus, preliminary comparisons with lions and wild dogs are consistent with what we observed in chimpanzees. Both, the tendency to hunt in groups and the level of cooperation within a population of hunters increase when the hunting conditions become more difficult.

Does cooperation favour the evolution of sociality in predators?

In recent years, the debate about the contribution of group hunting to the formation of social groups in animals has been revived by analyses at the individual level. New evaluations, especially of the Serengeti lions, have shown that despite the fact that groups of lions might increase their hunting success by capturing more prey (Schaller 1972), daily success is greater for single lions (Packer *et al.* 1990). Recent reviews have concluded that cooperation in hunting cannot be a factor influencing the formation of groups, because most observations did not show a resulting increase in individual food intake (Packer and Ruttan 1988; Caro 1994, Dugatkin 1997). The results presented in Table 8.15, which include the most recent published observations, indicate that care is needed before drawing broad conclusions, and that we really must distinguish between different populations in the same species. The observations of the Serengeti lions support the conclusion that cooperation is a by-product of sociality that possibly evolved for other reasons, but observations on Etosha and Uganda lions are quite different, and support the opposite conclusion, namely that individuals profit directly from group living through increased meat income. Trends are similar in other species. Wild dogs in the Serengeti do better in groups when hunting both wildebeest and Thompson gazelles, though in the case of the gazelles it is because they thereby prevent scavengers from stealing their prey. In the Selous Game Reserve, group hunting is even more advantageous for individual wild dogs. Of the three hyena populations studied, only in the Serengeti, individuals fare better when hunting in groups, and then only for wildebeest calves, their main prey target (Kruuk 1972). In cheetahs, adult males and mothers also fare better when hunting in groups, though this does not apply to subadult individuals (Caro 1994), which suggests a learning process like that found in chimpanzees. Thus, a cross-population comparison reveals that, depending upon the environmental conditions prevailing, individual benefit may increase when hunting in groups.

We may ask why previous studies have not reached the same conclusion. Creel and Creel (1995) stress that food intake curves can change shape depending on how they are measured, and that it is important to include the cost of hunting in the calculation. The only three studies that have done this found bell-shaped curves (Table 8.15). Thus, in other studies the results might change when the cost of hunting is included, dependent upon the hunting conditions. If a capture is easily achieved, we should not expect group hunting to evolve or individual success to increase with group size. When a capture is difficult to achieve, then individuals are forced to hunt together and cooperation may be the best way to keep hunting profitable. This is what we observed in chimpanzees, and this seems to be supported by the observations in lions and wild dogs. In conclusion, we could say that cooperation will be observed more frequently under difficult conditions and that it will then also become mutualistic, in the sense that all participants will benefit directly from it. Under such conditions, cooperative hunting could directly contribute to the evolution of group living.

Once group living has evolved, cheating by group members becomes a problem, which might account for the fact that foraging party size is often larger than the optimal hunting group size (Table 8.15). Larger foraging parties could be tolerated by hunters if a sharing system regulates meat access, as we observed in Taï chimpanzees (Fig. 8.3), and group living might also provide hunters with additional benefit. For example, the interests of cheaters and hunters may not always conflict, when cheaters trade their access to meat for other benefits for the hunters, such as sex, it would pay hunters to accept cheaters. This seems to be the situation in chimpanzees, where we know that females rarely hunt and may trade meat access for long-term sexual partnership and social relations. Similarly, cheaters may help hunters to protect their meat from scavengers and thus directly benefit the hunter, as has been observed in lions and hyenas (Cooper 1991; Kruuk 1972). Other benefits have been proposed such as cheaters helping to raise offspring or to defend territories (Packer *et al.* 1990). In such cases, cheaters are not really cheating: they are engaged in a quasi-economic exchange. We suggest that when mutual benefit has been demonstrated, they should not be called cheaters but bargainers. Hunters may simultaneously pursue several goals, and the fact that they cooperate for hunting does not prevent them from optimizing their strategy to find mates and protect their territory.

Taï chimpanzees hunt very regularly and have developed a sophisticated system of reciprocity in which hunters are rewarded for their contribution, not only for their participation in the hunt, but also for the type of contribution they make during the hunt. This indicates that they evaluate the contribution of different individuals and have a fine graded control of cheaters that pretend to hunt by adopting low cost tactics. Learning hunting tactics requires years and is fully acquired only by the older males within the community.

Theoretical models for the evolution of cooperation face the problem of finding ways of eliminating the advantages of cheating, as by definition they have less costs than cooperators and therefore can gain in many more circumstances. Two solutions have been proposed. The first is kin selection, where the cheaters' benefit is also considered as benefiting the cooperator through the high relatedness between individuals (Hamilton 1964). The second solution relies on game theoretical models, where under some specific assumptions, populations of pure cooperator individuals are evolving when the interactions are repeated many times (Axelrod and Hamilton 1981; Maynard-Smith 1982; Dugatkin 1997). Field observations with chimpanzees and lions point to some limitations in such approaches. First, kin selection may be less important than expected, since the genetic results show that relatedness between cooperative hunters might be absent, for example, Taï male chimpanzees are not more related to one another than females are (Gagneux *et al.* 1999). Second, cooperating and cheating can take many different forms and be complex and varied (Scheel and Packer 1991; Heinsohn *et al.* 1996; see above), which has the effect of producing varied rewards for each strategy. For hunting, costs and benefits for all strategies depend on the size of prey, the number of hunters, and the ease of capture (for Taï chimpanzees see Fig. 8.3). In game theoretical models, rewards of strategies are always simple and constant. In theory, as the number of participants and the number of strategies increase, this tends to make cooperation less likely (Axelrod and Dion 1988), but paradoxically it is in this situation that cooperation is stable in Taï chimpanzees.

Should we expect pure populations of cooperators in wild populations? Most theoretical models about the evolution of cooperation search for the conditions producing populations which contain only cooperators (Dugatkin 1997). Within structured populations, individuals interact more frequently with close kin and this allows cooperation to spread within the population. However, animal examples, like in the chimpanzee or the lion, show that even in small tight groups, cooperators have to face numerous cheaters, without any indication that they might be eliminated. In other words, animal populations may be stable but mixed, including both cooperators and cheaters. How can cooperation be stable when constantly confronted with cheaters? The Taï chimpanzee example shows that mutualism makes such a system viable and the disproportionate interest devoted to reciprocity might not reflect what happens in nature. Moreover, the cooperation game is not played in isolation from other aspects of an individual life, and cooperators themselves maintain cheating when they have other priorities and defect when a hunt starts. Lastly, a population is not homogeneous, and some individuals might rarely take part in the cooperative game (some female chimpanzees hardly ever hunt). Thus, the questions are: how do mixed populations of cooperators and cheaters coexist and why do individuals adopt mixed strategies?

Cooperation in hunting is stable among Taï chimpanzees thanks to an elaborate system based on individual recognition, temporary memory of actions in the recent past, attribution of value to those actions, and social enforcement of those values. The end effect is that cooperators gain more than cheaters, but for this to happen a complex social organization based on impressive cognitive abilities is required. If this example is representative of most cooperating systems, it makes the stability of cooperation dependent upon a certain degree of cognitive development, and these constraints will limit the animal species in which cooperation can exist. This has been illustrated by the apparent inability of baboons and capuchin monkeys to solve a task requiring coordination, whereas chimpanzees and orang-utans were successful (Chalmeau 1994; Chalmeau *et al.* 1997).

To conclude, all populations of chimpanzees have been observed to hunt small mammals. Their strategies are flexible and vary between the populations in many ways. We should expect more variability as more information from new populations becomes available, as environmental conditions seem to affect the hunting strategy used by a given population. Among primates, only human beings have been observed to hunt for meat as extensively and with such flexibility as chimpanzees. We return to this point in Chapter 11.

9 *Tool-use in wild chimpanzees*

Forest scene; April 1985: Early morning — Héra and her two sons Haschich and Eros are on their own quietly foraging. At 8.00, they arrive at the large *Parinari exselsa* tree that they have been visiting for the last few days. Its load of nuts covers the ground. Héra immediately gathers some nuts in her hands and settles down at the large anvil with a big granite stone. The $2\frac{1}{2}$-year-old Eros is already waiting there to be fed. He begs for nuts from the start, his mother is generous and he gets many of the nuts she pounds. Haschich, 6 years old, also settles down at the same anvil and stone he has used before. A heavy stone is a great help in opening the hard-shelled *Parinari* nuts. It would be possible to pound them with a wooden club, but it would take more effort. Haschich has two fingers of both hands paralysed, but is nevertheless a very keen nut-cracker. He is not begging at all from his mother. Cracking about eight to ten nuts in a row before gathering some more, the two of them are eating large amounts of this highly nutritious food. Despite his young age, Haschich cracks without interruption for five hours and fifteen minutes.

Some months later, we follow Héra when she finds a nest of sweat bees in a large freshly fallen branch. The nest is hidden inside the branch, its small entrance betrayed by some bees, but most must have already abandoned the nest. Héra quickly removes a stick from a sapling with her hand, approaches it to her face, and reduces its length to 25 cm by cutting it with her teeth. Then she dips one end of the twig deep into the hole, rapidly pulls it out with pieces of the drying honey sticking to it at the extremity. She licks the honey and dips again. Meanwhile Eros, 3 years old, sits nearby. When she removes the twig, he places his fingers tightly around it and she pulls it through his fist and he licks the delicacy that remains on his palm. She continues to dip, and soon she seems to collect more of the honey and Eros begs now directly for the loaded tool. Héra hands it over to Eros and rapidly makes a new one and now they exchange tools — empty for loaded. She dips while he licks. Héra shares every second dip and this goes on for about four minutes.

Tool-use in chimpanzees varies greatly, and we discuss it in this chapter to assess not only its diversity but also its complexity, its nutritional importance, and how youngsters learn different techniques.

Tool-use was for a long time considered in anthropology to be one of the abilities unique to man, and was thought to have evolved with *Homo habilis* some three million years ago (Leakey 1980; Isaac 1978; Washburn 1978). This belief was based not only on the evidence of tools associated with *Homo* fossils, but also on the expectation that the ability to modify the shape of an external and independent object intentionally, and to use it to fulfil a predefined function, were outside the range of non-human abilities. In support of this, reviews of such behaviours revealed that tool-use as observed in birds and most mammals is restricted to a few specific contexts and is always performed in a quite rigid manner. However, during the last three decades, evidence of tool use in wild and captive primates has accumulated, but only in great apes it is reported to exist in many different contexts and to be performed often in quite diverse ways (Beck 1980; McGrew 1992; Goodall 1970; Parker and Gibson 1977; Hall 1963; Sugiyama 1990). Intriguingly, tool-making seems to be much more limited than tool use, and has been convincingly observed only in great apes (Beck 1980). By tool-making we mean modification to the shape and size of an object to produce a tool. This ability has been observed regularly only in wild chimpanzees and orang-utans (Boesch and Boesch 1990; Goodall 1973, 1986; Nishida and Hiraiwa 1982; Teleki 1974; McGrew 1974, 1992; van Schaik *et al.* 1996). Knowledge of tool-use and tool-making in chimpanzees and orang-utans will increase our understanding of the abilities these animals possess in manipulating objects and their environment, and lead us to a better evaluation of their uniqueness in humans.

Tool-use in captive chimpanzees has been known since the beginning of the century, but was dismissed as being an artefact of captivity resulting from interactions with caretakers. The first observations of tool-use in wild chimpanzees made by Goodall in the early 1960s (1963) had a special impact, for it was the first time that chimpanzees demonstrated the ability to make and use tools without any human intervention. More observations rapidly revealed that chimpanzees not only use tools but do so in such a variety of ways and with so much sophistication that understanding the behavioural diversity of this species concerning tool use and manufacture became a goal in itself (Goodall 1986; Boesch and Boesch 1990; Nishida and Hiraiwa 1982; McGrew 1992; Sugiyama 1990). What is particularly impressive is their ability to use various tools in different ways and for different purposes, since this requires great flexibility in manipulating external objects. This ability can be population-specific, in the sense that different populations within one species use different sets of tools. This is still generally regarded as a characteristic of the human species and is often associated with the notion of culture and higher mental abilities in our species.

In the first part of this chapter, we describe for the first time all the tool-use observations in the Taï chimpanzees. Then we review and discuss tool-use in wild chimpanzees to assess the present state of knowledge. In the second part, we present a detailed analysis of nut-cracking behaviour, a characteristic of the most western populations of chimpanzees in West Africa (Boesch *et al.* 1994). Since the nut-cracking behaviour occurs very frequently in the Taï forest, we were able to follow most of its aspects in detail. Our observations show that tool-use may imply much more than simply using a tool, for it may represent a way of life, including elaborate planning of tool-use and a long learning phase, with important food-sharing interactions in mother–infant pairs that can persist for several years. In nut-cracking, tool-use and food-sharing intermingle in a complex web of interactions that affect many aspects of social life.

Tool-use and tool-making in Taï chimpanzees

Tools can be used in many different ways within one species. Studies of tool-use have tried to produce an exhaustive list of all tools used within one population. Whenever possible, information has been collected about the frequency of tool-use and about its contribution to the survival of the chimpanzees, since tool-use in animals was considered to be purely facultative, whereas in humans it was thought to be essential for survival (Isaac 1978; Leakey 1980). Table 9.1 presents these observations in Taï chimpanzees. Three points need to be stressed here. First, tool-use is a common behaviour in this population of chimpanzees. Even leaving aside nut-cracking, we observed on average one tool being used every second day, which is an underestimate, since we followed only a few chimpanzees per day. Including nut-cracking, we estimated that more than two tools were used per day. Whereas most tools are used for just a few minutes, nut-cracking is observed for hours each day during the nut-cracking season (see below). The tools listed in Table 9.1 have been observed as being used by individuals of all ages except for the very youngest ones. Thus, tool-use is part of the daily life of almost all chimpanzees in the Taï forest.

Table 9.1: List of tool use in Taï chimpanzees observed between 1979 and 1995

Tool-use activity Tool-use aim	Number of observations	Number of tools used (number of tools made)
(1) *Extracting*		
Ant-dipping	72	156 (141)
Wood-boring bees killing	24	43 (42)
Grub-extracting	20	44 (36)
Honey-fishing	70	119 (116)
Bone-marrow extracting	83	140 (138)
Brain-eating	1	1 (1)
Eye-eating	1	3 (3)
Nut-emptying	98	205 (181)
Mushroom	2	3 (3)
(2) *Probing*		
Wood-boring bee nests	8	16 (14)
Dead bodies	4	4 (0)
Wounds	6	7 (4)
Bark-trunk interstice	8	11 (7)
Other bodies	3	3 (1)
(3) *Cleaning*		
Sponging	75	84 (83)
Wounds	2	2
Dead body	6	7 (7)
Sperm	1	1
(4) *Displaying*		
Aimed throwing	11	27 (2)
Throwing	3	13 (1)
Dragging	16	31 (0)
Hitting	8	8 (0)
Play	11	13 (13)
Lever	1	1
Rake	1	2 (2)
(5) *Pounding*		
Nuts*	~1800	1037 (85)

* For nut-cracking, data strongly underestimate the number of tools used and made as they include only data collected with focal individuals between 1979 and 1988.

Second, tool-use can be very varied and flexible. In Table 9.1, we have listed 26 different types of tool use. This number will certainly increase further with more observations,
as chimpanzees are inventive and tend to try new solutions to commonly encountered
problems and these may include using tools. Tools are used mainly in the feeding
context, but also in social interactions and to improve personal comfort. We expect that
the different types of tools used by a population are influenced by the environment. For
example, insects are abundant in the Taï forest and chimpanzees feed on some of them
daily, as well as on insect products (for example, honey and bee-bread). These are often
hidden underground or under the bark of trees, situations in which tools are helpful.
However, as different species of insects live in different environmental conditions, different populations of chimpanzees will need different solutions when feeding on them.

Third, a large proportion of the tools used by chimpanzees in the Taï forest are made by
them (Table 9.1). Tool-making is common. Excluding hammers used for cracking nuts,
83% of all tools are manufactured by the chimpanzees before they use them. The reason
why only 8% of the hammers used to crack nuts are modified by the chimpanzees may lie
in the nature of the material itself; hammers are hard, whether they are made of wood or
particularly if they are made of stone. They can usually be used as they are and they last
for years. If there is need for modification, it is not frequent. Twigs and leaves, on the
other hand, are much more fragile materials, and as a rule the chimpanzees make such
tools whenever they need one.

Making a tool can indicate sophisticated cognitive abilities. Tool manufacture has been
presented for centuries as the hallmark of humanity, for it can require the production of
an object with a shape that might be very different from its original shape. This ability to
transform raw material into a fully fledged tool according to a mental representation held
by the tool-maker has been seen as the criterion of culture (Washburn 1978; Leakey
1980; Isaac 1978). Tool manufacture can be achieved in different ways, not all of which
demonstrate special cognitive abilities. A tool has to fulfil some specific requirements to
be of any use. For example, fishing insects from a depth of 10 cm through a hole can be
performed only with an elongated object at least 10 cm long with a diameter smaller than
that of the hole. Thicker or shorter tools would not work, and by trial-and-error manipulation of the object the tool user could produce tools specifically adapted for this task.
Standardization of the tools would be the result. But in this case, it might only be a
reflection of the requirement of the tasks performed and not a sign of anticipatory abilities
by the tool user. If, however, the complete modification procedure was done on the object
before its use, this process would reveal sophisticated anticipatory abilities, and this has
been considered by some to be the hallmark of culture. To distinguish between the two
processes, we need direct observations of the tool users from the time when they start to
handle the objects.

Many tools used by the Taï chimpanzees are manufactured before they use them (46%
of 1893 tools). How do chimpanzees make their tools? Tools are made from different
materials and this leads *per se* to tools having different shapes. But do chimpanzees
modify the raw material in such a way as to produce differently shaped tools for different
purposes? In other words, do chimpanzees standardize their tools according to their
future functions? In Fig. 9.1, we compare only tools that chimpanzees made out of sticks
for four different purposes. In all cases, the chimpanzee introduces a stick into a hole to

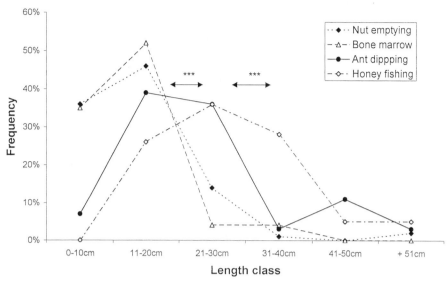

Fig. 9.1: Standardization of sticks made by Taï chimpanzees with statistical comparisons between length classes.

extract some food; it can be a pounded nut with some kernel remains in the shell, a broken bone with marrow inside, a hole in the ground leading to an ant nest, or a hole in a tree where bees have made their hive with honey in it. Tools produced by chimpanzees for these purposes clearly differ both in length and in diameter in three of the four tool types (only tools for bone marrow and emptying nuts are of similar size and shape) (Boesch and Boesch 1990). The length is directly determined by the modifications performed on the raw material, suggesting standardization in tool production by Taï chimpanzees.

The next question is whether this is the result of standardization made before use or the result of successive modifications performed during utilization. Taï chimpanzees modified their tool before they started to use it in 93.5% of the instances. In only 6.5% of the sticks used were some modifications made after they had already started to use them. Thus, standardization in tool production results from a process in which tool users rely on a mental representation of what an object should look like to become a tool, and also of what changes are required to produce such a result before use.

Table 9.2 lists the different procedures used by Taï chimpanzees to manufacture tools. Tool-making in Taï chimpanzees is common. In 81% of the instances more than one type of modification is made before using a tool, and this rises to 98.6% if we exclude hammers for nut-cracking which are mainly produced by altering the length once. In 76% of the observations, three or more types of modification were made before using the tool. Owing to the structure of most tools, which are elongated pieces of wood or twigs, the modifications are mainly of two kinds, reducing the length and changing the shape. Reducing the length is done in different ways depending upon the size and hardness of the material, and it can require considerable strength to break up long, hard wooden branches. Altering the shape includes the removal of leaves from the sticks, as well as sharpening the ends of some twigs, especially when trying to remove the remains of nuts from cracked shells.

Table 9.2: Tool-making observed in Taï chimpanzees between 1979 and 1988

Type of tool-making	Tool	Number of observations
(a) Detaching from substrate with teeth or hands	Stick + club	5
(b) a + cutting to a specific length with teeth or hands	Stick	18
(c) b + removing leaves or bark with teeth or hands	Stick	293
(d) c + sharpening the end with teeth	Stick	13
(e) c + modifying length with teeth or hands	Stick	23
(f) Breaking in two by hitting on hard surface	Hammer	72
(g) Breaking in two by pulling while standing on the branch	Hammer	6

(a)

(b)

(c)

(d)

Fig. 9.2:   (a) *Ant dipping*: Four mothers and their infants dip for driver ants using the short sticks typical for the Taï culture. (b) Cacao, a six-year-old male, dips efficiently for ants with one stick, while holding a second one in the other hand. (c) While his mother, Narcisse, dips for ants, Noureyev is reaching for her tool to get some ants just before she eats the soldiers which are biting the stick. (d) Vanille, a seven-year-old female, peels the bark of a stick with her teeth.

Tool-use and tool-making in wild chimpanzees compared

To assess the flexibility of tool behaviour in a species, we have to compare the sets of tool behaviour used in different populations, and we therefore have to classify the tools. Because agreement between authors on tool definitions is rare, we used Goodall's (1986) terminology to facilitate the comparison of chimpanzee technology. This classification is supported by the fact that Taï chimpanzees produce distinctly different tools for obtaining different types of food (Fig. 9.1). thus for the same tool activity (for example extracting), tools vary according to the aim (e.g. ants or honey). Table 9.3 reports all the 42 different types of tool-use observed in wild chimpanzee populations, classified in the following main categories: extract, probe, clean, display, pound, and some combinations of these. The description of their tool-use repertoires documents the great variability shown by chimpanzees in choice and use of tools for different activities. The frequency of use of the tools varies also between populations. All classifications underestimate the real variability existing in tool-use among wild chimpanzees, since different tools are pooled within one category. For example, inserting twigs in an ants nest to feed on driver ants at Bossou, Gombe, and Taï is done with tools of 25 cm or longer, whereas at Mahale the chimpanzees feed on arboreal ants with much shorter sticks. Moreover, Gombe chimpanzees dip the ants with a technique involving both hands, which is strikingly different from the one-handed technique used by Taï chimpanzees (Boesch and Boesch 1990). Similarly, Gombe chimpanzees insert twigs to fish for termites, whereas in Cameroon and Rio Muni they first perforate the termite mound to gain direct access to the termites and use larger and different tools than those used at Gombe. Tools for cracking nuts vary according to the type of nuts being opened, involving wooden clubs or stones with weights ranging from about 100 g to 40 kg. Still, this classification allows us to gain an impression of the wide variety of tools that wild chimpanzees have been seen to use spontaneously.

Some 51% of all the tools used by wild chimpanzees are used in a feeding context, 17% in aggressive contexts against conspecifics or against other species (mainly leopards), 12% for communication purposes, 11% to inspect the environment, and 9% to clean their own body.

This table also suggests several generalizations about tool-use in chimpanzees. First, tool-use is a regular behaviour observed in all long-term studied populations of chimpanzees. The size of the tool repertoire may vary, but all populations use at least a few different tools. Second, in some populations the toolkit is important (Taï chimpanzees use 26 different tool types), whereas in others it may be minimal (Kibale chimpanzees so far have been seen to use only 9 different tool types). None of the chimpanzee populations possesses the same tool repertoire. Taï chimpanzees share 57% of their tool types with Gombe chimpanzees, who share 54% with the Mahale chimpanzees. Clearly, tool-use is a flexible and regular activity in chimpanzees.

It is difficult, however, to provide a simple explanation for the differences observed. We expect the environment to affect tool-use in chimpanzees, and accordingly a small tool repertoire might reflect an environment in which few problems need or can be solved with the help of tools, or an environment in which there is enough nutritious food accessible without the need of tools. It is striking that the distribution of tool-use types does

Table 9.3: Tool-use observed in wild chimpanzee populations. Short-term study results are listed under Other. + means the tool-use has been observed, ++ means it has been observed more than 20 times, and – indicates that the tool-use has not been observed in that population

| Tool-use activity | Populations | | | | | |
Tool-use aim	Gombe	Mahale	Taï	Bossou	Kibale	Other
Extracting						
Termites	++	(+)[1]	–	–	–	Assirik Cameroon Rio Muni
Ants	++	++	++	+	–	
Bees	+	–	+	–	–	Liberia
Honey	+	–	++	–	–	
Bone marrow	++	–	++	–	–	
Brain	–	–	+	–	–	
Eyes	–	–	+	–	–	
Mushroom	–	–	+	–	–	
Grub	–	–	++	–	–	
Nuts	–	–	++	–	–	
Gum	–	–	–	+	–	
Algae	–	–	–	+	–	
Probing						
Termite nest	+	+	–	–	–	
Ant nest	–	++	–	–	–	
Bee nest	–	–	+	–	–	
Tree hole	+	+	+	–	–	
Bodies	+	–	+	–	–	
Bark	–	–	++	–	–	
Wounds	+	–	+	–	+	
Hook	–	–	–	+	–	
Cleaning						
Sponge	++	+	++	+	++	
Dirt	++	+	+	–	+	
Ants	–	+	–	–	–	
Brush	+	–	–	–	–	
Catch	+	–	–	–	–	
Self-tickle	+	–	–	–	–	
Leaf-napkin	+	–	–	–	+	
Foot-stick	–	–	–	–	–	Sierra Leone
Displaying						
Aimed throw	++	++	++	+	+	
Throw	++	++	++	+	–	
Drag	++	++	++	+	++	
Hit	++	++	++	+	–	
Play	++	++	+	+	++	
Weapon	–	–	+	–	–	
Leaf clip	–	++	++	++	++	
Leaf groom	++	++	–	–	++	
Fly whisk	+	–	+	+	–	
Pounding						
Nuts	–	–	++	++	–	Liberia
Pestle	–	–	–	+	–	
Secondary[2]	–	–	–	+	–	

Table 9.3: *continued*

Tool-use activity	Populations					
Tool-use aim	Gombe	Mahale	Taï	Bossou	Kibale	Other
Combination						
Pound + Insert[3]	–	–	++	–	–	
Probe + Insert[4]	–	–	–	–	–	Liberia
Total	22	14(1)	26	15	9	

[1] One community in the north has been observed to use such tools, but not the main study communities (Collins and McGrew 1987).
[2] Secondary tools at Bossou were stones placed under a stone anvil to stabilize it (Matsuzawa 1996).
[3] Taï chimpanzees often use small sticks to extract remaining pieces of kernel in the cracked shells of nuts they opened with a hammer.
[4] Chimpanzees were using three types of branches to perforate, enlarge the hole, and extract honey from arboreal bee nests.
Gombe data are from Goodall (1973, 1986), McGrew (1992). Mahale data are from Nishida and Hiraiwa (1982), Nishida and Uehara (1980). Bossou from Sugiyama and Koman (1979), Sugiyama (1994a). Kibale from Whiten *et al.* (1999). Sierra Leone from Alp (1997). Mt Nimba from Kortlandt and Holzhaus (1987). Liberia from Hannah and McGrew (1987). Rio Muni from Sabater Pi (1974). Cameroon from Sugiyama (1993).

not follow a clear pattern. With few exceptions, the distribution among the different chimpanzee populations differs for each tool-use listed. That means it will be difficult to find a simple explanation for these differences. Previous explanations of regional differences in tool-uses (McGrew 1974; Teleki 1974) have not been confirmed by more recent observations of additional types of tool-use in new populations under study.

Tools are made in various ways; in Table 9.4 we list all the techniques observed so far in wild chimpanzees. Here again, we pool some methods that are in fact quite distinct. For example, cutting a twig with one hand is rather different from breaking a large branch by pulling forcefully with both arms, but both result in shortening the object to the right length. Likewise, shaping a twig by removing leaves with one hand is quite different from removing the bark by biting it away with the teeth. Nevertheless, the classification demonstrates the variety of methods used by chimpanzees to modify an object to make it a tool. In most of the cases, the chimpanzees performed more than one modification on an object (81% of the cases at Taï, 72% of the tools used to fish for ants at Mahale [Nishida and Hiraiwa 1982]).

Table 9.4: Tool manufacture techniques observed in three chimpanzee populations

Type of tool-making	Gombe	Mahale	Taï
I) *Cutting* To the right length (grass, twig, stick, stone)	(1) Breaking with the hands (2) Cutting with the teeth	(1) Breaking with the hands (2) Cutting with the teeth	(1) Breaking with the hands (2) Cutting with the teeth (3) Pulling while standing on it (4) Hitting against a hard surface
II) *Shaping* (twig, stick)	(1) Removing leaves or bark	(1) Removing leaves or bark	(1) Removing leaves or bark (2) Sharpening ends with the teeth

Fig. 9.3:   (a) *Honey dip*: Agathe extracts honey of a wood-boring bee nest (*Xylocopa sp.*) with a stick. (b) While she eats some honey, her infant, Aphrodite, looks at her and another infant inspects the nest entrance. (c) Agathe gives her tool to Aphrodite who licks the honey, while she dips further with another stick.

The main modification procedures used by wild chimpanzees are very similar in the different populations studied, reflecting the similarity in the type of tools used: mainly twigs, leaves or grass. The fact that Taï chimpanzees have more ways of making tools than the other populations studied results from the fact that they use hard material to crack nuts, and this demands other methods than those used to modify the length of twigs, leaves or grass. In addition, because they eat the nuts they open, Taï chimpanzees have an additional way of sharpening the ends of twigs to extract the remains of the kernel, which would be impossible to reach with the finger or the teeth. Thus, tool-making is influenced by the environment in which the animals live, as this determines what kind of tools are required to get access to certain food sources.

Tool-use and tool-making has been seen in all chimpanzee populations which have been studied in detail. Each population possesses a diverse tool repertoire, stressing the flexibility of this species when using tools. Tool-use allows wild chimpanzees to modify some part of their environment and directly expand their access to certain food sources, modifying their environment daily to their own benefit. We shall see that, in the case of nut-cracking, this effect is far from being trivial and may directly affect their survival.

The nut-cracking behaviour of Taï chimpanzees

Nut-cracking behaviour is very common in Taï chimpanzees. We studied the planning of actions it requires, described sex differences, made a cost–benefit analysis, and did a detailed analysis of apprenticeship. Some of these results have already been published elsewhere, and we summarize them here. However the learning of nut-cracking behaviour is presented here for the first time.

Five species of nuts of different shapes and degrees of hardness are eaten by chimpanzees in the Taï forest; the most abundant one, *Coula edulis*, is the softest, *Panda oleosa* is the hardest, whereas the large trees with huge fruit production of *Parinari excelsa*, *Detarium senegalense* (irregularly cracked at our site), and *Sacoglottis gabonensis* (very rarely cracked at our site) produce nuts of intermediate hardness (Boesch and Boesch 1983). *Panda* nuts require a weight of 1600 kg to be cracked without pounding (Peters 1987). These nuts are rich in protein, sugar, fat, and amino acids. Most of these trees are widely distributed throughout West and Central Africa, where the majority of the remaining wild chimpanzee populations live. The distribution of nut-cracking behaviour, however, is much more limited than the tree distribution. Nut-cracking has been known for some time to be confined to West Africa, but a recent survey of nut-cracking showed that its distribution is surprisingly limited even within the West African forest region (Boesch *et al.* 1994). The Sassandra river in the western part of Côte d'Ivoire is the eastern border of this behaviour (see Fig 1.1). None of the sampled chimpanzee populations living east of this river in forests with Coula and Panda trees crack nuts. When comparing chimpanzee density, tree density, and the presence of nuts and hammers between sites east and west of the river, we could find no ecological factor that differed between the two sides of the river. Even when separated by only 30 km, only the chimpanzees on the west side pound these nuts. This strongly suggests a cultural difference.

To crack nuts, the chimpanzees bring together three different objects: a hard tool to pound the nuts, the nuts, and a hard substrate as an anvil, on which to place the nuts

(Boesch and Boesch 1983; Sugiyama and Koman 1979*b*). These materials rarely occur together naturally, and the chimpanzees need to select and transport them to a given place. The anvils used by the chimpanzees to stabilize the nuts include emerging roots, the base of large trees, rocky outcrops and suitable branches in a tree. At Bossou, chimpanzees have been seen to use mobile stones as anvils, a behaviour never seen throughout Côte d'Ivoire (Boesch *et al.* 1994). The reason is most probably that Bossou chimpanzees pound palm fruits near the oil-palm trees within young secondary forests, and exposed roots in such forests are far too small to be suitable as an anvil (Sugiyama 1994*a*; personal communication). Most hammers used to pound the nuts are fallen branches of various shapes, sizes, and degrees of hardness. Compared with branches, stones are rare in the forest, and the chimpanzees use them consistently if they are available. By using a stone hammer for the soft *Coula* nuts, a chimpanzee invests 30% less pounding energy than would be expended using a wooden one. This gain increases to 43% when cracking the harder *Panda* nuts (Boesch and Boesch 1983). Chimpanzees select more stones to crack the hard *Panda* nuts, while they content themselves with the more common wooden hammers for the *Coula* nuts.

Nut-cracking in the Taï forest constitutes a very important aspect of the chimpanzees' life and diet. On average, during the four months of the *Coula* seasons, the chimpanzees crack nuts for 2 hours 15 minutes per day. This represents 700 grams or 270 nuts. Chimpanzees obtain an average net gain of 3450 kcal per day from this activity. During the nut season, the chimpanzees obtain the large majority of their calorific intake and a large portion of their protein intake from nuts and through tool use (Boesch and Boesch 1984*b*; Günther and Boesch 1993). Without tools, the Taï chimpanzees would have much less access to a very rich food source. Thus, we can say that they depend on tool-use during the nut-cracking season.

Transport of tools and nuts

Taï chimpanzees transport nuts very frequently. The nuts are carried to an anvil in the hands, the mouth, and even the feet (Boesch and Boesch 1983). Chimpanzees crack on average about 270 nuts per day; this represents more than 20 transports. Hammers are the factor limiting nut-cracking, since good quality hammers (of a convenient size, weight, and hardness) are not abundant. Chimpanzees adopt an 'energy saving' strategy and transport heavier hammers for longer distances for the harder nuts (Boesch and Boesch 1983). The larger the gain, the longer the transport distances. For *Coula* nuts, 85% of the transports are less than 20m long, whereas for the harder *Panda* nuts only 59% of them are made for less than 20 m. This is combined with a selection of the hardness of the hammer, so that for softer nuts 62% of the hammers are wooden clubs, whereas for harder nuts 99% of the hammers are stones. These observations suggest that hammer transport is far from random and that the chimpanzees select them depending upon the nut species they want to crack and by considering the energy expended while pounding.

There are two main questions concerning hammer transport. How are hammers chosen and how are they transported? As visibility in the forest is limited to about 20 metres, long transports are of special interest. Both qualitative observations and the quantitative analysis of the stone-hammer transports for the hard *Panda* nuts indicate that chimpanzees anticipate their need of a hammer and search for one before transporting it to the nut-cracking site (Boesch and Boesch 1984a). The hard *Panda* nuts are produced by isolated trees that are relatively rare in this forest, and the stones used to crack them are rare as well. Hammers are transported only when no hammer is available at the nut-cracking site; and in the few cases where we found two hammers at a tree, two individuals were cracking at the same time. For the abundant *Coula* nuts, less anticipation is required as most hammer transports are done within sight of the anvil where the chimpanzees are going to crack the nuts, and a few are done before actually seeing the anvils but within a region in which there are many *Coula* trees. More demanding are the hammer transports for cracking *Coula* nuts in the trees, as the chimpanzees always pick up a hammer before climbing the tree. Chimpanzees could adopt a 'play safe' strategy by always carrying a hammer before going to pound nuts. This is, however, observed only for the abundant *Coula* nuts, which is often a social activity, done within foraging parties and in areas with many *Coula* trees. Good quality hammers tend to be quickly monopolized by those who find them first.

The situation in the case of the *Panda* nuts is ideal for studying the mental map of the nut-cracker. Stones are rare in the forest, but are necessary to crack the hard nuts. The trees of this species are widely dispersed, and chimpanzees have to transport the stones over long distances between nut-cracking sites (average transport distance = 120 m). This represents a special challenge due to the limited visibility of about 20 m in the forest. How do chimpanzees select the hammers under such extreme conditions? In a given region of 30 hectares, we marked all the fruiting *Panda* trees, measured the distance between them, marked and weighed all the available stones, and kept a daily record of their location. Analysing 76 of the stones transported to crack nuts at a *Panda* tree out of sight showed that the chimpanzees choose the nearest stone to that tree within a given class of weight in 48 cases, the least energy-demanding one in 40 cases, and the lightest stone in 26 cases (Table 9.5). Almost all (95%) of the energy-minimizing transports were also transport of the nearest stone to the *Panda* tree, while only 65% of the nearest stone transports were energy-minimizing ones. This indicates that the chimpanzees follow a minimal distance strategy when transporting hammers for long distances (Boesch and Boesch 1984a).

We compared the decision rules of the chimpanzees for stone hammers that were at a same distance to a given *Panda* tree, when the distance is either smaller than 20 metres (46 cases) or larger than 40 metres (22 cases). It appeared that weight also plays a role, in

Table 9.5: Stone transports of more than 20 m classified according to three possible strategies when Taï chimpanzees cracked *Panda* nuts

Minimal distance	Minimal weight	Minimal energy	Number of transports
+	+	+	17
+	−	+	21
+	−	−	10
−	+	+	2
−	+	−	7
−	−	−	19

A plus indicates that the transport was minimized for the strategy considered, otherwise they are classified with a dash.

the sense that chimpanzees transport the heavier stones for small distances but the lighter ones for long distances (Boesch and Boesch 1984*a*).

Thus, chimpanzees perform frequent hammer transports, and they select the hammers they are going to transport after an evaluation that includes the weight of the hammer and the transport distance. This is made possible by a mental map that allows flexible positioning and relocation of stone hammers and nut-producing trees (see Chapter 10 for further discussion of the cognitive abilities required). The underlying rule on which the chimpanzees in the Taï forest base their decisions about the transport of nuts and hammers is an energy-optimization rule: minimal energy investment in hammer transports for those that allow the nut-cracker to reach optimum efficiency.

Sex differences in frequency and performance of techniques

At Taï, we observed two main techniques of nut-cracking; the ground technique, used for all five species of nuts, and the tree technique, used only for nuts of *Coula edulis*. Most commonly, chimpanzees pound the nuts on the ground by selecting a surface root or an outcrop rock as an anvil. They place the collected nuts on the ground, then place and pound one on the anvil. When eating it, they place the hammer on the ground. *Coula* nuts are ripe before they fall to the ground, and in November and December, the chimpanzees crack them directly in the trees (Boesch and Boesch 1981). Cracking in a tree requires particular dexterity: the nut-cracker must carry both the nuts and the tool in the tree and handle them during the pounding so that neither falls to the ground. This is especially impressive when a mother, supporting her baby against her belly, holding spare nuts in one foot and in her mouth, supports the nut on the anvil with one hand while pounding it with the hammer she holds in the other hand. Later in the season, the *Coula* nuts are to be found mostly on the ground, and the ground technique is then preferred.

Thus, two nut-cracking techniques used at Taï are particularly demanding: *Coula* cracking directly in the tree (requiring the anticipation of transporting tools and dexterity in handling tool and nuts), and cracking the very hard *Panda* nuts (requiring transport of stone hammers and technical skill due to the hardness of these nuts). Extracting kernels of *Panda* represents a further difficulty. *Panda* nuts have three kernels independently embedded in a hard wooden shell. To extract them without smashing the kernels, very powerful blows must be given at the beginning to open the nut. Then, to gain access to all the kernels, the nut has to be precisely repositioned two or three times, and a careful administering of very gentle taps is needed to extract them. The tree technique, as well as pounding *Panda* nuts, are both used more frequently by females, and their performance (number of hits used to open the nut and number of nuts eaten per minute) is superior to that of males (Boesch and Boesch 1981, 1984*b*). We found that the reason for this sex difference is social, for males favour social contact whenever there is a conflict between cracking more nuts or remaining with other group members. The rather solitary activity of these two particular techniques, one taking place high up in the trees, usually out of sight of others, the other happening at the few isolated trees, being further restricted to one or two individuals due to the lack of stones, are thus performed less frequently by males. One could speculate that in early humans there may already have been such a sex difference. The first tools, most likely wooden ones, were possibly used for gathering and food-processing activities, such as cracking nuts. The first users and makers of these

(a)

(b)

(c)

(d)

Fig. 9.4:   (a) *Nut-cracking*: Xérès cracks the *Coula* nuts with a wooden hammer and her daughter Xindra eats some while watching carefully. (b) Perla uses a heavy granite stone on a stone anvil to crack the hard Panda nuts, while her daughter Pandora takes some pieces from the anvil. (c) Dilly pounds a soft *Detarium* nut with a large wooden hammer.(d) To extract more of the *Detarium* nuts she opened with a hammer, Dilly uses a small stick.

tools might well have been females, in contrast to the common theory that the first tools were invented by male hunters.

This female superiority in nut-cracking performance is mirrored by a male superiority in hunting (Chapter 8). Both may be precursors of the division of labour that has been proposed to be a key feature of early hunter–gatherers in human evolution (Isaac 1978).

The learning of nut-cracking

Infant chimpanzees, who remain mostly in close contact with their mothers, show very early a strong interest in manipulating hammers and in learning to open nuts by themselves. At the same time, they are eager to eat nuts, although they do not yet manage to open them. Mothers share the nuts they pound with their infants for many years. This leads to a situation in which the learning attempts and food-sharing occur simultaneously, and our aim was to study how the attitude of the mothers affects the performance of the infants. We shall first quantify the progress of the infants in cracking nuts, then include the nut-sharing interactions, and finally try to see how these effect the progress of the youngsters.

The analysis is based on net energy gain when cracking nuts, so that we could compare it to the gain when begging for nuts. For this we calculated both the cost and the benefit of nut cracking.[1] Figures 9.5 and 9.6 show the performance of young chimpanzees when cracking the two species of nuts, as well as the performance of the adults. Reflecting the difference in the hardness between the two nut species, from the age of 4 years onwards, efficiency is clearly higher for the soft *Coula* nuts than for the hard *Panda* nuts. Youngsters start to pound *Panda* and *Coula* nuts during the first two years of life, and they succeed in opening their first *Coula* nuts when three years old, but not until two years later they are able to open their first *Panda* nuts. After one year of practice on the *Coula* nuts, the youngsters reach their first positive gain when four years old (Fig. 9.5). For the hard *Panda* nuts, the youngsters achieve the first positive gain when seven years old. They

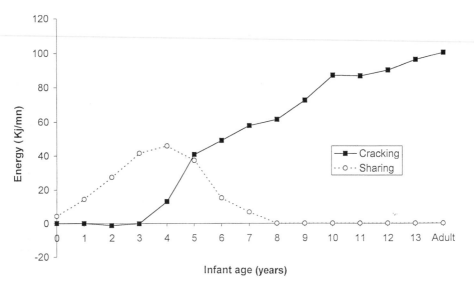

Fig. 9.5:   Youngsters net benefit when cracking *Coula* nuts or through sharing by the mother.

1   The costs include the sum of the energy spent to collect the nuts, to pound and eat them, as well as the energy required to transport the hammer to another anvil. We based our quantification of the costs of nut-cracking on a film analysis of a young adult male cracking nuts (Günther and Boesch 1993), but included in the calculation for each age class the specific physiological value for the basic metabolic rate, the body and arm weight, and the observed proportion of time spent hitting and eating the nuts, as well as collecting the nuts and the hammer. The benefit of nut-cracking is the gain in energy resulting from eating the nuts that were cracked successfully.

The nut-cracking and nut-sharing data for *Coula* nuts are based on 412 hours of observations on 23 male and 30 female infants observed from birth to 13 years old, with some individuals followed for one nut season and some followed for up to 8 years. The same data for *Panda* nuts are based on 154 hours of observations on 16 male and 14 female infants observed from birth to 13 years with some individuals followed for up to 4 years. Mother–infant pairs were selected *ad libitum* in the forest, and whenever a pair stopped cracking nuts, we switched to another one to increase our sample size. For each nut-cracking pair, we recorded all nut-cracking attempts and the performances (hits/nut and nuts/minute), as well as all nut and hammer sharing events. The observations were concentrated during the nut seasons from 1982 to 1990.

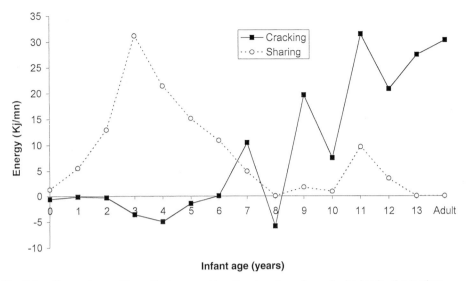

Fig. 9.6:   Youngsters net benefit when cracking *Panda* nuts or through sharing by the mother.

benefit from nut-cracking three to four years later for *Panda* than for *Coula*. Once having achieved the first benefit, progress at first is quite rapid, as they double their gain for *Coula* nuts within a year (between four and five years). For *Panda* nuts, the progress is slower, for it takes four years to double the benefit of the seven year olds (Fig. 9.6).

The learning of nut-cracking seems to proceed through four distinct phases. First, the youngsters make unsuccessful attempts by hitting the nuts. Typically during this phase, youngsters do not understand fully how to achieve success and regularly make mistakes, such as hitting the nuts directly with the hand or with another nut, without using a hammer, or without placing the nut on an anvil. The second phase is reached at the age of three years when they understand the technical problems and crack nuts only when the three elements are present, but they are limited by their lack of muscular power for opening a nut. The third phase starts when they have enough strength and manage to pound the nuts correctly. In spite of some success, the balance of energy remains negative. Through practice, progress is quite rapid, and youngsters achieve 42% of the adult efficiency for the *Coula* nuts within two seasons. For *Panda*, this phase starts much later, when eight years old, and youngsters achieve adult efficiency within four seasons. Hence, the third phase is characterized by rapid progress in the nut-cracking behaviour of the youngsters. Then the fourth phase starts in which the youngsters progress slowly and reach adult efficiency only once they are adult themselves. This slow progress reflects the fact that subadult individuals have problems in gaining access to good hammers and must often content themselves with less than optimal tools. This effect is less marked for *Panda* nuts for two reasons. First, it is useless to try to crack them with wooden clubs, and they rapidly stop doing so at a certain age. Second, pounding *Panda* nuts is a much more solitary activity, and infants can regularly gain access to their mother's hammer once she has stopped pounding. Mothers most often wait for their juvenile offspring to crack some nuts before going elsewhere, and regular training with easy access to a good hammer more than compensates for the fact that *Panda* nuts are harder to crack. Thus, although young chimpanzees demonstrate an early interest in cracking nuts, it takes years to achieve a net benefit. Once the first benefits are achieved, progress is relatively quick.

For years, the infants spend most of their time with their mother, including the time she is cracking nuts. When they themselves start to crack the nuts, they do so for very short bouts (Table 9.6). The older they get, the more time they spend cracking both species of nuts (*Coula*: $r_s = 1.0$, $N = 9$, $p < 0.0001$; *Panda*: $r_s = 0.87$, $N = 14$, $p < 0.002$). The increase in nut-cracking time correlates between the two species of nuts ($r_s = 0.81$, $N = 9$, $p < 0.01$). For *Coula* nuts, the energy gain increases both with age ($r_s = 0.85$, $N = 9$, $p < 0.002$) and with the time spent cracking nuts ($r_s = 0.85$, $N = 9$, $p < 0.002$). However, for *Panda* neither age nor time spent cracking the nuts correlates with the energy gain. Control of strength, which is more important for the *Panda* nuts, cannot be understood by watching; it requires personal, direct practice. Nut-cracking progress does not correlate with the time spent watching the mother for either of the two species of nuts ($p > 0.7$). Young infants tend to wait until their mothers stop pounding nuts, so that they can use her hammer and crack as long as she is willing to wait for them. Five-year-old chimpanzees still crack nuts for only a little time, although this is the period when they are being weaned (see Chapter 3). That infants spend more time watching the mother crack *Panda* nuts is mainly because *Panda* cracking, in contrast to *Coula*, is a mostly solitary activity and the youngsters have no playmates around.

The slow and difficult start of the nut-cracking learning process makes us wonder why, after many years of unsuccessful trials, youngsters keep on trying. The answer to this question probably lies in the mothers' generosity in sharing nuts with them for many years. Without this sort of permanent incentive, they might give up this struggle. We could not test this idea as all mothers of the study community shared nuts with their youngsters. This brings us to the question of nut-sharing in Taï chimpanzees.

Table 9.6: Proportion of time the infants spend cracking nuts or watching the mother cracking nuts (percent).

Age (years)	*Coula* nuts			*Panda* nuts		
	N	Watch	Crack	N	Watch	Crack
0	10	3	0	3	26	0.6
1	20	19	0.5	11	45	2
2	15	21	3	6	63	33
3	17	31	8	4	64	14
4	15	31	20	7	85	20
5	13	17	47	4	80	17
6	10	11	79	4	78	22
7	8	27	368	4	75	86
8	8	0	∞	1	0	100
9	8	–	–	3	70	464
10	8	–	–	3	1	39
11	6	–	–	2	15	216
12	8	–	–	2	13	111
13	9	–	–	2	30	482

Values higher than 100% result from a bias in our method, as for juveniles the mother's activity was not always recorded because she was sometimes out of sight.

Sharing the nuts and tools in Taï chimpanzees

Tool-acquired food has rarely been observed to be shared in chimpanzees, even between mother–infant pairs, compared to other food types (Goodall 1986; Nishida and Hiraiwa 1982; Silk 1978). This is in marked contrast to the meat-sharing that is common between group members (Chapter 8). Nut-cracking is an exception, for in Taï chimpanzees nuts are shared abundantly by the mothers. Figures 9.5 and 9.6 show the nut-sharing of *Coula* and *Panda* nuts provided to youngster by the mother. Nut-sharing by the mother is observed right from the first year of life onwards and lasts for many years: 8 years for *Coula* and 12 years for *Panda* nuts. Maternal investment in terms of nuts extends well beyond the weaning period and, for *Panda* nuts, extends into adolescence. The investment provided during infancy would be predicted, for it complements the maternal milk investment and infants eating nuts might become less dependent upon milk. However, why do mothers invest in their infants for such extended periods after weaning? Nuts are a very rich food source available to the Taï chimpanzees during several months of the year. Hence, it is important that infants learn to crack nuts, and while they are struggling to do so, the mothers support them by sharing. Sharing level is tuned to the difficulties the youngster encounters. Mothers reduce their *Coula*-nut sharing when their infants are about six years old, when they have reached about half the efficiency of adults. For *Pandas*, the situation is complicated by the hardness of the nuts and youngsters' progress is less rapid. Accordingly, the mothers invest more in their youngsters for this species. They appear to ascertain that their infants receive enough energy from nuts during the learning phase.

How much do the mothers share during this investment period? Figure 9.7 shows that sharing is very costly. They share more than 20% of the nuts they pound for five years for *Panda*, and four years for *Coula*. Nut-sharing reaches its maximum around the age

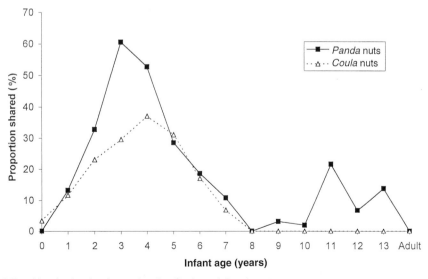

Fig. 9.7:   Nut sharing by the mother for *Coula* and *Panda* nuts.

of four to five years and decreases slowly with the increasing age of the infant. Remembering that mothers crack *Coula* nuts for about 2 hours and 15 minutes, infants of four and five years still obtain from their mother more than 1750 kJ of nuts per day, and mothers expend about 252 kJ to pound them. This represents 13% of the mother's total daily nut-pounding. Due to the rarer and more irregular occurrence of *Panda* cracking, we were unable to evaluate the amount of *Panda* nuts obtained through nut-sharing as precisely.

The amount of *Coula* nuts shared by the mother correlates positively with the time spent by the infant watching her crack the nuts ($r_s = 0.81$, $df = 7$, $p < 0.01$) (Table 9.6). The older the infants grow, the more time they spend cracking *Coula* nuts and the more efficient they become, so that they reduce the time looking at and begging from the mother. However, whenever they do so, they are rewarded proportionally to the time they are present. On the other hand, for *Panda* nuts, the infants also spend more time cracking nuts as they grow older, but their progress is not proportional to the time practising, and they go on watching their mother cracking nuts throughout the juvenile period. However, when they watch their mother crack *Panda* nuts, they are not rewarded proportionally to the time they are present, despite the fact that mothers share a larger amount of *Panda* nuts with their infants and they support the learning process for a longer period of time than they do for *Coula*. Qualitative observations showed that mothers adapt their behaviour to the species of nuts even in situations when they crack the two species within one hour.

Maternal investment in terms of opened nuts gives the impression of covering both a nutritional need of the infant, as well as stimulating its learning process. Nut-sharing complements nut-cracking, so that infants tend to eat a fairly constant amount of nuts from four years onwards. To start with, this amount is provided by the mother, and as the infant improves its technique it is progressively able to cover this amount by itself.

Mother–offspring conflicts in nut cracking

Trivers (1974) was the first to present the idea that conflicts between parents and offspring have to be expected, for they will disagree about the amount of investment the infant should receive. Typically after a certain age, parents may profit more by investing in a new infant than by investing further in the previous one, whereas for the infant the opposite is true, as it profits more from the investment it receives than from that provided to one of its siblings. This results in weaning conflicts (Trivers 1974). Parental investment is costly, for it has been shown to increase mortality rate and decrease fertility in parents that have reproduced compared to parents that have not (Clutton-Brock 1991). Therefore, we should expect conflicts between parents and offspring not to be restricted only to weaning (Trivers 1974; Godfray 1995; Parker 1985). Empirical studies have provided frustratingly little support for these ideas, mainly because in a natural habitat it is extremely difficult to measure conflicts over investment precisely: how much food would an infant like to receive and how much less does it get (Clutton-Brock 1991)?

From this point of view, nut-sharing might be a unique opportunity because it is possible to see from the begging gesture what the infant would like to receive and compare it with what the infant actually gets. The infant can beg with its outstretched hand to receive the freshly opened nut before the mother takes it from the anvil (begging for a whole nut), or it may beg for a part of the nut she is holding in her hands while she is eating, or for pieces that remain on the anvil, or for scraps that fall to the ground while she is eating. These were all considered as begging gestures because mothers were seen to refuse access to all of them. The mother, on her part, can refuse by not giving anything after such a begging gesture, or she can give a smaller piece than was wanted (for example, she will take the nut from the anvil despite the begging gesture, leaving just a small piece to be taken, or she might refuse anything from her hand by turning away but will allow to take scraps from the ground). We have to specify here that mothers were seen to share only with their own infants, even for the tiny scraps that can be found on the ground near the anvil. Invariably other infants would be pushed away if they tried to take some of them (this applies only to mothers, as infantless females were seen to share small amounts with orphans or infants of nearby mothers). We recorded all begging gestures by the infant and noted whenever there was a discrepancy with the amount it actually received (which is what we call a conflict). To quantify the conflicts, we estimated the size of the nuts required and received in the situations in which the visibility was perfect and extrapolated it to the begging gestures (for example, begging for a piece of nut from the hand of the mother represented 0.4 of the whole nut, whereas scraps from the ground were estimated to represent 0.06 of a nut).

The first question is whether there are any reasons why we should expect a conflict between mother and infant concerning the nuts. It seems obvious that infants want to suckle for longer than the mother would like, as milk is of a general profit to them. We likewise need to know whether infants would still derive benefit from nut-sharing from their mother after they have started to crack nuts on their own. Figures 9.5 and 9.6 compare directly the energy gain obtained by cracking nuts and by sharing them for *Coula* and *Panda* nuts. It appears that obtaining nuts from the mother remains the most profitable way of eating nuts for youngsters under six years old for *Coula* nuts and under nine years old for *Panda* nuts. For both species of nuts, the offspring still benefit from nut-sharing well after weaning, when the mother has given birth to another infant. Thus, the preconditions for a conflict are fulfilled.

Figure 9.8 shows the variation of the conflicts in relation to the amount of nuts shared for *Panda* and *Coula* nuts. Conflicts are indeed very much part of the sharing relation-ship, for they are observed whenever the mother shares nuts with her offspring. For *Panda* and *Coula*, the amount of conflict increases according to the number of nuts that are shared (*Panda*: $r_s = 0.74$, $N = 14$, $p < 0.001$; *Coula*: $r_s = 0.69$, $N = 8$, $p = 0.06$). Contrary to expectation, conflicts decrease with age for *Panda* and remain stable for *Coula* (*Panda*: $r_s = -0.61$, $N = 14$, $p = 0.02$; *Coula*: $r_s = -0.30$, $N = 8$, $p < 0.5$). From Figure 9.8, we see that conflicts increase very sharply at the start of the sharing period and then tend to decrease, despite the fact that the sharing increases for two more years. Strikingly, conflicts are observed for most occurrences of sharing during the first three years, but later a large proportion of sharing events occur without any visible conflict. Hence, conflicts between mothers and infants in Taï chimpanzees do not support theoret-ical expectations (Trivers 1974), but rather give the impression of a constant and fine tuning in the investment between the pair. The end of the maternal investment looks more like a consensus than a fight. Taï mothers do not perceive a conflict between present and future reproduction, but seem able to optimize both goals at the same time, as we

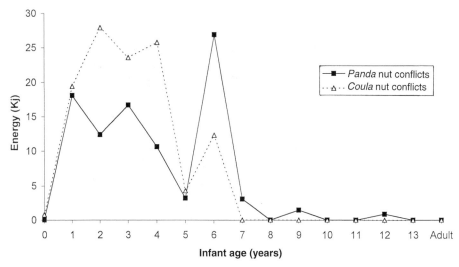

Fig. 9.8: Mother–offspring conflicts: quantity contested by mothers when sharing *Panda* and *Coula* nuts.

witnessed in those mothers who shared *Panda* nuts with two or three of their offspring at the same time. A similar tendency has been found at Taï in mother–offspring conflicts over milk, in which conflicts were concentrated during the first six months of life and were very limited afterwards (Pfluger *et al.*, in press).

Much theoretical work has been devoted to the question of whether the mother or the infant should win the conflicts (Trivers 1974; Alexander 1974; Parker and MacNair 1979; Parker 1985). If we interpret the conflicts over the amount of nuts shared as a sign of the desire of the mother to give less than the infant wants, and we look at the amount shared between the two as the compromise they have reached, we can estimate who wins the conflict by looking at how much of the initial request of the infant has been granted.

Infants tend to win all conflicts, for they obtain on average 90 to 92% of all the requests the mother initially contested (Fig. 9.10). For *Panda* nuts, the success of the infant is proportional to the amount shared ($r_s = 0.76$, $N = 14$, $p < 0.001$): The more that is shared by the mother, the more conflicts there will be, and the more of them will be won by the infant. With time, the amount shared, as well as the conflicts, decrease. This does not apply to *Coula* nuts ($r_s = 0.45$, $N = 8$, $p = 0.26$). The more time the infant spends near its mother, the more *Coula* nuts will be shared by the mother, but the number of conflicts and their outcomes show no trend with either the amount shared or the age of the infant.

It is difficult to claim that infants win all the conflicts over nuts, for they might be very sensitive to the mother's refusal gestures even if she gives in afterwards. The infants would thereafter simply avoid a similar pressure (which is not measured with our present method) and ask for less, or try it another way. For example, when infants are about four years old, mothers start to deny them access to whole *Coula* nuts from their anvils, and the infants rapidly shift to begging exclusively for nuts from their mother's hands. About two years later, mothers start to deny them access to *Coula* nuts from their hands, and the

Fig. 9.9: Loukoum, cracking the hard *Panda* nuts, shares them with her two sons at the same time (note the hands of different sizes).

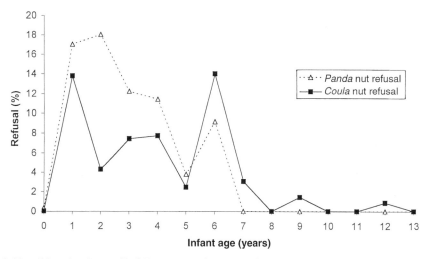

Fig. 9.10: Who wins the conflict? Frequency of sharing refused by the mother for *Coula* and *Panda* nuts.

infants rapidly learn to beg only for nut remains from the anvil after the mother has taken her share. Such changes in sharing practice often occurred quite rapidly, and were all made on the initiative of the mothers.

The species of nut that is shared makes a difference to the conflicts observed between mothers and infants. Conflicts are over larger amount of nuts for *Coula* nuts than for *Panda* nuts (Wilcoxon test for 0 to 7 years: $T = 4$, $N = 8$, $p < 0.05$), and infants win more conflicts for *Panda* nuts than for *Coula* nuts (Wilcoxon test: $T = 4$, $N = 8$, $p < 0.05$). Mothers try harder to restrict their sharing for *Coula* nuts, which infants learn more quickly to crack on their own. This is in support of our proposition that mothers not only invest food but also support their infants' own nut-cracking attempts through the way they share nuts.

Teaching and stimulation by the mother

A mother who is concerned that her infant should learn nut-cracking behaviour as precisely as possible may not only be supportive in the way she shares nuts, she can also directly guide her offspring's attempts to crack nuts. In humans, such actions by the parents or older group members may be of central importance for the transfer of knowledge and skill between generations (Boesch and Tomasello 1998; Greenfield 1984; Rogoff *et al.* 1993; Whiten and Milner 1984; Tomasello *et al.* 1993). Such pedagogical actions are often presented as a 'scaffolding process' (Wood *et al.* 1976), whereby the teacher's selective interventions provide support to the learner, extending his or her skills to allow the successful accomplishment of a task not otherwise possible. This means that the teacher helps the learner to produce new skill components that are often understood but yet not performed. This includes not just teaching but all the ways used by parents to stimulate and facilitate their offspring's attempts on a given task. Teaching is considered to be the most elaborate form of pedagogy, but is often less frequently used for learning a task than attention-fixing or observation (Rogoff *et al.* 1993; Whiten and Milner 1984; Greenfield 1984; Greenfield *et al.* 1989). The existence of pedagogy in animals has often been denied, because such behaviour was rarely observed. But animals should be expected to use a given behaviour only when needed and not otherwise, and we may have been looking at the wrong situations. The nut-cracking behaviour is an appropriate situation for several reasons: acquiring the task is very important, because of the nutritional value of the nuts; it is a particularly difficult task to learn; and mothers invest a large amount of time and energy in helping their infants to acquire their daily share of nuts. Thus, any acceleration in the learning process would be to their direct benefit.

At Taï, chimpanzee mothers rely on many forms of pedagogy to help and stimulate their offsprings acquisition of the nut-cracking technique (Boesch 1991*a*). We distinguished three different ways:

Stimulation: The mother may stimulate the nut-cracking attempts of her offspring by leaving the hammer behind on the anvil while she collects more nuts. The mother may, in addition, leave some intact nuts near the anvil during her nut collecting. In this way, the mother provides her offspring with the opportunity to learn what a good nut and a good hammer look like, as well as giving it the chance to practice with a convenient tool. In some cases, the mother even places the hammer on the root used as an anvil after having placed an intact nut on it, ready to be pounded. All the infant has to do is lift the hammer and pound the nut. In half of the instances of stimulation, the offspring made some hits with the hammer. Chimpanzees without infants were never seen to leave a hammer behind or to collect more nuts before having opened all of the previous collection. Stimulation became significantly more frequent for three year olds, who had started to use the hammer (Fig. 9.11), occurring once every 15 minutes. However, the mother may go further than just temporarily lending her tools.

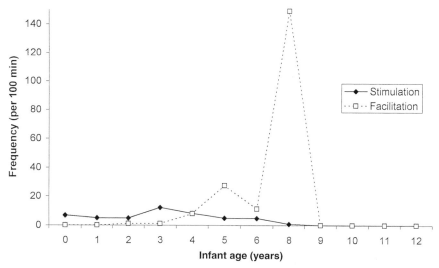

Fig. 9.11:   Pedagogy in Taï chimpanzees: frequency of stimulation and facilitation in nut-cracking.

Facilitation: The mothers may facilitate the task of her offspring, when it is trying to open nuts, by providing it with a better hammer or intact nuts that she collects for it. Facilitation is costly for mothers, for they have to search for the nuts, and when they give away their own hammer they have to content themselves with a less optimal tool, which reduces their own performance. Facilitation, like stimulation, is more frequent when the infants have made some progress. While stimulation occurs most frequently for three-year-old infants, facilitation starts when four to five-year-old infants have begun to be successful. The mothers' acts are adjusted to the level of skill attained by their infants ($r_s = -0.48$, $p < 0.05$) (Fig. 9.11). Facilitation by the mothers occurred on average once in 7 minutes and peaked at more than one case per minute for eight year olds. The offspring always took advantage of this help, and their nut-cracking performance always improved.

Teaching: Despite their efforts and the use of convenient tools, the offspring may face technical difficulties in nut-cracking that they are unable to overcome. In two cases, the mother, resting near her nut-cracking offspring, noticed its technical difficulties and was seen to make a clear demonstration of how to solve them (see description in Boesch 1991*a*). In the first example, after her six-year-old son incorrectly positioned a piece of *Panda* nut on the anvil, the mother, who was resting nearby, corrected the positioning of the piece of nut. In the second example, the mother, seeing the difficulties of her four-year-old daughter, demonstrated how to hold the irregularly shaped hammer by turning it in a very slow and conspicuous manner into the correct grip which the mother kept while cracking a few nuts for her. Thereafter her daughter showed she had learned the lesson, since she maintained the grip demonstrated to her despite varying her own position and that of the nut. In these two cases, the mothers were not cracking nuts at the time but interrupted what they were doing, that is resting, and walked to their offspring. They both resumed their siesta after teaching, while the offspring went on cracking.

Pedagogy in Taï mothers includes stimulation and facilitation of the nut-cracking attempts of their infants. If necessary they may further complement their actions by actively teaching specific technical points. These pedagogical interventions by the mothers are frequent (on average 12 times per hour of nut cracking) and all bring specific aspects of this technique to the attention of their offspring. Through sharing nuts, mothers support the nut-cracking attempts of their offspring and they choose interventions appropriate to their offspring's level of ability when improving its performance. In some human societies, transmission of technical skills relies mainly on pedagogical processes other than active teaching, such as attention fixing and facilitation (Rogoff *et al.* 1993; Whiten and Milner 1984; Greenfield 1984; Greenfield *et al.* 1989), which is quite similar to the way chimpanzees facilitate the learning of nut-cracking.

Infant–mother cooperation in nut cracking

Conflicts in nut-cracking can be viewed as a method of adjustment between mothers and infants during the learning of this task. Nuts obtained through sharing remains by far the most profitable way of eating nuts for youngsters, hence we should expect them to try different ways of increasing their mother's willingness to share for a longer period. However, we should expect the infant to influence the mother only for a limited time, while infants benefit from it and while the relationship with the mother still permits such manipulations. We have observed such behaviour patterns in some detail in two mother–infant pairs, Salomé–Sartre and Héra–Eros. We detail them here to show how elaborate interactions can become within the context of tool-using behaviour. These two pairs were not the only ones showing such behaviour, but they were observed at a time when both tolerated our presence.

A case study: Salomé and Sartre

When we began our observations, Sartre was a five-year-old male with a generous high-ranking mother, Salomé. As can be seen from Table 9.7, Sartre at that time received 90% of the nuts he begged for. Normally, at this age, mothers start to restrict their sharing (begging success decreases, and the frequency of refusal increases) and infants react to this new situation. Unlike two other males of the same age (Kummer and Ali) that responded by cracking more on their own, Sartre responded by starting to cooperate with his mother in two different ways. First, while she cracked, he collected nuts on the ground and brought them to her (called 'Collect' in Table 9.7). This represented a gain for his mother, as she could save the time and energy needed to collect the nuts and go on cracking. In this way, Sartre was reducing some of the cost his mother had when pounding nuts for him. Sartre provided his mother with 151 nuts in 52 collections. Most of these nuts were eaten by the mother herself. In a second type of collaboration, he cleaned his mother's anvil of the nut shells and placed a new nut on it for her to crack (called 'Place'). This way of cooperating with her was more directly beneficial to him, for he ate 84% of 61 nuts he placed and she pounded. Qualitative observations indicate that this was perceived by Salomé almost as an obligation to share that nut with the youngster, as we saw her leave the anvil where Sartre had just placed a nut and go to crack another one on another anvil three times, and once she prevented Sartre from placing a nut for her.

When six years old, Sartre cracked nuts much more frequently and begged less (Table 9.7). He still cooperated with his mother by collecting nuts and cleaning the anvil for her, but he was then more interested in obtaining her optimal hammers. Sartre could have looked for

Table 9.7: Cooperation for nut-cracking by Sartre and Eros with their mothers. Data for each *Coula* nut season are separated into three periods in order to follow the interactions more precisely.

Age	Nov.–Dec.	January	Feb.–March
Sartre 5 years	did not crack	Beg: 90% Refusal: 6% Collect: 0 Place: 0 Fac: 0	Beg: 87% Refusal: 9% Collect: 62n (54%) Place: 12n (92%) Fac: 3n, 5h
Sartre 6 years	Beg: 100% Refusal: 0 Collect: 0 Place: 0 Fac: 11n, 2h	Beg: 58% Refusal: 25% Collect: 34n (47%) Place: 20n (90%) Fac: 25n, 21h	Beg: 67% Refusal: 9% Collect: 55n (36%) Place: 29n (76%) Fac: 0n, 11h
Eros 4 years	Beg: 42% Refusal: 6% Collect: 1n (0%) Place: 0 Fac: 0	Beg: 43% Refusal: 1% Collect: 0 Place: 0 Fac: 0	Beg: 65% Refusal: 8% Collect: 4n (25%) Place: 19n (95%) Fac: 7n, 2h

Beg: Infant's begging success.
Refusal: Proportion of sharing refused by the mother.
Collect: Number of nuts collected by the infant for his mother (proportion received by the infant after the mother had opened them).
Place: Number of instances the infant cleaned the anvil and placed a nut for the mother (proportion received by the infant after the mother had opened them).
Fac: Facilitation by the mother: n = number of intact nuts given, h = number of hammers given to the infant.

hammers himself, but he never did, and he usually obtained his mother's tool. Salomé then had to look for a new one for herself, and very often it was not as good as the one she let Sartre have. When Sartre was seven years old, he cracked most of his *Coula* nuts on his own. However, the cooperation between the two persisted for two more years for the *Panda* nuts, and we saw them happily share the same stone hammer for hours, one pounding while the other was eating, sitting face to face, the anvil between them.

A case study: Héra and Eros

Eros was the second son of Héra, a middle-ranking mother and a skilled nut-cracker. She shared 48% of the nuts she opened with Eros when he was watching her (Table 9.7), and Eros successfully resisted most of her few refusals to share. In February 1987, she was clearly pregnant and started to refuse to share nuts. Eros' response was similar to that seen with Sartre: he started cooperating, collecting nuts for her or placing nuts for her to pound. As with Sartre, collecting nuts led to limited success, while placing nuts was almost always successful. Héra was even seen once to put aside a nut placed by Eros and place one herself that she then cracked and ate. During Eros' fifth year, Héra had a new baby and Eros was never again seen to beg for *Coula* nuts.

These two examples illustrate how complex the interactions between a mother and her infant can become in the nut-cracking context, when the infant is at an age when it can start to crack nuts itself and at the same time the mother reduces her investment. Both the mother and the offspring react specifically to the changes they observe in each other.

Sex differences in learning to crack nuts

Taï females more frequently crack *Panda* nuts and *Coula* nuts in the trees than males, and they are more efficient when they crack *Coula* nuts on the ground (Boesch and Boesch 1981, 1984*b*). How do such sex differences emerge? We suggested earlier that they result from the higher social inclination of males, who prefer social contact to nut-cracking when there is a conflict between the two. This explains why males crack *Panda* and *Coula* nuts directly in the trees less often than females do; in both activities contact with other group members is hindered. However, this does not explain why females should be better nut-crackers than males, for both perform the technique regularly. A sex difference during the learning process of this technique might lead to such a contrast (Boesch and Boesch 1984*b*), a hypothesis we now test.

Figure 9.12 presents the efficiency curves for males and females for *Coula* nuts; we see that age had a clear effect on the efficiency of nut-cracking (ANOVA, Type III SS: data between 4 and 12 years are normally distributed: $F = 17.82$, $df = 8$, $p < 0.0001$), and neither sex nor the interactions between age and sex have an effect. It is only in late adolescence that the adult sex difference starts to develop. Mothers support the acquisition of the technique differently in sons and daughters. Sons are provided with more nuts and fight constantly to obtain even more with great success (85%) (Fig. 9.13(a) and 9.13(b)). In the first year, daughters obtain a similar amount to sons, but face stronger opposition right from the start, and lose a larger proportion of the conflicts during the first two years. Thereafter, daughters seem to be satisfied with a smaller share and fight with less determination, despite the fact that they win almost all conflicts (93%), while mothers become harder with their sons, so that they lose more conflicts than daughters from the second year onwards. Thus, in general, sons rely more on their mothers for nut-eating, while daughters become independent earlier and try harder to open nuts by themselves. During the first four years of nut-cracking, daughters crack on average 25% longer than sons. It

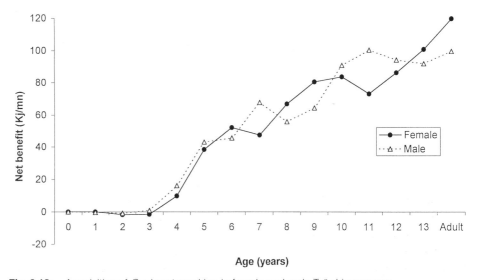

Fig. 9.12:   Acquisition of *Coula* nut cracking in female and male Taï chimpanzees.

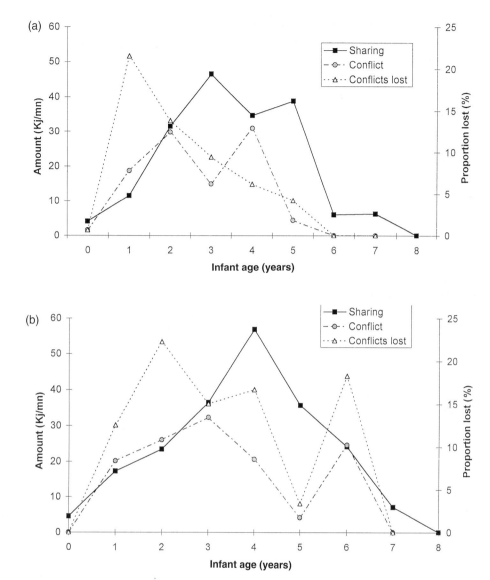

Fig. 9.13: (a) Female chimpanzees' begging success and conflicts for *Coula* nuts. (b) Male chimpanzees' begging success and conflicts for Coula nuts. (*Fig. 9.13(c) and (d) on next page*)

seems possible that the females' earlier and longer practice at cracking nuts is reflected in their adult performance and in their nut-cracking technique. Females are significantly better at cracking nuts on the ground than males, and males do so with a more rigid pounding movement (Boesch and Boesch 1984*b*).

For *Panda* nuts, age has an important effect ($F = 11.51$, $df = 10$, $p < 0.0001$) and sex also explains some of the difference ($F = 5.2$, $df = 10$, $p < 0.05$): between 4 and 13 years males are more efficient than females (Fig. 9.14). This is surprising for we know that

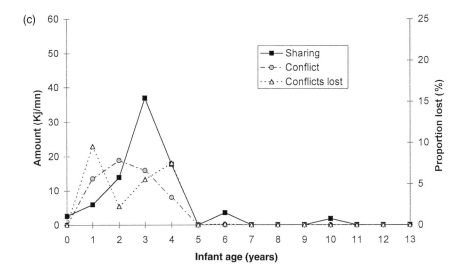

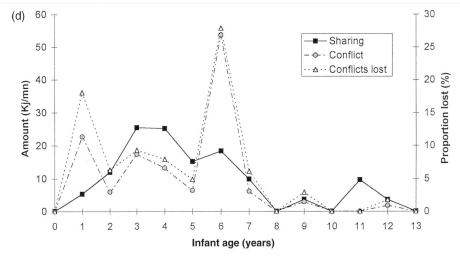

Fig. 9.13:   (c) Female chimpanzees' begging success and conflicts for *Panda* nuts. (d) Male chimpanzees' begging success and conflicts for the *Panda* nuts.

adult females are more efficient crackers of *Panda* nuts than adult males. A comparison of the sharing pattern between the two sexes reveals that sons from four years on obtain more nuts from their mother for all the remaining six years of sharing (ANOVA planned comparison: $F = 14.96$, $p < 0.002$), despite the fact that conflicts between mothers and sons tend to be more abundant and that sons lose more conflicts than daughters (Fig. 9.13(c) and 9.13(d)). Thus, the sharing pattern for *Panda* nuts is very similar to that for *Coula* nuts, and the consequence seems to be the same, namely that sons spend 47% less time cracking nuts than daughters during the first ten years of life. The argument for *Coula* nuts could apply to *Panda* nuts: through more practice when young, females achieve greater efficiency as adults.

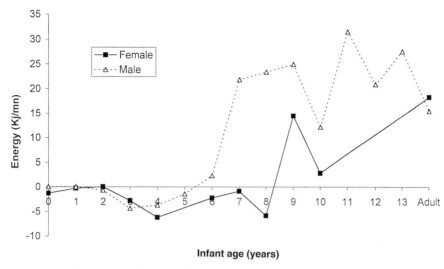

Infant age (years)

Fig. 9.14:   Acquisition of *Panda*-nut-cracking in female and male Taï chimpanzees.

Why are then males more efficient than females while learning to crack Panda nuts (Fig. 9.14)? The *Panda* comparison relies on a smaller sample than the *Coula* one and the most elaborate forms of cooperation we observed were seen between mother–son pairs, while during the same period the females in our sample were mostly cracking on their own. For example, Salomé shared her optimal granite tools with her son Sartre for extended periods of time and Sartre achieved surprisingly high gains (during his sixth year he cracked nuts with a positive gain of 11.3 kJ/mn and during his seventh year of 23.7 kJ/mn). Ali, the other male in our sample, was adopted by Brutus, the alpha male, when six years old, and he too had access to Brutus' optimal hammers. This limited sample suggest the hypothesis that sons have better access to the mothers' tools and progress more quickly without having to overcome the technical difficulties of using a bad hammer, while daughters have less support from their mothers and have to learn on their own to overcome such difficulties. This could have the effect that females learn earlier to overcome technical problems related to varying quality of hammers which makes them more efficient as adults.

Both for *Coula* and *Panda* nuts, the mothers' style seems to be sex-dependent. Sons tend to receive more nuts and tools than daughters for several years, and as a consequence daughters practice nut-cracking more frequently from a very early age. This difference in the learning process could explain why, once adult, females generally do better than males. We saw in Chapter 3 that dominant females invested more in sons than in daughters, which might reflect what we observed in the nut-cracking apprenticeship. However, although some dominant mothers, like Salomé and Ella, are part of the sample, many more low-ranking females, like Héra, behaved with their offspring in exactly the same way, suggesting the maternal investment is sex-dependent whatever the mother's rank.

In conclusion, nut-cracking behaviour is an elaborate tool-use that needs many years to be fully acquired. The outstanding feature of this behaviour are the nut-sharing interactions between mothers and offspring. Mothers share for hours per day and for years with their offspring. This maternal investment adds considerably to nursing, and lasts much longer. The collaboration of the mothers with their offspring in foraging also seems to function as a help in learning a difficult technique more quickly. Taï mothers not only share nuts with their offspring, but also regularly actively stimulate and facilitate their nut-pounding. The offspring can also directly influence their mothers' behaviour by cooperating in an attempt to obtain more nuts. Growing up in such a generous social environment may have influenced the adult chimpanzees' attitude towards the unusually important meat-sharing interactions in the Taï community.

A comparative approach: is chimpanzee tool-use special?

Tool-use and tool-making has been observed regularly and in many different contexts in all studied chimpanzee populations throughout their range in Africa. The technological reper-toire of each population is specific and seems to reflect a technological tradition, as there is no possible simple explanation based on genetic or ecological differences. On the other hand, tool-use has been observed in a large variety of animal species, ranging from different bird species to many primates (Beck 1980; Goodall 1973; Parker and Gibson 1977; Hall 1963; Hunt 1996). In many instances, the types of tool-use observed can be very similar to that seen in chimpanzees, but the repertoire within each species is much more limited and is observed only in a limited number of populations. For example, only northern Californian otters have been observed to use stones to open mussels, or only Galapagos finches were seen to use twigs to extract grubs from hollow branches. Caledonia crows have been pro-posed to make three different kinds of tools to fish for insects, suggesting that tool-use and tool-making is probably within the possibilities of many animal species (Hunt 1996). Thus, the question should not so much be whether a given species is able to use tools, but why most animal species possess such a limited repertoire of tool-use.

How common is tool-use in monkeys? Tool-use in monkeys stands out as never being widespread in the few wild populations known, and as being observed more proficiently in captive individuals of the same species. One adult male brown capuchin (*Cebus apella*) used a piece of oyster shell to pound on other oysters (Fernandez 1991) and a few instances of dropping objects on intruders, of nut-cracking, and of using sticks in aggres-sive encounters with conspecifics have been reported (Chevalier-Skolnikoff 1988; Boinski, personal communications). This contrasts with captive capuchins who have been found to use tools quite proficiently in a wide variety of skills, such as nut-cracking, sponging liquids, using rakes, probing crevices, and inserting sticks in tubes (Visalberghi and Trinca 1989). Similarly, there are only a few observations of wild baboons using a twig to extract a small stone fragment from the ground or to pound on scorpions (Beck 1980). In captivity, baboons use tools with reasonable proficiency to rake food that is out of reach, soak up liquids, or pound objects with stones (Beck 1980; Tomasello and Call 1997). Observations on macaques are restricted in the wild to a few tool-uses in a few individuals and to a comparably higher variety of tool uses in captivity (see review in Tomasello and Call 1997).

How common is tool-use in great apes? So far, wild bonobos have been seen to drag branches and make rain hats (Kano 1992; Ingmanson 1996), while wild gorillas do not apparently use any tools. However both species do so regularly in many varied ways in captivity (see review in Tomasello and Call 1997). In wild orang-utans, one single population in South Aceh, in Sumatra, manufactures and uses feeding tools on a regular and population-wide basis (van Schaik *et al.* 1996), while all captive orang-utans are considered to be very adept tool-users and use them in a large variety of circumstances (Lethmate 1982; Russon and Galdikas 1995). Chimpanzees are the only non-human primates in which both captive and wild populations routinely manufacture and use tools on a population-wide basis.

How can we explain these differences in primates? We have already mentioned the difficulty of evaluating why an animal species does not use tools. Is it because it is unable to do so, or because there is no need for it? The bird and otter examples indicate that probably most primate species possess the minimal cognitive requirements to use tools. However, it would be wrong to say that all types of tool-use are equal: some tool-uses are more demanding than others, and a same tool-use can be executed in a more or a less rigid way. A twig can be held more precisely with a hand than with a beak or a paw, and it can thus be also manipulated more precisely. A bird cannot do anything but insert a twig in a hole in a straight line or pull it out, whereas, for example, a Gombe chimpanzee can insert a twig in long and irregular termite-mound tunnels and wiggle it softly with a delicate movement of its wrist to incite the soldiers to bite in it. Thus, it is the precise voluntary motor control of the hand movements that enables chimpanzees to make more out of a simple twig and an insert movement. Similarly, bi-manual dexterity is not possible for many animal species, which also limits the kind of actions that can be done with a tool and the resulting benefit. We readily see that for an animal species to develop a large repertoire of tool-use requires both some cognitive capacities (that we think are widespread in primates) and some manipulative dexterity.

These abilities are, however, present in all primate species and especially in great apes, and the total or relative absence of tool-use in wild primate populations is the puzzle we need to explain. Most of these species live in tropical forests where insects such as termites, ants, or beetles are abundant, and it seems difficult to suggest that the absence of tool-use is due to the absence of food sources to exploit. It has been proposed for gorillas and orang-utans that their strength enables them to reach all necessary food sources and they thus have no need to use tools (Chevalier-Skolnikoff 1988). We are sceptical of such arguments. First, Taï chimpanzees can open the nuts directly with their teeth, but they almost invariably use a tool instead. With the help of a tool, the Taï chimpanzees eat 6–10 times more nuts in a given time span than they would without one (personal observation). Second, the recent observations of tool-use in wild orang-utans (van Schaik *et al.* 1996) directly contradict this explanation. The puzzle may disappear if we remember that gorillas, bonobos, and many monkeys regularly use tools in captivity. In captivity, they can use tools because they are provided with the opportunity to do so, while in the wild they do not spontaneously 'see' where an opportunity for tool-use may arise and don't always comprehend the benefit of using one. This may result from a limited understanding of the possible action a tool, an external object, can have on the environment, another external object; they can solve problems with their own body, but do not see how the problems could be solved with an external object. At Taï, for generations, mangabey monkeys have observed chimpanzees using hammers to feed on nuts, and they eagerly eat the bits and pieces that remain once the chimpanzees have left the site, but we never saw them use a hammer. They simply don't 'see' how a piece of wood or stone could help them to reach the nuts, despite watching the chimpanzees doing it for hours.

We suggest that the propensity of humans and chimpanzees to use tools spontaneously relies on a better understanding of how causality applies to external objects. Similar to the notion of 'intelligent tool use' (Parker and Gibson 1977), we need the concept of an 'intelligent understanding of a tool', implying a flexible anticipation of how to use a new tool, and an evaluation of the potential benefit of such a use. Most tool-uses observed in captivity are about either 'cleaning or absorbing a liquid' or 'bringing a visible object within reach', while in nature tool-uses are more complex to understand for they are mainly about 'reaching invisible objects or food', (for example insects in nests and trees, or embedded food like nuts) or 'increase the efficiency of a technique' (for example, touching or hitting dangerous or feared social partners or predators). This would explain why an ability, tool-using, is observed in captivity in many primate species, but is so rarely spontaneously used in wild populations. Chimpanzees and humans, thanks to a better understanding of causality, use them in a wide range of contexts, be it for food or for reasons of social or personal comfort. In Chapter 10, we shall present a model to explain how such a difference might have evolved.

10 *Intelligence in wild chimpanzees*

Forest scene: Panda *nut cracking and cooperation*
Once in the early days of the study, when habituation was close to zero, we had searched for chimpanzees all day long without success, when suddenly nearby we heard the typical sound of nuts being pounded with a heavy tool. Very carefully, we approached the chimpanzee, expecting it would disappear before we could get a glimpse of it. However, thanks only to a rustling sound, we saw the shadow of a chimpanzee that was obviously collecting nuts. Surprisingly, at the same time, we heard the pounding go on. Then the spell broke, and in a second the two animals had disappeared. As expected, we found nuts and shells of *Panda oleosa* but only one stone hammer. We were left with the intriguing question of what exactly these chimpanzees had been doing. Had they been sharing the hammer, the nuts, or both, or were we just imagining a nice story?

Two years later, in 1981, in a similar situation, but by then with partly habituated chimpanzees, we saw Snoopy, a young male, cracking *Panda* nuts with a heavy granite stone. We knew this stone was previously elsewhere and wondered how Snoopy had dealt with this fact. Visibility being restricted to about 20 m in the forest, he could not have seen it from where he was. We wondered whether he knew about its location and had picked it up before arriving at the cracking site or had he walked round searching for a tool after having found the nuts? Why did he take this stone rather than any other? We then saw that Snoopy was with another larger male we had not yet identified and who fed on the remains of kernels near Snoopy's anvil. To our surprise, this male then collected more nuts, put them next to Snoopy, who continued his cracking session, and then sat and ate more of the bits and pieces. Snoopy very naturally opened the nuts that were brought to him. This went on for more than thirty minutes.

Apparently, the impressions we gained by traces and glimpses of this set of behaviour in the very early days of the study were correct; the chimpanzees do collaborate when nut-cracking. With knowledge, came questions and doubts: why should two grown males collaborate in cracking *Panda* nuts and why would the stone-owner and user let somebody else eat the fruit of its work? We named the unknown male Rousseau, after the philosopher, as a tribute to the many new questions his behaviour brought up.

'Intelligence' is a term used to classify a large array of faculties all related to the ability to solve a problem. We shall use the term 'intelligence' to review what contributes to how wild chimpanzees 'see and manipulate their world' (see Cheney and Seyfarth 1990). Traditionally, studies on animal intelligence have been done in experimental settings with captive subjects. This enables subjects to be tested under controlled situations where the impact of most factors on performance can be evaluated. It has the disadvantage of placing the subjects in unnatural settings, which is of limited value when the aim is to understand the mental faculties an animal uses to survive in the wild which have resulted from evolution. Captive studies inform us about the potential of some individuals within the special situation of their upbringing and social interactions. Studies of wild individuals inform us about the abilities that contribute to their biological success. These may overlap, but probably only partly. Wild studies face the challenge that there is little chance of controlling for all the factors affecting the performances of an individual, and thus careful analysis and planning of the observations are needed. The value of wild studies has been shown by Cheney and Seyfarth (1990) and Whiten and Byrne (1988), who discovered many unsuspected facets of primate intelligence. In this chapter we attempt to describe the level of intelligence used by Taï chimpanzees in their daily activities.

In the best studied animal species, *Homo sapiens*, cognitive abilities vary, for there are differences between individuals belonging to distinct populations, and the cognitive capacities of the adults within a single population also differ (Segall *et al.* 1990). For example, hunter–gatherer populations, like the Eskimos, possess very elaborate mental maps using coordinates, but have limited numerical ability, whereas traditional farmers, like the Baoulés in West Africa, have more restricted mental maps that rely mainly on topological landmarks, but have sophisticated numerical abilities (Dasen 1975, 1982; Dasen *et al.* 1978). Cognitive abilities can also develop to varying degrees in the two sexes, for example, in traditional farming populations, where the sexes perform very different tasks from an early age, more women gave non-operational answers to tasks assessing concepts of time and speed than men (Bovet and Othenin-Girard 1975, in Dasen and Heron 1981). These and similar observations demonstrate that intelligence is a complex notion, that we should not expect all individuals to achieve the same level, and that experience may influence its development.

From an evolutionary perspective, we expect individuals to develop the cognitive abilities they need to survive and reproduce within the niche they occupy in a given environment. Not all chimpanzees living in the rainforest occupy the same niche; for example, only some populations crack nuts, a specially rich food source. Different environments present different challenges, and more demanding habitats select for more complex cognitive abilities. We should expect different levels of intelligence in populations that encounter different environmental challenges. The results on humans illustrate this point. However, history sets limits on the variability of cognitive faculties observed within a species, for each individual is born with physical and physiological features that are the result of a long evolutionary history. Individuals inherit their brain size and structure from their ancestors, and this limits what they can do. Mammals tend to have larger brains for their body size than other vertebrates, and primates have larger relative brain size than other mammals (Byrne, 1995).

Some general hypotheses have been proposed to explain the evolution of intelligence. The social intelligence hypothesis (Jolly 1966; Humphrey 1976; Whiten and Byrne 1988) argues that living in groups requires sophisticated mental abilities in order to keep track of social complexities. Some tests of this hypothesis show that the ratio of the volume of the neocortex to the rest of the brain correlates well with average group size and with the frequency of use of tactical deception (Dunbar 1992; Byrne 1995). The ecological intelligence hypothesis (Parker and Gibson 1979; Milton 1981), the arboreal clambering hypothesis (Chevalier-Skolnikoff 1988; Povinelli and Cant 1995) and the technical intelligence hypothesis (Byrne 1997) propose that different aspects of surviving in a complex environment require sophisticated mental abilities. These last hypotheses attempt to explain differences seen in the cognitive development between monkeys and great apes (Byrne 1995, 1997).

Captive and wild chimpanzees should therefore be expected to possess different abilities. Thus, what may hold for Taï chimpanzees does not have to apply to other populations and vice versa. We are still far from a complete comprehension of intelligence in chimpanzees. At present only four or five wild chimpanzee communities are followed regularly, not enough for a full understanding of an animal species. The effect of captivity has been rarely studied as such, but we know from studies with surrogates or no mothers at all that the separation from the mother has devastating effects on the rhesus monkey infants (Mason 1978). Up to twenty years ago, all captive chimpanzees were 'wild-born'. This euphemism meant that after their mother was killed or injured, the very young infants were forcefully separated from her, tied with ropes, or pushed into sacks, carried through the jungle, kept in poorly made cages with inadequate food, exposed to villagers, and eventually shipped to Europe or the USA. It has been estimated that for every survivor, nine died during this process (Teleki 1989). Once chimpanzees arrived in Europe or the USA, they were kept in isolation or in small groups in crude and bare cages — not much impetus for cognitive development. Such conditions still prevail for many chimpanzees kept in captivity. What are the effects of such traumatic experiences on cognitive development? It is probably safe to assume that they are not positive. Psychologists testing children for cognitive faculties would certainly never think of relying solely on individuals that had, say, survived the traumatic experience of a civil war.

Recently, a small proportion of the captive chimpanzees have been kept in larger enclosures with enriched conditions and more social partners. This is an improvement, but captivity still means boredom, limited social partners, and limited space. How do captive conditions compare with wild ones? Most people doing studies on captive animals have not studied the species in the wild and know about natural conditions only from publications. Because we are field workers, we are not impartial judges either, but we do plead for more care when scientists state that captive conditions compare with natural ones, that captive animals are semi-free, or that a species does not possess an ability when a few captive individuals have failed to show it.

Studying intelligence in the field is a rather frustrating enterprise. Sharing the life of another animal for years gives us a feeling about what they do and how they do it, and some knowledge of their world, which makes us aware of abilities we would not have noticed otherwise. But how to prove them? We often have no way of proving something because we find no way of setting up a control. To take one example, a striking feature of chimpanzee movements is the straight line they take between feeding points, which only seems possible with a precise mental map that allows them to take short cuts to a given point from all possible directions. To study this systematically is not easy. With time, we could recognize the individual trees they fed on and the route we took with them when going from one such tree to another, during periods when they fed on them regularly. But we were always surprised when they joined these same trees in a straight line from a totally different direction. How could we demonstrate that they know the territory so well that they can make these straight lines? And, more importantly, how could we show before they arrive at a given tree that it is the one they are actually aiming at? Be aware then that what we are describing and analysing in this chapter as evidence of intelligent behaviour in wild chimpanzees is only a small fraction of all the intelligent behaviours we observed, and contains none of those we saw but did not understand.

Field studies have sometimes been judged as producing nice anecdotes, but these have been considered to be of little scientific value, because only experimental work is thought of as contributing to scientific progress (Bernstein 1988; Heyes 1993, 1998). This overly restrictive view should not keep us from thinking about the value of anecdotes compared with data that are obtained in controlled situations and that can if necessary be replicated. The difficulties of the controlled experiment lie (a) in the question of the biological relevance of the experimental variable, and (b) in the uncontrolled connected variables in the experimental situation. Its main advantage lies in its repeatability, and consequently the possibility of statistical analysis, the value of which, though, is limited by (a) and (b). The advantage of anecdotal observations stems from their occurrence in biologically relevant situations. Multiple observations allow for a checking of consistency, which to some extent can compensate for the lack of statistical analysis, and constitute a qualitative validity test. If we want to know the chimpanzees' reaction to members of a stranger community, observations with captive individuals will simply tell us nothing about intergroup interactions in chimpanzees. Here, a few anecdotes are of great value. The comparative method in behavioural ecology, in which one compares field data from different populations, species, or individuals while controlling for as many confounding variables as possible, has proved its value now for two decades and has greatly contributed to our present understanding of animal behaviour (Krebs and Davies 1993). Similarly field experiments (see Kummer *et al.* 1974; Cheney and Seyfarth 1990) have proved their value in the understanding of complex social mechanisms. This shows that there are ways to control for confounding variables that can also be used in natural settings. When the goal is to understand a wild animal species, a flexible approach is needed. Let us not forget that in everyday life as well as in clinical experience most of our knowledge of human beings is based on anecdotes.

Interpreting data in the cognitive domain of a non-speaking animal species is far from being straightforward, as exemplified by the long unresolved debates about the possible presence of imitation, theory of mind, or empathy in any animal species (see Carruthers and Smith 1996; Byrne 1995; Heyes 1994, 1998; Tomasello and Call 1997). The observations we present here will not settle the debate, but should add to the spectrum of what can be observed within a species. Our interpretations will stress the complexity of the behaviour of wild chimpanzees in this domain and favour the view that intelligence is domain-specific in chimpanzees. We present here data on the intelligent behaviour of wild chimpanzees, first in contact with their physical environment and then with members of their social group. Our questions are moulded by an evolutionary framework. What are the cognitive faculties used by wild chimpanzees living in the Taï forest, a typical tropical rainforest environment? How do these compare with what we know from other populations of chimpanzees or from their captive counterparts? This may help us to understand the factors which favour the evolution of intelligence in chimpanzees.

The physical world

Wild animals live in a world where they have to solve problems daily and are continuously confronted with new and complex external information. Taï chimpanzees live within a home range of up to 25 square kilometres, which contains many patchy food resources that vary in quality and in quantity over time. They have to adapt to these changes in a flexible and efficient way. In the Taï forest, the visibility is at most about 20 m on the ground. This is an additional challenge, as they never see more than a tiny fraction of their home range and of the resources they use. We concentrate here mainly on aspects of tool-use that allow chimpanzees to exploit resources that would not be available to them without a tool. One decisive advantage of tool-use for our present purpose is that the goal of the chimpanzee behaviour is obvious to the observer, and features of the behaviour, such as tool availability, choice, and transport distance, can be quantified.

Cognitive mapping

Cognitive maps are clearly important for the survival of some animal species. The astounding orientation abilities of migratory birds, salmons, eels, and turtles are well known. Squirrels and some birds have been shown to be able to relocate hundreds of food caches during the winter. Having a mental map means having mental images, but beyond that it is an imprecise term, for it encompasses many different ways of memorizing and manipulating spatial representations, each of which allows different sorts of performances (Piaget and Inhelder 1947). It is thus important to analyse and describe what operations can be performed with the cognitive maps of given individuals.

Studies of the mental maps of chimpanzees in captivity have revealed that they can memorize many locations of hidden food and revisit them by adopting a 'short route' strategy independent of the routes they followed when they were shown the locations (Menzel 1974). At Taï, for nut-cracking, hammers of wood and stone as well as nuts are regularly transported, the anvils being fixed objects such as outcrop rocks or roots (Boesch and Boesch 1983). We never saw them transport hammers or nuts in any other context apart from nut-cracking behaviour. Stone materials are rare in the territory of the study community, and transport distances for stones can be quite long, well exceeding the roughly 20 m visibility distance in this forest. This makes transport of hammers especially convenient for studying mental maps.

Object permanence

Object permanence is required for using a flexible mental map, for, without it, landmarks could not be used as reference points, and one could not refer to or select objects that are not seen directly. Object permanence appears progressively in humans. Children of less than 14 months do not know that an object that disappears behind a curtain is still there. Only when some months older do they get the idea of searching for the object behind a second curtain, if they do not find it behind the first one (Piaget 1935). But object permanence is not only knowing where an object is that cannot be seen any more, but also being able to conserve its volume and weight despite physical transformation. Some of these elaborate forms of object permanence are performed only by children older than six to ten years (Piaget 1945). Some primate species apparently never fully acquire the first level of object permanence: baboons or macaques do not search for an object that has disappeared behind a hide (Chevalier-Skolnikoff 1977; Parker 1977). Chimpanzees may perform better: captive chimpanzees searched for objects that had disappeared, and Sarah, a female chimpanzee, was able to understand that a quantity of liquid presented in differently-shaped glasses conserves its quantity despite its different appearance (Woodruff *et al.* 1978).

In the Taï forest, 38% of all hammer transports are made over distances where the goal tree is not in view from the starting-point of the transport. Such transports would prove object permanence only if the individual had an invisible goal in mind when starting to transport the stone, so that we can say that it somehow mentally remembers its existence and location. The results of the analysis of the long transports of hammers for *Panda* nuts (Chapter 9) show that these were not done randomly, and that hammers were selected with a 'goal tree' in mind. Object permanence is also needed if the choice between different hammers is made before transporting one. On average, chimpanzees compared the distance of five stones to select the closest one to a given goal tree when in the presence of at most one of them (Boesch and Boesch 1994*a*). Thus, the Taï chimpanzees demonstrated object permanence for hammers in their long distance transports. Chimpanzees do more than just remember that hammers are around. They associate with each of these hammers a direction, a distance to the goal tree, a hardness, and a weight. The properties of hammers, hardness and weight, remain constant whatever the distance and the position of the chimpanzee in relation to them. Comparison of distances were made only for hammers of a certain class of weight and hardness; to make such comparisons, distances have to be mentally compared while remaining permanent. The elaborate notion of object permanence found in Taï chimpanzees is one that might be expected to exist in the forest, for a low-visibility environment selects for a more elaborate concept of object permanence than other habitats.

Mental maps

Mental maps have been studied in various organisms. During ontogeny, humans first use a topographic map in which landmarks are used for orientation. Later, a projective map develops that enables the use of angles and directions. Eventually, a Euclidean map appears in which coordinates are used to relate landmarks one with another, allowing the use of shortcuts and the measuring of the relative distances between objects (Piaget and Inhelder 1947). Humans living in different cultures may use different kinds of mental maps. For example, traditional agriculturists and taxi drivers in Paris use primarily topographic maps, whereas hunter–gatherers, like Eskimos or Australian aborigines, rely for their orientation primarily on Euclidean maps (Dasen 1975, 1982; Dasen and Heron 1981). The mental rotation of a map has been tested in a classical Piagetian test with five-year-old captive chimpanzees. The chimpanzees failed to transpose landmarks on a landscape that had been rotated (Premack and Premack 1983), which is one criterion for a Euclidean map. However, these abilities appear in humans only at the age of six or seven years, and it is thus unrealistic to expect young chimpanzees of five to possess them.

In Chapter 9, we presented the results of an analysis of stone-hammer transports to crack *Panda* nuts (Table 9.5), in which the chimpanzees relied on object permanence and performed mental operations to select the best hammer to crack nuts at a given tree. This not only implies an evaluation of the suitability of the tool, but also of its distance to the cracking site. On the average, the chimpanzee chose between five stones lying within a distance of about 300 m of a target tree. To compare distances oriented in different directions in space, they have to somehow mentally align them. Furthermore, as transports were made to different target trees, and the position of the stones changed regularly because they had been transported by chimpanzees to crack nuts at different trees, hammer selection had to be done anew each time (Boesch and Boesch 1984*a*). According to our analysis of choosing hammers to crack *Panda* nuts, Taï chimpanzees demonstrated a mental representation of space allowing them to perform four specific mental operations:

1. Evaluate and conserve distances between objects which are out of sight (hammers and trees).
2. Compare distances orientated in different directions in space.
3. Permute objects (hammers) within the map and assign them new distances to the reference point (tree).
4. Permute the reference point (nut-bearing trees).

In Piagetian terms, the simultaneous presence of these four operations are characteristic of a Euclidean mental map observed in the representation of space by nine-year old children (Piaget and Inhelder 1947). The hammer carriers we saw were all adult females. The fact that young captive chimpanzees of five to six years old were proposed to be unable to permute objects in a map (Premack and Premack 1983) may likewise indicate that the age of acquisition of mental representations in chimpanzees is comparable to that observed in humans.

The elaborate conception of space shown in Taï chimpanzees is not a surprise, for wild chimpanzees live in a habitat where resources are often out of view and where searching for resources randomly would be extremely time consuming. Selection should have improved this process, and both mental maps and permanence of invisible objects are tools that improve performance for such problems.

Thus, wild chimpanzees possess sophisticated abilities in the domain of object permanence and mental mapping. That most studies carried out on captive chimpanzees fail to demonstrate similar faculties reflects the fact that individuals only acquire the faculties they need in their daily forays to solve the challenges they encounter. Captive chimpanzees and other primates have rarely been tested for conservation of distance in space, space being sparse in captivity, but have mainly been tested for numerical abilities or conservation of liquid quantities. As human studies have demonstrated, numerical abilities develop mainly when used daily, and hunter–gatherers have more developed spatial abilities than numerical ones (Dasen 1982). The demands for foraging experienced by wild chimpanzees are comparable to those of hunter–gatherers that select for spatial competence. Therefore, they need more developed abilities for understanding and predicting distance in space, rather than numerical abilities. It appears that by testing chimpanzees and other primates for abilities they rarely need or do not need in their daily lives, captive studies have been mostly interested in studying the presence of human abilities in other species.

Object manipulation

Objects are by definition external to the body. The manipulation of objects requires different skills from those needed to manipulate one's own body. The manipulator has to realize that objects are not simply a part of its body, that they react differently and need to be manipulated accordingly. For this reason, much attention has always been devoted to tool-use, which has been observed extensively throughout the animal kingdom, with primates standing out as the most proficient tool users (Beck 1980; Goodall 1970). Limits on tool-using abilities have been shown in the capuchin monkeys, once presented as 'ape-like' in their intelligence and tool-using skills, as they do not seem to understand cause and effect. For example, they do not understand what length or width an object needs to have to be a good tool (Visalberghi and Trinca 1989). Chimpanzees stand out in primates as using very diverse kinds of tools, particularly so in their natural environment. How much of this is based on elaborate cognitive abilities?

Tool-making

Tool-making has always had a special status. To make a tool, the individual needs to possess a mental image of the tool as well as of the transformation needed to produce the tool from the raw material. In Chapter 9 we saw that chimpanzees make tools very regularly, and commonly perform up to four modifications on the raw materials before they use it as a tool. The data on the modifications applied to objects to turn them into tools (Table 9.2), combined with the fact that for over 93% of the tools all modifications were done *before* use, reveals that chimpanzees possess a precise mental image of the tool and that this image guides them when making it. The capuchins probably lacked such mental images when they were using sticks much thicker than the tube out of which they wanted to push a peanut (Visalberghi and Trinca 1989). It is true that these mental images of a tool might have been acquired through trial-and-error learning and much practice, but the point is that they have become so independent of practice that chimpanzees can use them in new situations to guide their shaping of raw objects, without having to rely on actual trials to know if their shaping is appropriate for the task.

We can get an idea of the accuracy of these mental images by measuring the various tools produced for one particular function: the more precise these images are, the more the tools should be alike if they are made before use. At Taï, the mental image is sufficiently precise to produce a strong standardization of tools, even for tools fulfilling very similar functions (Fig. 9.1).

Cultural traditions also seem to affect these mental images, the size of the tools produced not only being related to the function, but also reflecting the social norms in force in a given social group. For example, ant-dipping in chimpanzees is performed with two types of dipping twigs: at Taï a small one of about 24 cm, and at Gombe a longer one of about 66 cm (Boesch and Boesch 1990; McGrew 1974). With the Gombe tools, driver ants are dipped for more efficiently than with the Taï tools; but nevertheless, all chimpanzees at Taï produce smaller tools and use the Taï technique to dip for the ants (Boesch and Boesch 1990). The mental images used by chimpanzees to make tools are apparently detailed and precise enough to achieve a certain standardization in tool productions.

Causality

Notions of cause and effect could notably improve the performance of tool-use, particularly when selecting the raw material and making a tool. Appropriate choice of hammers allows a 30 to 43% energy gain in nut-cracking. It is thus not surprising that chimpanzees select suitable hammers in anticipation of the nut species to be pounded (Boesch and Boesch 1983; 1984*a*). Even when the gain is less obvious, Taï chimpanzees select the thickness of the twigs in accordance with their intended use (Fig. 9.1). Thus, the understanding of cause and effect allows chimpanzees to mentally anticipate their actions and to choose tools adapted to specific purposes.

The understanding of causality is the ability to understand the dynamic relations between objects when external forces affect them. Some limitations on primate understanding of causality were proposed when capuchin monkeys proved unable to predict the length or the diameter of a tool required to push a peanut out of a transparent tube. Chimpanzees and humans were able to solve such tests, although some chimpanzees faced problems (Visalberghi and Limongelli 1996; Visalberghi and Trinca 1989). These results suggested to some that the understanding of causality is a uniquely human skill (Tomasello 1998; Tomasello and Call 1997). This proposition underestimates the fact that the comprehension of causality has many aspects and that a species may possess it to different extents in relation to different physical principles or in different domains (Piaget 1935; Visalberghi and Limongelli 1996). The question is not, do chimpanzee comprehend causality, but rather, what kind of causal relations do they understand?

Two aspects of causality need to be considered: first the comprehension of causality, when one is involved oneself in the relation, and second, when external relationships are concerned in which one is able to consider external objects as not being part of or related to oneself. In selecting and making tools, chimpanzees demonstrate an understanding of a cause–effect relationship (certain properties of the tools are required to perform the task), as well as a force–effect relationship (the force applied to a tool is dependent on the property of the tool and the strength needed to execute the task optimally). In Taï chimpanzees, such understanding of causality is often made before performance, by mentally anticipating the requirements for the future action. This is the case when choosing tools that will be transported over long distances before they will be used. In these instances, the chimpanzees have to understand the causality of a relation in which they are the actors (they are using the tool), and their optimal selection and use of tools demonstrates that they understand such causality perfectly.

More demanding is the comprehension of causality in situations where the relation between objects is totally independent of the individual, that is the causality of external relations. This is the case when hunting cooperatively. When hunting, chimpanzees sometimes anticipate the response of the prey to other hunters (Chapter 8): 24% of all hunting movements performed by the chimpanzees at Taï anticipated the movements of the prey resulting from the pressure of another hunter. To be able to do so, the individual has to appreciate that other hunters perform movements different from its own, that the prey reacts specifically to these movements, and that its own action can increase the likelihood of a capture. Thus, when hunting cooperatively, chimpanzees understand causality both when it involves themselves and for external tertiary relations, and they can combine the two to plan their actions. We suggest in Chapter 9 that this broad understanding of causality allows chimpanzees to anticipate the benefit of tool-use in many new situations and explains why chimpanzees in the wild have been seen to use so many more kinds of tools and so more often than any other great ape and primate species, except humans. The causal understanding necessary to build coalitions is partly available to other primates (see discussion of coalitions below), but extending such knowledge to other animal species (as in hunting), and to inanimate objects (as in tool-use) is not an obvious step. Such understanding might not yet represent a full understanding of causality, but it might have led to the evolution of tool-use in chimpanzees and humans.

In their interactions with tools, wild chimpanzees show well-developed mental abilities, particularly in the domain of object permanence, causality, and spatial representation. These are faculties needed for their survival, particularly in a low-visibility environment.

The social world

The key difference when interacting with the social world rather than the physical one is that social partners can move at their own initiative. The more one knows about social partners, the better one can react. Obviously, group size will play a role: the more partners there are, the more combinations are possible. These considerations motivated the 'social intelligence' hypothesis (Humphrey 1976; Whiten and Byrne 1988), and are partly supported by the increase in the relative size of the cortex in animals living in larger groups (Dunbar 1992). However, the number in the group is only one dimension of the problem, because with the same number of individuals one may interact in many different ways. Aggregations of fishes or ungulates interact, but without distinguishing individuals. The first increase in complexity comes with individual recognition, which allows differentiated relationships among group members. A second step is reached when individuals consider others as behaving as individuals, allowing them to predict and anticipate the actions of others, and to manipulate them. A third step is achieved when one attributes to others a mind different from one's own, for this can be used to further elaborate social relationships with group members. Providing evidence of such social faculties is complicated. Preliminary results converge in showing that within primates not all abilities are equally distributed, and great apes are better at the two last levels (Byrne 1995, Povinelli *et al.* 1997). Studies with human children have shown that it is only when they are three and a half years old that significant changes appear, allowing them to understand desire and belief in others; while the attribution of knowledge to others, based on inference or informative perception as a source of knowledge starts to be understood first in six-year-old children (Gopnik and Meltzoff 1994, Roth and Leslie 1998, Wimmer *et al.* 1988). More observations of wild chimpanzees are badly needed to learn how much they know about social partners and how they use such skills to coordinate movements, plan actions, and anticipate those of others.

Social knowledge
Individual recognition and voice identity

While we followed chimpanzees daily for years in their natural habitat, it became progressively clear that their vocalizations contain more information than just expressing their mood. To understand this, we had to undergo a period of training, during which we learned to identify the chimpanzees' voices. This learning process required years, but once acquired we used it to find individual chimpanzees, and it proved to be reliable, for their voice intonation and rhythmic properties (especially during the pant-hoots) remained stable for years. Male chimpanzees seem aware of the importance of having clear individual calls. When Le Chinois disappeared in February 1984, Macho changed his formerly rather atypical pant-hoot sequence, so that within three weeks he used exactly the same 'hooaa' sound that Le Chinois had used. We were so confused that for many weeks we checked whether Le Chinois had come back. At the same time in 1984, shortly after four adult males had disappeared, Brutus stopped producing his typical long series of rapid hoots, and instead made shorter series of slower hoots. Thereafter, Falstaff, the oldest male of the community, started to imitate Brutus' long series of rapid hoots exactly, and he used them until his death in 1987 (Boesch 1991*b*).

This ability of the chimpanzees to notice the quality of their own voice and to imitate perfectly the intonation and rhythms of other voices resembles their ability to recognize themselves in the reflection of a mirror (Gallup 1970; Hart and Karmel 1996; Povinelli *et al.* 1997). Self-awareness entails many different abilities, and chimpanzees use it to recognize themselves visually in a mirror and to recognize and modify their own voices.

Communication and symbolic drumming

Many studies of captive chimpanzees have revealed surprisingly developed symbolic abilities with sign language (Fouts 1983; Gardner and Gardner 1971), plastic symbols (Premack and Premack 1983), and artificial idiograms (Savage-Rumbaugh *et al.* 1978). Although there is still debate about how much of the human language is demonstrated by these chimpanzees, there is agreement that they show the ability to use symbols and combine them (Ristau and Robbins 1982; Luce and Wilder 1983). How this relates to what they do in their natural environment remains unclear. Wild chimpanzees may be using symbolic-like abilities in the wild, but field workers have not yet either seen them or recognized them. The example below illustrates the difficulty of documenting such abilities, while suggesting that they do probably exist in wild chimpanzees.

Chimpanzees typically forage in parties of 7–12 individuals remaining permanently in auditory contact with the majority of the community (80 chimpanzees at the time of analysis), and they follow a constant direction for hours even if they remain totally silent. Normally the community splits in at least three major parties that may communicate with one another by vocalizing and drumming. Buttressed trees are abundant in this forest, and adult males, after pant-hooting loudly, hit these buttresses powerfully and rapidly with their hands, or feet, or both. Drumming is a way for males to communicate their position to other group members, and it indicates the direction in which the drummer progresses. We suspected that some of these drummings were more than just an indication of an individual's position, for we tended to loose contact with them just after they were heard. It seemed that the whole chimpanzee community had abruptly and often silently changed direction following such drumming. In early 1982, three years after we had started the study, we began to realize that it was only after Brutus, the alpha male, drummed that the community reacted by abruptly changing the direction of travel. On some occasions, Brutus' drumming sequence appeared to transmit a specific message. There was no

Table 10.1: Symbolic communication in Taï chimpanzees: Brutus' communication system with the frequency of drumming containing information about travel direction and resting duration (Boesch 1991a)

Number of drumming incidents	Location of emission	No. of cases (No. of exceptions)	Group response
2	same	8 (1)	1 hour rest
2	different	8	Change direction
3	same	0	–
3	different	6	1 hour rest + change direction
4	same	1	2 hours rest
4	different	0	–

audible difference between sequences that did or did not have such a message; rather this message was indicated by their spatial and numerical combination. During a sixteen month period (January 1983 to May 1984), we studied the information conveyed in Brutus' drumming and could identify three messages (Table 10.1).

Change in travel direction.   Brutus, by drumming twice at two different trees, proposed a new travel direction, which corresponded to the direction between the two trees he used for drumming, and this was understood by the other chimpanzees. Such drumming always occurred within a time interval not exceeding two minutes. Individuals that were not part of Brutus' party apparently inferred the direction proposed by mentally visualizing Brutus' displacement between the two trees and then transposing it to their own travel direction.

Indication of resting period.   On other occasions, Brutus, by drumming twice at the same tree within two minutes, proposed a resting period of a specific duration that the community would follow. We were able to identify this message from Brutus in 14 cases when the community activity stopped for an average of 60 minutes ($N = 12$, range = 55 to 65 minutes). Community activity was judged to be resting by the absence of foraging movement of the chimpanzees we followed and by absence of vocalizations from parties out of view. After a rest, parties started to move without giving any vocal indication. There was one exception of a party not resting at all (Table 10.1).

Direction and resting time combined.   By combining both messages, Brutus could propose both a change of direction and an hour's rest; in such a case he would drum once at a first tree and then twice at another tree in the direction he was proposing within a short period of time. Alternatively, Brutus could drum twice at a first tree (on the axis) and then once further in the proposed direction. In all cases, the information about resting time had an immediate effect, whereas that about direction applied later. If Brutus were simply adding information about direction and time, he would have drummed four times (twice for each piece of information). In fact, he combined them and drummed only three times; one drumming contained information on both direction and time.

  Brutus stopped using this code rather abruptly, when several of the prime males suddenly disappeared in 1984, probably due to poaching, and the number of travel parties diminished, making the benefit of such a communication system much less obvious

(Boesch 1991*a*). This symbolic communication mean has so far to our knowledge only been observed in the Taï chimpanzees, and in this community its use was also limited to a short period, which emphasizes its arbitrariness. The symbolic communication system based on drumming presented here shows that wild chimpanzees can spontaneously develop such a system and the symbolic communication is more developed in this species than is suspected at present. Such a communication system functions only if the signallers intend to communicate information to others, and if the receivers understand that they are informed about something they should understand. As both are out of sight, there is little room for associative learning, and individuals must know the communicative value of the signals.

Coalition tactics and social tools

The use of 'Social tools' refers to the phenomenon where an individual uses members of the group to attain a goal he would not be able to reach on his own. For this, an individual must be aware that others behave on their own, that their behaviour can be anticipated and manipulated, and that the effect of their behaviour on a third party can be predicted as well. If, for example, A supports the alpha male in an attack against B who is dominant over A, this gives A an easy chance to win against one of the dominant group members, and the alpha male might reward A for his support. Such winner-support coalitions, however, do not benefit the one supported, since the winner might have won anyway. A loser-support coalition can benefit the one supported, if the coalition is stronger than the opponent. However, in both types of coalition, the individual needs to have a way of evaluating the different forces present in order to make the correct choice. In loser-support coalitions, an understanding of the strength relationship within the group would be helpful. Despite the fact that A is stronger than B, and A is stronger than C, the combined force of B + C can be stronger than A. This could be experienced over time, but since coalitions are not stable, C should be able to make such evaluations with individuals A, B, D, E, and so on. Furthermore, social status varies, so that B + C might not always dominate A, and C has to keep track of the dominance relations within the group. Keeping track of all these changes and reacting appropriately makes loser-support coalitions more demanding than winner-support ones, for which one only has to know that A (the attacker) dominates B. Since the costs of being mistaken are paid in hits and bites, the risk of being wrong should not be underestimated.

Compared with collaborative hunts, coalitions are simple to perform, since the particular behaviour of the partners is usually similar and the coordination between the two individuals is only in time and space. The difficulties of coalition lie in the evaluation of the forces present. Coalitions have regularly been observed in various primate species, but they are mostly of a winner-support type (for a review see Harcourt and de Waal 1992). Loser-support coalitions have so far only been proposed in baboons for access to oestrus females (Bercovitch 1988; Packer 1977; Noé 1992) and in chimpanzees (Goodall 1986). One special aspect of the coalitions observed in chimpanzees is their ability to take revenge against group members, which shows a score-keeping ability and a value system for judging some actions as requiring revenge, a feature not yet observed in macaques (de Waal and Luttrell 1988; de Waal 1996). Coalitions in chimpanzees are regularly found, and at Taï 60% of them are of a loser-support type, compared with Mahale with 17% (see

Table 6.14). The fact that most loser-support coalitions were successful shows that they were not simply support given to a friend, or the emotional reaction when seeing an individual being attacked, but that the support provided followed a mental evaluation, such as supporting only when B + C are stronger than A, despite the fact that A is dominant over B and C taken separately.

Cooperative hunting

The mechanism of cooperation is discussed here to see how it is performed and what abilities are needed to perform it. We discuss two distinct aspects. What abilities do chimpanzees demonstrate when they perform a cooperative action (the performance question)? And second, how do chimpanzees learn this behaviour (the ontogeny question)?

The performance question

Cooperation is defined as the joint actions of at least two individuals to achieve a given task. If the joint actions are different and complementary, it requires considering another group member as an independent individual whose actions contribute to team success. Collaborative hunting involves a specialization in the actions of the hunters and precise timing, for the actions often have to be performed in a specific order. The social knowledge of what others may be able to contribute is also necessary. In humans, young children of 18 months rarely or never cooperate, whereas two-year-old children coordinate their behaviour quickly and effectively (Brownell and Carriger 1990). It has been shown in these children that those who cooperate more effectively are also those that were better able to represent others as being independent of themselves. Thus, some aspects of a theory of mind are required for cooperation in humans. In captivity, chimpanzees have successfully passed tests in which they need to coordinate their actions (Chalmeau 1994) and were obliged to collaborate (Savage-Rumbaugh *et al.* 1978). Similarly, orang-utans successfully performed such coordinating tasks, whereas capuchin monkeys succeeded only rarely, and macaques and baboons failed (Chalmeau *et al.* 1997). At closer inspection it became apparent that capuchins learned the test individually, but performed it so frequently that if two of them did it at the same time they were successful. However, they never understood the social dimension of the task and were never seen to wait specifically for a partner nor to synchronize their actions with those of the other. This, plus the comprehension that one's own action is successful only thanks to the joint action of another, is what is required for cooperation.

Cooperative hunting in wild chimpanzees is remarkable in two respects. First, it is a joint action of many individuals. In our hunting sample at Taï, a chimpanzee coordinates its actions in 81% of the hunts with two or more hunters (average number of hunters = 3.07; N = 248 hunts). This coordination is done under conditions of very low visibility; only rarely all hunters are in view of each other and they have to somehow 'read' the others' intentions, so as to anticipate their future actions. Because the prey is moving to escape, the planning of the actions has to be projected spatially and in time for the co-ordination of different actions to be effective (we often had to run to keep up with the chimpanzees during a hunt). Hunters must proceed while making a continuous evaluation of the positions of all the participants to know which position needs to be filled. In other words, hunters must be able to judge the causality of external tertiary relations (relations between other individuals than themselves). For example, if chimp A is driving the prey in a given direction, and chimp B is blocking that direction, it has to be understood that they will move in a new direction and that an ambusher is needed over there and that you, chimp C, might fill this gap.

Hunters must perceive other hunters as independent agents with their own specific competence and intentions. For example, imagine the situation in which Darwin (a young and not very experienced hunter) and Brutus (the most experienced hunter) are in some trees with monkeys, driving them towards Kendo, who climbs a tree to block some of them. If Kendo now chases the monkeys away from Brutus and Darwin, he is hunting on his own. This situation happens at Taï with inexperienced hunters. If, however, Kendo remembers that Darwin and Brutus are in the trees from which the colobus came, he might push them back towards Darwin. This would be an error, for Darwin does not have the competence to anticipate that far ahead and would have already gone to the ground to follow the hunt. Kendo should remember where Brutus is and push the monkeys in his direction, for Brutus can anticipate such a move and wait for Kendo to chase the prey back in his direction. Collaboration requires of an individual not only to perceive another entity as an independent agent, but also to attribute to this entity abilities that may not be within its own range. Accurate attribution of competence here makes the difference between the success or failure of the hunt. If the hunt is of a synchronous or coordination type, with all hunters performing the same hunting movements but coordinating them in time and/or space, there would be no need to attribute competence to others. They would only need to differentiate between hunters and non-hunters. Only once hunters perform different types of hunting movements with different efficiency it is important to be able to attribute competence to them. Consequently, the ability to 'read' the intentions and actions of non-visible hunters and to attribute competence to them accurately makes hunts successful.

Second, the goal of the joint action is to capture an intelligent prey that fights for its life and does all it can to escape. This requires permanent and flexible adjustments of the movements of each hunter, that not only depend upon the actions of the other hunters, but also depend upon the rapid movements of the prey. This is a special challenge as the prey is another animal with its own understanding of the situation and its own physical skills. Hunts occur in a three dimensional space, and colobus monkeys have different physical abilities from those of chimpanzees. For example, blockers regularly place themselves so that monkeys cannot escape into trees above the canopy. No chimpanzees could ever enter such a tree as a monkey would — with one great leap. Here, hunters show that they are aware that colobus have abilities (jumping over large gaps) and attributes (lesser weight that thin branches can support) unlike those of adult chimpanzees. Hunting successfully requires attributing to the prey abilities other than those of chimpanzee hunters. Ambushing with full anticipation of prey movements is mastered only by chimpanzees over 30 years old (see Chapter 8); this reflects the difficulty of the task. Alternatively, one might argue that chimpanzees learn through trial-and-error what colobus can do, and progressively adapt to their abilities. This should be reflected in the number of errors made by hunters when attempting to anticipate colobus movements. Of the 262 anticipations we observed, only 4 (1.5%) were erroneous, in the sense that hunters selected trees in which the colobus could not or did not come. Thus, in 98% of the anticipations, the hunters anticipated the prey movements correctly. This supports the hypothesis that chimpanzees can attribute abilities not only to other conspecifics but also to other species.

How do chimpanzees anticipate where the prey is going to flee? Old males may be able to make such advanced anticipations because they remember the possible escape routes of colobus in the same trees, which would require a tremendous memory. Alternatively, they could know in a generalized way how the colobus monkey reacts when facing certain situations. Humans would do this by a theory of mind. In support of this, we see that comprehension of some hunting roles seems to precede the ability to perform them. This excludes memory as an explanation for the observations. Hunters regularly showed that they comprehended that an anticipation is needed (27% of 959 hunting movements were anticipations) to achieve success, but only a small proportion of the actions were performed with full anticipation of the prey's movement (32% of 262 anticipation movements).

The question of ontogeny

How is cooperative hunting learned? As a cooperative action produces a benefit only when performed in concert with at least another individual, learning it is complicated because key elements of the behaviour cannot be learned alone. Coordination may result from the intentional synchronization of only one of the two partners, but collaboration requires synchronization and coordination by both partners. Youngsters may start to take an active part in a hunt by synchronizing their behaviour with other hunters, which provides them with real situations to learn the cooperative task. At this level, cooperative learning implies that something is learned through the interactions with another individual, as it is the joint action that leads to the result. The easiest way seems to attribute to older hunters better hunting skills and to base one's actions on them. The reciprocal and synchronous learning of a cooperative task by two youngsters is possible (Tomasello *et al.* 1993), but this would require much more time, like reinventing the wheel. How do youngsters learn to hunt?

According to the individual learning hypothesis, cooperative hunting would be explained by the fact that each individual hunter increases its own success by hunting in groups. In the Taï forest, the conditions are such that a lone hunter is not likely to capture a colobus monkey (Chapter 8), and it makes sense for an individual to hunt with others. However, it is not all that simple. First, the hunting success per individual does not increase for all group sizes; we expect them to hunt only in groups of three to five hunters (see Fig. 8.3). However, only 52% of the hunts are carried out by groups of these sizes. This means that in 48% of the hunts, individuals are not hunting under the optimal conditions, which is unexpected if we follow the individual learning hypothesis. Second, rewards for the different types of hunting movements are quite diverse (see Table 8.13), and according to this hypothesis we should expect all of them to learn the most rewarding tactic. In other words, according to the individual learning hypothesis, all chimpanzees would follow the hunt from the ground to make a capture, since 81% of the hunters on the ground do make a capture. However, monkeys do not fall simply from the tree like fruit, and hunting would simply disappear if the hunters just waited on the ground. In the Taï forest, the conditions are such that, chance aside, a capture is only possible through co-operative group actions. In addition, some hunting tactics look like altruistic roles, in the sense that they contribute to the success of the group but not to an individual success. Drivers achieve a capture in 1% of the instances, whereas ambushers have a success of 11%. Drivers and ambushers seem to be altruists, especially compared with ground hunters, making themselves vulnerable to a possible selfish reaction from the actual captors. This is not the case at Taï, but it is at Gombe. Nevertheless, Gombe chimpanzees still cooperate in some hunts. Thus, cooperative hunting is learned, despite a decrease in the individual success for some hunters, which contradicts the individual learning hypothesis.

The cooperative learning hypothesis predicts that it is through the observation of others that the youngster learns the hunting tactics and sees the result of their combined actions on the hunt. At the beginning, it was not clear at all to us what would be the best way to hunt and what contribution the different tactics had on the outcome. When we saw the colobus moving away from a chimpanzees which was blocking its way, we also thought at first that the chimpanzee should go down on the ground and reposition himself under the prey. That is indeed what most of the chimpanzees do. We were very surprised in some situations to see that Brutus or Falstaff would, on the contrary, remain there totally silent and quiet. We learned with time that they knew better, and that the colobus might come back under pressure from another hunter. In other situations, we saw Brutus sud-denly run away from a hunt in a direction where we could see no monkey. It was only by following him that we realized that he was simply thinking further ahead, predicting pre-cisely the further movements of the prey, estimating the time he would need to climb and very cautiously so as not to move any branches that could betray him, position himself high enough, and dart up in a perfect surprise attack when the prey entered his tree. We learned how such movements could actually contribute to a capture by watching the chimpanzees, which is exactly what we expect the youngsters to do under the cooperative learning hypothesis. Had we simply followed the chimpanzees that were pressing the colobus monkeys, we would have missed some of the most important aspects of hunting. At Taï, youngsters clearly observe good hunters precisely, follow them in their hunting movements, and adapt their own movements to theirs. This is especially clear when more than one hunter is pursuing the prey, for they then orientate their movements to the best hunter. So it looks as if the apprentices were attributing more competence to some males than to others, which makes sense since they are not all equally gifted.

Testing how the learning process takes place for cooperative hunting is difficult. But for one infant at Taï we were especially lucky. When the young male Ali was just five years old, his mother died, and he was almost immediately adopted by Brutus, the alpha male. We could follow Ali's hunting progress until his death when he was fourteen years old. Thus, in a test-like situation, contrary to his age mates, Gipsy, Sartre, and Marius, Ali had the best hunter of the community as a 'surrogate mother', from the age of five. He was associated 100% of his time with Brutus, following him everywhere, copying his behaviour for nut-cracking and other behavioural patterns. And, from the age of five onwards he followed Brutus in all the hunts he took part in. As Brutus was the male making most full anticipation of the movements of the prey and of other hunters, Ali had the best model possible in front of his eyes at an age when other youngsters hardly participated in a hunt, and the result showed. When Ali was fourteen, 6% of his hunting movements included full anticipations ($N = 140$ movements), whereas his age mates performed only one, 0.5% ($N = 177$ movements) ($X^2 = 5.81$, $df = 1$, $p < 0.02$). Ali's achievements were exemplified in one hunt in 1990, involving Brutus, Macho (23 years old), and Ali (12 years). After an unsuccessful movement, Brutus was searching for some more colobus toward the west. Both Ali and Macho had blocked an escape route in two different trees to the east. Macho climbed down and placed himself further west on the ground, while Ali remained in the trees but moved into the one Macho had left. Brutus manoeuvred the colobus in such a way that he drove them towards the trees where he knew Macho and Ali to be — he was out of view and could not have noticed that Macho had gone down. Noticing Brutus' manoeuvre, Macho rushed back on the ground to climb up his previous tree — too late — but Ali had in the meantime captured an adult monkey in that tree. Ali, 11 years younger than Macho, had anticipated Brutus' actions on the colobus' moves, as he saw Brutus do so many times, whereas Macho needed to see the colobus' move before he realized what was going to happen. This supports the cooperative learning hypothesis. Other males with a more normal youth with their mother, also learn these sophisticated hunting techniques, but it requires about six more years to reach the same level of proficiency. An additional factor may lie in the fact that it was natural for Ali to copy Brutus, his adoptive mother, without having to make a choice, whereas others would need to select good models — to whom they need to attribute competence correctly.

In conclusion, the performance of collaborative hunting in a three-dimensional space with low visibility requires that the hunters perceive other hunters as independent agents with their own intentions and competence, attribute abilities to the prey that differ from those of conspecifics, and understand the causality of the external relation between prey and other hunters. It takes an elaborate understanding of causality to make collaboration between different hunters possible. Cooperative hunts can be performed with less elaboration, as seen in the open African plains performed by lions, wild dogs, wolves, and hyenas (see Table 8.14). However, only a few percent of the group hunts in lions and wolves are collaborative hunts with the performance of different and complementary roles. Probably only some of these carnivores were able to attribute competence to other group members. At Taï, collaborative hunts are the majority, suggesting that more individual chimpanzees are able to make such attribution.

Theory of mind

To possess a theory of mind means to be aware that other individuals may have feelings, knowledge, beliefs, and competence, and that these might differ from yours. This ability to put oneself in the mind of somebody else is the essence of the theory of mind. Thereafter, one can use this knowledge to guide ones behaviour with them. This contrasts with what we could call a 'theory of behaviour', in which one knows that other individuals perform behaviours and that these behaviours might be different from yours without referring to what the other individuals' perspective might be. Individuals might have the ability to manipulate and anticipate behaviour of other individuals. Recently, these two types of theory have been contrasted, and humans have been proposed to be the only species to possess a theory of mind (Heyes 1998; Tomasello 1998; Tomasello and Call 1997; Povinelli 1996; Povinelli *et al.* 1990, 1992; Povinelli and Eddy 1996 in Heyes 1998). However, a theory of mind was initially proposed to account for the behaviour of captive chimpanzees (Premack and Woodruff 1978). In their classical experiment, Premack and Woodruff showed that a chimpanzee can differentiate between two human observers, of which only one had seen where a fruit had been baited, by considering only the indications of the one that knew. For this one needs to attribute different knowledge to the two observers, which was proposed to be the base of the theory of mind. There is presently a debate with opinions ranging from those opposing anything like a theory of mind in animals (Heyes 1998; Smith 1996) to those presenting evidence of imitation, deception, or teaching as signs of a theory of mind in great apes (Boesch 1991*a*; 1993; Whiten and Ham 1992; Byrne 1995).

In this context, the observations of cooperative hunting in chimpanzees are important. They show that abilities typical of a theory of mind are important for good hunters. Attribution of intentions and competence is decisive to perform actions such as those of a blocker or ambusher properly, and for apprentices to select a proficient hunter model. We are not saying that all chimpanzees possess all these abilities, but it seems difficult to see how gifted hunters could perform collaborative tactics without aspects of a theory of mind. Collaborative hunting also shows how a theory of mind might be beneficial in nature. Such situations differ strikingly from those tested in captivity, in which young chimpanzees have to recognize themselves in mirrors, care about small food items hidden for example under cups behind a plexiglass wall, with human observers disguised in strange laboratory clothes, and wearing various hats on their heads (including in some cases a bag or a bucket) (Provinelli 1996; Gallup 1970). In other words, no captive study has so far attempted to study the chimpanzee's theory of mind, but all have confronted the chimpanzees with totally new situations to pass tests to show the human's theory of mind. This may address the question of chimpanzee potential, but does not answer questions about the theory of mind that chimpanzees use in their daily lives. If some of these tests did not demonstrate a theory of mind in captive chimpanzees, we should not be surprised but rather ask ourselves 'What kind of a theory of mind is adaptive for chimpanzees to acquire?' and 'When do they use it?'. We now present some contexts, apart from cooperative hunting, where we feel chimpanzees in the Taï forest demonstrate abilities that are normally considered as being part of a theory of mind.

Imitation and teaching

Imitation is a cognitively demanding task, as it requires perceiving another individual as an independent agent with its own intentions. It is through this perspective-taking approach that the naive individual can perceive the benefit of copying another's action, as well as understand which aspect of the action needs to be copied. For example in nut-cracking, the fact that the nut-cracker sits or stands, or that it sits to the right or the left of the anvil is not relevant, but the way it hits the nut with the hammer should be imitated. But you may copy a behaviour without thinking about the intentions of the model (some would call it then mimicking). Similarly, for teaching to be effective it is required that the teacher evaluates the abilities of the naive individuals, so that it can adapt its actions to their level of proficiency. Thus, teachers have to consider others as independent agents with their own level of competence and knowledge. This is essential, since it is this ability that will allow teachers to determine which information is appropriate for the naive individual.

In nut-cracking, some evidence for imitation could be found (Boesch 1996*a*). To discriminate whether a behaviour is learned individually or copied from group members, it should be acquired in different ways or have different final forms once required. An attempt to introduce nut-cracking to a group of naive chimpanzees in the zoo of Zürich (Funk 1985) allowed us to answer one question about the individual learning strategy. The chimpanzees in the Zürich zoo were offered nuts and hammers and were observed for two weeks while they manipulated the objects and tried to open the nuts. None of them succeeded and no enforcement from other group members was present. If we compare the methods used by the Zürich and the Taï chimpanzees, we see that Taï chimpanzees tried fewer methods; of the 14 methods used by the Zürich chimpanzees, only 7 were seen in Taï chimpanzees (Table 10.2). This is intriguing because some of the rarely or never used methods described in the Zürich study were actually observed in Taï chimpanzees in other contexts. Stabbing with a stick was observed being used against a leopard, and rubbing is regularly observed as being used when feeding on other kinds of fruit. Thus, in comparison with the Zürich chimpanzees, a social canalization of individual learning potential is at work in Taï youngsters, and it is strongly influenced by the behaviour observed in the model: five of the seven methods used are behavioural movements commonly observed in adults cracking nuts. The object used by the models is less of a guidance, since young Taï chimpanzees used as hammers large hard-shelled fruits,

Table 10.2: Social canalization in the learning of the nut-cracking behaviour in chimpanzees. List of all methods used to attempt to open nuts by two populations of chimpanzees. The study on the captive chimpanzees in the Zürich zoo was performed by Martina Funk (1985). + indicates that the method was used in the population, whereas – indicates that it was never observed.

Method	Zürich chimpanzees	Taï chimpanzees
Hit with a hammer	+	Often
Bite the nut	+	Often
Pound the nut against hard surface[1]	+	Regular
Hit the nut with hand	+	Regular
Hit with an object[2]	+	Rare
Rub the nut against hard surface[1]	+	Rare
Throw the hammer on nut	–	Regular
Throw the nut against hard surface[1]	+	–
Hit with other body part[3]	+	–
Shake the nut	+	–
Press nut against teeth	+	–
Sit on nut	+	–
Scratch the nut with fingers	+	–
Press on the nut	+	–
Stab with a stick	+	–

[1] Chimpanzees can rub, pound, or throw the nut directly by hand against the ground, a stone, a tree trunk, or a root.
[2] As objects were included materials that could not make a hammer, such as a piece of cloth, small twigs, or, in the Taï forest, another nut, a piece of termite mound, or a hard-shelled fruit.
[3] Other body parts are understood as being the back of the hand or the elbow.

pieces of termite mounds, and rotten branches. In nut-cracking behaviour, social canalization through imitation is at work and individual learning is confined to the different types of objects that can be used to pound the nuts. The models imitated were their mothers.

Taï chimpanzees mothers have been seen to teach their offspring in such a way that the acquisition of the nut-cracking technique by their youngsters is accelerated (Chapter 9, Fig. 9.11). Teaching included stimulation, facilitation, and active teaching, and is a common action of female chimpanzees in the nut-cracking technique. The first two types increase the infants' attention to the nut-cracking technique and make the learning process more attractive, and, when possible, easier. Such techniques, called 'scaffolding' in psychology, were also seen in the process of learning many human techniques (Greenfield 1984). Once youngsters start to crack seriously, the mothers are ready to share their tools, which is often to their own disadvantage as they must find another tool for themselves. By adjusting such pedagogical acts to the level achieved by the youngsters, mothers demonstrate the ability to attribute competence to other individuals. In some rare cases, mothers actively taught their infants by intervening only for a special point that presented a problem.

Case study 1: Ricci teaching Nina

On 18 February 1987, Ricci's daughter, Nina, tried to crack the not very hard nuts of *Coula edulis* with the only available hammer, a wooden club of an irregular shape. As she struggled unsuccessfully with this tool, alternating her position sitting by the anvil (14 times), her grip holding the hammer (about 40 times), and the position of the nut or the nut itself, Ricci was resting on the ground some 10 meters away. Eventually, after 8 minutes of this struggle, Ricci joined her, and Nina immediately gave her the hammer. Then, with Nina sitting in front of her, Ricci, in a very deliberate manner, slowly rotated the hammer into the best position for pounding the nut efficiently. As if to emphasize the meaning of this movement, it took her a full minute to perform this simple rotation. Ricci did nothing else, which means she did not change her position, the position of the nut, or the nut itself. She thereby demonstrated her precise judgement of the problem Nina had. With Nina watching her, she then cracked ten nuts (of which Nina ate 6 entire kernels and portions of the 4 others). Then, Ricci returned to the same place she was before. Nina immediately resumed cracking, holding the hammer exactly in the position her mother did. She succeeded now in opening 4 nuts in 15 minutes, but she still had difficulties and she regularly modified her position (18 times) and the position of the nut, but never that of the hammer or her grip. Nina demonstrated that she had understood what Ricci had shown her and copied faithfully the grip throughout the difficulties she encountered.

Deception

Whiten and Byrne (1988) collected anecdotes on wild primates and proposed that apes deceive others by inducing false beliefs in them with manipulated information. Alternatively, it has been suggested that these anecdotes might merely be intelligent social strategies in which an individual learns through time that a given action might affect another's behaviour in its favour (Heyes 1998). This sounds like an argument of last resort, for many of those examples were unique occurrences, which do not give much room for associative learning. The question is, are they intended to change somebody else's belief or knowledge?

In some hunting attempts, chimpanzees try to make the red colobus monkeys move by pretending to be going one way when in fact they will move another way, where they expect the colobus to flee. For example, the 26 September 1990, Ulysse was hunting a group of black-and-white colobus with only Macho following his attempts from the ground. The colobus were waiting high up in a tree and it was clear that they would jump out of that tree if Ulysse climbed into it. From below, Ulysse started three times to climb into this tree, then immediately jumped over to the tree in which he expected the colobus would try to escape. Each time he immediately checked if his stratagem worked, and as it did not, repeated the whole set. The fourth time he climbed a bit higher, and then the monkeys did jump into the predicted tree. Ulysse immediately followed but did not catch up with them. It is difficult to quantify this kind of deception, but we saw it being used on twelve occasions by five different hunters. These twelve movements came from a sample of 959 hunting movements, giving two occasions in five years per individual to learn by association how monkeys react. This shows the limited scope existing for associative learning and the need for a complementary mental process to understand the monkeys' reaction in different situations and to predict how to deceive them.

Another routine deception used by chimpanzees is hiding. During displays by a dominant male if an individual anticipates that it is going to be the target of that male, it may hide instead of running away. We regularly observed individuals hiding from a displaying dominant. Hiding in chimpanzees does not mean simply sitting behind a tree opposite to the dominant, it means monitoring the position of the dominant carefully and moving accordingly so as not to be seen. Byrne (1995) said that deception may be comparatively rare in apes because of the large number of possible counter-strategies by the deceived. This is exactly what happens in this situation. Dominants regularly anticipate that others are hiding and search for them. Once, in 1995, Macho, the alpha male, was displaying against Brutus, who tried to avoid him by hiding behind a fallen tree. Macho immediately noticed it and circled around the log with Brutus moving too, so as to remain on the opposite side. After two turns, Brutus deceived him a second time by running away quickly and Macho circled once more before he realized that Brutus had gone and then chased after him. This monitoring of the dominant indicates that the deceiver is aware that hiding is a perspective-taking problem.

It is worth noting here that the Mahale chimpanzees have never been observed to hide from others in such a way (Nishida, personal communication).

Compassion and empathy

If an individual is wounded another may feel compassion, because it knows the other is in pain (attribution of a mental state). This is only possible if one adopts the other's perspective. In empathy, one should not only realize that because of its wound the other one is in pain, but that it is now no longer able to do things it was doing some minutes before (Cheney and Seyfarth 1990; Hart and Karmel 1996). Hence, by such a definition, compassion requires attributing feelings to others, while empathy requires attributing competence to others. Observations in nature about such questions in primates are very rare. In the course of our study, several individuals were wounded or killed by leopards (Chapter 2). Although most of these injuries were superficial, some of them were severe, and the wounded individuals appeared to us obviously to be in pain and handicapped. Were chimpanzees able to perceive it? Monkeys are said to be incapable of compassion because they cannot understand others' needs and feelings. They have not been seen to care for injured individuals, but look at disabilities as anomalies or as objects of interest rather than as handicaps (Cheney and Seyfarth 1990). Gombe chimpanzees have been described as showing compassion to close kin tending their wounds and even bringing them food, but the reaction to unrelated wounded group members was one of fear and sometimes disgust (Goodall 1986). How do Taï chimpanzees react to wounded individuals?

Not all wounded individuals have the same needs or the same handicaps. Small wounds resulting from bites with little bleeding happened about once a month within the study community as a result of social fights, and they were not seen to elicit much reaction from other group members, apart from some inspecting of the wounds. Larger wounds can result from territorial fights ($N = 1$) and from leopard attacks ($N = 8$). In all these instances, most adult group members attended the wounded chimpanzees, taking special care in cleaning their wounds with saliva, removing dirt, and chasing away flies. Tending was very common, provided by and to all group members, and not limited to close kin; during the three and a quarter hours following his attack by a leopard, Falstaff, an old male, was groomed all the time without interruption by nine individuals and his wounds were licked for 2 hours and 55 minutes (see case study in Chapter 2). The day after, when resting, Falstaff was tended during 84% of the observation time. Similarly, when Ella was wounded by a leopard, she was constantly surrounded for the rest of the day by other chimpanzees and her wounds were licked for 4 hours and 20 minutes (see example 3 in Boesch 1991*c*). This tending was provided as long as needed, some wounds being cared for for longer periods than others. For example, Gitane had many cuts in her back that she was unable to reach herself and other group members were seen to lick her wounds for as long as one month after the attack. Similarly, Falstaff's wounds, although hardly visible, were tended for two months following the leopard attack that had probably perforated his lung under his right arm. Rousseau had the scrotum of his left testis

Fig. 10.1. Fitz licks the wound of Darwin's left foot. Intriguingly, the common tending of the wounded seen in Taï chimpanzees is rare in other populations, except between close kin.

removed by the claws of a leopard and other individuals were seen to lick him as much as two years later. His testis never healed, and he may have died eventually of an infection. For other wounds, mainly on arms and legs, the tending stopped much more quickly. This suggests that tending was proportional to the severity of the wounds and the need of the wounded, and not just to the external aspects of the wounds.

Wounds tend to be painful and other chimpanzees seem to be aware of this. Ella's son Gérald and Gitane's daughter Goyave were still dependent upon their mothers for transport when they were wounded, but the two of them adapted their position so as not to touch the wounds. Freshly wounded individuals are shocked. Dominant group members waited for more than three hours for Falstaff to be able to move, and they waited for Ella at the site of the attack for more than two hours. Was this behaviour a reaction to the fact that the wounded did not move or were group members aware that they were in pain and handicapped? Kendo, Ella's adult son, chased away all infants that came in her vicinity during the first few hours after the attack and he was regularly supported in this by other males. Similarly, after a leopard attack, Fossey, a young immigrant adolescent female, was quite afraid of the resident females, and two males took turn to protect her of their approaches, so that she could rest quietly during the hours following the attack. Waiting for long periods of time and chasing infants away from an adult are not part of the normal life of the chimpanzee. These observations show that chimpanzees not only attribute special needs to wounded individuals, but that they attribute different needs to different wounds and that different wounds elicit different feelings, some of them being alleviated by rest and calm.

Case study 2: Tina's death resulting from a leopard attack

This anecdote is so extraordinary that we wish to give it here in detail. Tina was a 10-year-old juvenile female who had lost her mother some four months earlier. Together with her 5-year-old brother, Tarzan, she was since that time seen to be regularly associated with the dominant male, Brutus. Obviously, Tarzan wanted to be adopted by Brutus and he was even seen to share his night nest. Tina was more careful with Brutus and followed him at a certain distance. Regularly, Brutus was seen in the forest followed by Tarzan, then Ali who Brutus had adopted some five years before, and finally by Tina.

On the 8 March 1989, at 7.45, Grégoire Nohon, a field assistant, was following the female named Bijou when he heard unusual calls nearby. Rushing to the spot with Bijou and the alpha male, Macho, he saw Brutus encircling the body of motionless Tina, some of whose viscera were visible through a cut she had on the belly. We can only guess how the attack had happened: the hidden leopard possibly let Brutus pass, followed by Tarzan and Ali, and attacked Tina who was lagging behind. Brutus immediately reacted and the leopard fled. However within the first seconds of the attack, the leopard had killed her by biting through her neck, breaking the second cervical vertebra, as we could confirm later when examining the body. When Grégoire arrived, already four males and several females had gathered around Tina and there were loud calls.

We arrived at 8.17 and found six males and six females sitting silently near the body. The males showed some aggressive behaviour by displaying nearby and by dragging the corpse over short distances. Ulysse hauled it over two metres and Brutus pulled it back to where it had been before, about 5 metres away from the place where the attack had taken place. Kiri, Poupée, and Ondine, all high-ranking females, were nearby as well, and they smelled Tina's wounds and some leaves on the ground. Ulysse rapidly inspected one of Tina's hands, holding it. Four females arrived and very carefully approached the body, which was now guarded by the males and Ondine, the alpha female. Malibu smelled the body, while the infant Lychee was chased away as she approached. Malibu, as had done all the others, smelled the body near the wounds, but did not lick them. At 8.30, Macho lay down and started to groom Tina for the first time. Brutus did the same from the other side. Ricci, a low-ranking female, smelled the body, but Ondine and Brutus chased her away. During a period of 1 hour and 20 minutes, Ulysse, Macho, and Brutus groomed Tina's body for 55 minutes. This was unusual because neither Ulysse nor Macho were ever seen to groom Tina alive and other males seldomly did so for a few seconds. Salomé, the beta female, came and smelled the wounds and the genitals of the body.

Nearby, subadults and low-ranking females inspected with great intensity the place where the attack had taken place and where the ground showed clear traces of a fight with traces of blood. In contrast to what had happened when Ella was wounded by a leopard, not a single drop of blood was licked. Goma and Héra approached and they were allowed to smell the wounds, whereas their two infants were chased away by Ondine and Brutus. At 9.07, Brutus gently tapped Tina's chin while looking at her face. Macho and Ulysse later softly shook one of her hands and legs, looking at her face. It looked as if they were testing for some kind of reaction. Also for the first time, Brutus and Ulysse started to play briefly together with their hands, showing a distinct play face, and then groomed her. Brutus played with Ulysse, Macho, and Ondine in this way several times, always very close to Tina and generally grooming her afterwards. It looked as if some tension among the guardians of Tina's body had to be released and this could explain the short duration of the play along with strong play face and laughter. Xérès wanted to approach, but Brutus chased her away and she fled screaming. Following some alarm calls by Kiri in the north, they moved on a bit and for the first time no chimpanzee was within two metres of the body.

From 10.10 onwards, the flies on the body were numerous and started to be a nuisance for the chimpanzees. They waved them away frequently and removed the eggs laid in the nose, eyes, and wounds of the neck. Two hours thirty-eight minutes after Tina's death, Tarzan came to smell gently over different parts of the body and he inspected her genitals. He was the only infant allowed to do this. Then, Tarzan groomed her for a few seconds and pulled her hand gently many times, looking at her. At the same time, Brutus chased away Xérès and her juvenile daughter Xindra. 11.45: Most chimpanzees stayed at 5 m from the body due to the impressive number of flies, the males coming close intermittently to wave them away. Many chimpanzees left the site for a while to feed and came back later. All females of the community came back to look at the body, the males stayed generally for longer and Brutus remained without interruption 4 hours and 50 minutes, except for 7 minutes. In all, there were chimpanzees constantly with the body for 6 hours and 15 minutes.

After having inspected and weighed the body, we left it at the site and checked regularly for the passage of a leopard or other scavengers. Two days after the attack, on the 10 March in the afternoon, a leopard had come and eaten part of the body. That day, the flies that had been covering the body, were gone. The leopard had cut the body in two through the lower back region and carried the lower part (with no open wound and relatively free of maggots) 15 m away and eaten it.

Leopard attacks were fatal in several cases ($N = 4$) (Chapter 2) and the chimpanzees' reaction to dead individuals was strikingly similar and different from their reactions to wounded individuals. They were similar in that the adult dominant group members seemed to required respect towards the dead of other group members and particularly so of youngsters. They were chased away whenever they came close. The interesting exception was Tarzan, Tina's younger brother (see case study 2). In contrast to injured individuals, none of the wounds of the dead individuals were ever licked nor was dirt removed by any group member. Many inspected the place of Tina's attack, but whereas for wounded individuals they always licked all the blood on the ground, on the wounds or the running blood on the body, not a single drop of Tina's blood was licked. Chimpanzees differentiate between injured and dead individuals: the injured need to be tended, but dead ones do not.

Case study 3: A notion of death in wild chimpanzees? Bambou's death

The 23 March 1991, at 10.45, Bambou, a two-year-old male, died, probably by breaking his neck when he fell from a tree. His mother, Bijou, immediately carried him against her chest making loud alarm calls for 10 minutes. She climbed into a tree to avoid the arriving display-ing males, who all climbed behind her to smell the motionless body. Bambou presented no visible wound. After 30 minutes, when the situation had calmed down, Bijou went to the ground and followed the group. Kendo, the new alpha male, guarded her for the next 2 hours and 50 minutes. One hour after Bambou's death, she started to lay him on the ground for short periods of time, but continued to carry him along all day long. On 24 March, the body of Bambou was swollen and smelling, but Bijou still carried him against her chest.

The next morning, she was eating in a tree, having left Bambou on a branch, when Kendo started warming up for a display against her. Brutus, in anticipation, came and took the dead Bambou in his arms. Bijou did not hurry to retrieve him and Brutus brought him to the ground, where he rested with him near Macho and Ulysse, who both smelled at the body.

At 9.20, the body was so swollen that the skin tore in several places and flies started to swarm. At 14.30, after a long rest, the group started to move and Bijou, hesitatingly, looked alternately at the group and at Bambou's body. Then, Mystère, Goma, Belle, Agathe, and Ondine came back to Bambou. Mystère and Sirène, Ondine's infant and preferred playmate of Bambou, climbed a small tree above the body, looking down at it. Ondine, Mystère, Goma, and especially Sirène made a few soft 'hou' calls. Then, they all left silently. At 14.56, leaving Bambou behind, Bijou started to catch up with the group. Brutus, Macho, Kendo, Fitz, and Ali were silently waiting for her. After 8 minutes, Bijou, alone, came back to Bambou and carried him over 20 metres. She hesitated in this way for another 80 minutes, until she left him definitively behind.

It is difficult not to wonder why these females came back and made these soft calls to Bambou. Were they aware they would not see him again? If they have a notion of death, this behaviour makes sense. If they don't, it is puzzling. It was certainly a very touching scene. The Taï chimpanzees suffer high predation from leopards, and empathy could improve the chance of survival of the wounded individuals. The important care given to them partly explains the incredibly rapid healing of most injuries.

Are the observations of tending and pain-relieving due to the fact that chimpanzees have compassion and empathy, or is it just that they associate wounds with some types of behaviour? An association explanation would have to take in account that the tending was provided only for certain kinds of wounds, when individuals could profit from it, and for behaviour the wounded could not perform any more. The rarity of important wounds, 10 to 20 individuals wounded by leopards in 15 years, makes it very unlikely that a pure association process could produce such a precise response of group members. In addition, it was directed to all group members and not only to close kin, excluding a kin-based explanation. We are aware that for such matters it will be difficult to provide conclusive evidence and that only the accumulation of such examples can make a case. We feel that attributing compassion and empathy to chimpanzees makes it easier to explain the chimpanzee varied responses according to dead, lightly or badly wounded companions.

There is another context in which empathy would be adaptive: poaching. On 20 June 1985, Véra, a young female, had her left wrist caught in a snare trap. Spinning wildly on the ground she was able to break the cable but not to remove the snare that was by then cutting deep into her flesh. Rapidly, Schubert, the beta male, approached, and while she held her arm towards him, he removed the cable with his canine. Snares represent a fatal danger for the chimpanzees, and associative learning is probably not possible as it remains, fortunately, too rare at the individual level: during the fifteen years of the study, nine individuals were trapped, including five subadults. Thus, it is more likely that Schubert felt empathy and reacted accurately and immediately. In a way it is comparable to what Little Bee did at Gombe, who was seen to help her dying mother by collecting food in trees and bringing it down to her (Goodall 1986).

To conclude, we have seen many examples suggesting that chimpanzees evaluate their companions' state, anticipate their actions, and adjust their own behaviour to these social evaluations. It might be time, therefore, to attribute to them a knowledge which has all the qualities of a theory of mind with the exception of reflective conceptualization. In other words, chimpanzees seem to be aware of the others' competence, knowledge, feelings, and are aware, too, that they may differ from their own. We cannot know whether their knowledge implies other aspects of the properties of a theory, that is hypothesis- and alternative-formations. Such a theory is much more complex than the knowledge that other individuals behave in their own way, without referring to their individual views, intentions, and feelings.

Parsimonious explanations based on associative learning will be presented to counter our interpretation. It seems, however, difficult to find a single straightforward non-mentalistic interpretation for all the observations in the social domain we have presented. The occasions allowing associative learning are simply too rare to make such a process realistic. Youngsters, who are still associated with their mothers, will see a hunting movement with full anticipation at most six times per year (mothers tend to join hunting parties only once the capture has been done). Likewise, in 15 years of observations, only 10 to 20 individuals were wounded by leopards (Boesch 1991*c*). An explanation based on a theory of mind seems the simplest to account for the diverse phenomena we have reported on the behaviour of chimpanzees in their social world.

Is associative learning really excluded from the acquisition process of a theory of mind? How does an individual acquire a theory of mind? Studies show that human infants progressively acquire the abilities necessary for the theory of mind, and that many aspects are affected by social interactions (Gopnik and Graf 1988; Hala and Chandler 1996; Meltzoff 1995, 1996; Pivonelli 1996; Wimmer *et al.* 1998; Roth and Leslie 1998).

Social experience is central for the acquisition of a theory of mind, and some sort of associative learning is presumably part of the process. We should expect youngsters to start to attribute competence to somebody else only once they have become very familiar with that person, in most cases the mother, once they have learned that the person possesses different special skills, and once they can actually reliably predict this person's performance (see Meltzoff 1996 for a similar argument). From that point on, infants can start to generalize and to expect competence in other individuals. In our opinion, such an ability can only build on extended and refined social experiences and does not come out of the blue. If that is true, associative learning ability is part of the process. Thus, we should not deny the existence of a theory of mind only because associative learning is involved, but rather try to understand how much has been built upon such a simple and efficient process to generalize attribution of competence in social partners.

The fact that aspects of a theory of mind were not obvious in studies of some captive chimpanzees may have to do with their particular experience and upbringing, which made it difficult to develop intimate and stable social partners. Human infants, like wild chimpanzees, usually spend their entire youth with their natural mother, and acquire social intimacy much more easily in the presence of stable social partners. Can we expect social intimacy to develop across the species barrier? Can we expect captive chimpanzees to develop such an ability with a human caretaker? Do chimpanzees attribute a mind to us? These are the basic assumptions of all captive studies. They may seem far-fetched if we consider that the majority of human beings, among them many scientists, are reluctant to attribute a mind or a notion of self to chimpanzees.

What might be a chimpanzee theory of mind? First, chimpanzees make judgements about other chimpanzees or prey species. Second, chimpanzees judge physical abilities achieved by other group members or prey species, be these competence at hunting or nut-cracking, the abilities of the prey or the feelings of wounded individuals. All these are adaptive abilities important for survival. An evolutionary approach to intelligence predicts such observations, and studies of a theory of mind in chimpanzees should orientate their investigation towards attribution of competence and abilities in the physical domain of social partners. Whether some aspects of the human theory of mind, such as the attribution of beliefs and emotions to others, or the intentional manipulation of such knowledge in a malicious way, are part of the chimpanzee's domain, is still open. More data are clearly needed to understand in which sense and to what extent the theory of mind in chimpanzees differs from the one found in most humans.

Reflections about human and chimpanzee cognitive uniqueness

The recent confirmation of the very close genetic similarity between *Homo sapiens* and *Pan troglodytes* has renewed interest in understanding what differentiates or unites the two species. Although interest in the difference between human and animal species has been around for ages, Aristotle, Plato, Rousseau, and many others had no choice other than to guess what chimpanzees and other animals might be capable of. Since the early 1960s, many detailed studies have revealed facts about the behaviour of chimpanzees, but still, on the whole, only about 300 wild chimpanzees have been followed for a reasonable length of time. This remains an exceedingly small sample compared with the number of humans that have been studied, and demands caution when affirming the absence of an ability. Owing to this situation, discussions about the differences between chimpanzees and humans may reflect personal and cultural preconceptions.

Before embarking on the task of comparing the two species, three points must be considered. First, are all adults within a species cognitively equal? The technology used daily by a worker at an aerospace factory and that used by a Tasmanian traditional hunter are extremely different. This stresses the point that in very recent history, the industrial revolution is only 200 years old, humans have made tremendous changes in some aspects of their lives. As members of the industrial world, some of us tend to forget how recent these changes are. The separation of human and the chimpanzee took place some five million years ago, and *Homo sapiens* appeared some 180 000 years ago. We lived as hunter–gatherers for 96% of our history, and agricultural and industrial history represents a mere 4%. Being an agriculturist or an industrialist is the exception for a human being, but many more cognitive studies have been done on individuals belonging to these exceptional ways of life. Sorting out the differences between human beings living such different lives is mostly guesswork. Cross-cultural psychology is a science devoted to understand some of these differences, and we have mentioned some important cognitive differences found between individuals living different lifestyles (Dasen 1975; Segall *et al.* 1990). The important differences observed between chimpanzees living within different environments indicate a problem that may be similar.

Second, a comparison between two species should try to compare what is comparable. No study of the cognitive abilities of humans kept under captive conditions from their early childhood onwards is available. This is a major limitation, as most studies on chimpanzees have been carried out on individuals that lived under captive conditions for all their lives, and most comparisons have been made between free-living humans and captive chimpanzees. However, if one compared captive leopards with wild lions, everybody would be sceptical about generalizing from such a comparison. Cognitive abilities have evolved over thousands of generations of our predecessors, and have enabled them to survive in their natural habitat. These abilities are acquired during ontogeny in the course of solving daily problems. For example, children succeed in understanding false-belief tests much earlier if they have previously been given the opportunity of participating in mental-state manipulation of conspecifics (Hala and Chandler 1996). Individuals living in poorer social environments should thus be expected to be less advanced.

The final point to consider in comparing different species is that all individuals develop from birth to adulthood, and so do their cognitive abilities. The most studied animal species, the human, shows that cognitive abilities develop progressively. A baby, a child of four years, and an adolescent of 12 years all possess different cognitive abilities. So, when comparing two species, what stage in the development should we consider? Comparing them at the same age does not take into account the differences in life-span or the developmental trajectories of each species. For example, human babies a few days old are especially retarded compared to babies of wildebeest. Most people would respond that at a later age humans will develop quicker and for a much longer period. This means that we are interested in the whole individual development and in what level of abilities is reached. For most animal species, we have no detailed information about cognitive development, and we should thus compare the cognitive abilities of adults of different species. There are a few studies that have investigated the ontogeny of cognitive development in chimpanzees, and all concur in showing that similar activities tend to appear later in captive chimpanzees than in humans (Langer 1996; Povinelli 1994; Chevalier-Skolnikoff 1988). Most studies of captive chimpanzees were carried out on young individuals, and comparisons were often made between free-living adult humans and captive young chimpanzees.

It will remain difficult to study all aspects of the intelligence of animals in the wild, and captive studies are essential for a complete understanding of an animal species. However, we plead that this should be done by taking into account the biology of the species under study. Many studies have only been interested in studying human abilities in other species, without taking into account the relevance of the test for, or the effect of captivity on that species. The unique advantage of being able to control for many factors in captivity does not mean such studies have no limitations. Some captive studies (such as those of Köhler, Hayes, or Yerkes) have had a decisive influence on fieldwork, and valid results from captive studies are most important for guiding field observations. Ideally, the results of fieldwork should be used to judge the relevance of the abilities tested and the choice of the subjects used.

Factors affecting the development of intelligence

We presented different hypotheses about the development of intelligence at the start of this chapter. The social-intelligence hypothesis (Humphrey 1976) expects that social challenges will favour the abilities to solve them. The observations on the Taï chimpanzees clearly support this expectation, for we saw them use their expert abilities to attribute competence to other group members in many social domains. Similarly, the ecological- or technical-intelligence hypothesis (Parker and Gibson 1979; Piaget 1935, 1945; Byrne 1997) is supported by the elaborate faculties chimpanzees have developed in the domains of spatial representation and comprehending causality. The fact that this involves different domains of intelligence shows that these hypotheses are not exclusive, but that they should be viewed as having a complementary role in explaining different aspects of the development of intelligence. We could actually pool them under one heading: 'the environmental intelligence hypothesis', environment being understood in its full complexity, to include the physical and the social environment. The physical environment includes technical skills (Byrne 1997), feeding techniques (Parker and Gibson 1977), as well as cognitive maps (Milton 1981). This would allow us to add a fourth important dimension located at the border between the physical and the social environment, the 'hunting intelligence domain'. The analysis of the hunting behaviour of the Taï chimpanzee stresses the special abilities required for hunting, for they go beyond what is needed in the intra-specific social domain (for example predicting actions and abilities of another animal species). Group hunting requires both these abilities as well as the more social abilities of coordinating actions with other group members. The hunting-intelligence hypothesis might also explain some of the especially elaborate abilities shown in social carnivores (Schaller and Lowther 1969). This would support the idea that hunting might have played a special role in the development of human intelligence, as has been proposed in a less specific way before (Leakey 1961).

The observations on the Taï chimpanzees confirm the very important role of both a natural social life and a life in a rich and demanding environment. The same has been shown in humans (Segall *et al.* 1990). For example, in humans, the impact of the environment is not a simple one and some aspects of it have more influence than others. Consider spatial representation. Traditional farmers, who move from their village to their fields following predetermined paths, going from landmark to landmark, do not develop a Euclidean concept of space, while Eskimos and Aborigines, who move in an open space with very few landmarks do develop one. So it is not the richness of the environment *per se* that favours a particular ability, but the combination of the environmental properties and the challenges an individual faces that favours the development of different abilities (Eskimos have to orientate themselves in an icy world to survive). Similarly, traditional farmers possess better numerical abilities than Eskimos and Aborigines (Dasen 1982). So, within one physical environment, some abilities might be favoured, while others are not. This specifies how the environment-intelligence hypothesis can work and how careful we need to be when making predictions about it.

The environment-intelligence hypothesis also provides us with a theoretical tool for understanding some of the differences that have been found between captive and wild chimpanzees. Not only do captive chimpanzees live in an environment that is much simpler and remains entirely stable for years, but in addition they face no challenge on which their survival depends. Wild chimpanzees live in an environment that is more complicated and changes regularly and frequently over time and space. They are faced with constant challenges to solve, and their survival depends directly upon how well they solve them. Most aspects of the environment-intelligence hypothesis would predict that such contrasting environments will select for markedly different levels of intelligence. Captive chimpanzees with especially intense interactions with humans might present a special case (Call and Tomasello 1996), as both the social and the physical environments have been improved. Thus, some improvements in their performance in some domains would be expected.

According to the environment-intelligence hypothesis we would expect wide differences in the levels reached by members of the same species living under very different conditions. Domain-specific intelligence would be the result, for in some domains individuals may need more advanced abilities than in others, as it was shown recently in studies on human intelligence (Gigerenzer 1997; Hala and Chandler 1997). We expect that under some conditions one species surpasses another, whereas under other conditions the opposite might be true. This seems to be the case in the domain of spatial representation, in which some chimpanzees possess a Euclidean mental map, whereas this has not been observed in all human populations. This shows that both species are flexible and react adaptively to environmental conditions. However, clear differences in brain volume, structure, and encephalization exist between the two species (Passingham 1982), and, although not all these physical differences will have a direct effect on cognitive abilities, there must be important constraints in some domains. Much more work will be needed on chimpanzees before we can answer such questions.

Discontinuity humans–chimpanzees?

The present approach based on the adaptive evolution of intelligence does not predict a basic discontinuity between human and other animal species. In evolutionary terms, differences should be expected where individuals of different populations or species have lived for long enough under different conditions. However, we may ask in which sense these closely related species differ one from the other in their abilities related to intelligence. In two important domains, some differences have been proposed to exist between the chimpanzee and human.

Evolution of intelligence

Despite a limited knowledge of the intellectual abilities of many animal species, recent reviews of the evidence have suggested the following evolutionary scenario of intelligence in primates. Primates differ from other mammals in having more rapid learning abilities and larger brains (Byrne 1995), as well as in understanding categories of relations between objects and individuals (Tomasello and Call 1997). Apes differ from monkeys in that they possess insightful abilities (Byrne 1995). Humans differ from apes in having linguistic abilities and a theory of mind, originating in the human understanding of causality and intentionality (Byrne 1995; Tomasello and Call 1997). As we have seen in the present chapter, wild chimpanzees show in many domains a more developed understanding of causality than has been shown in captive chimpanzees. The attribution of competence and feelings seen in Taï chimpanzees strongly suggests that here again there might be a difference in the level of intelligence achieved in individuals living under very different conditions. More studies of wild chimpanzees are needed to understand more precisely the breadth of their theory of mind and their understanding of causality, before we can compare it with that observed in humans. Eventually, language may be the only ability that differentiates the two species. Classically, emphasis has been placed on the abilities required for language to evolve (intentionality, symbolic abilities, theory of mind, a notion of self, and so on). However, most of these have been demonstrated in one or the other form in chimpanzees. We should not neglect the positive effect that language itself has had on cognitive abilities once it evolved. It probably had a decisive influence on the evolution of culture (Boesch and Tomasello 1998), allowing more detailed attribution of knowledge, and opening the way for more flexible and precise manipulation of beliefs, feelings, and knowledge.

Evolution of culture

Psychologists take the view that social learning processes are at the root of culture (Galef and Heyes 1996). Some psychologists have insisted that only imitation can produce culture, based on the assumption that otherwise the variation within the behaviour learned would be too large (Galef 1990; Tomasello 1990). However, depending upon the conditions prevailing when a task is learned, this assumption might be wrong, and different social learning processes allow culture to emerge (Whiten and Ham 1992; Boesch and Tomasello 1998). The question is, do behaviour patterns seen in wild chimpanzee populations qualify as cultural behaviour, if at least parts of them were learned by imitation? Recent analysis of the distribution of population-specific behaviour in wild chimpanzees, as well as of the transmission mechanisms of behaviour, support the notion of a culture in chimpanzees (Boesch 1996*a*, *b*; Boesch and Tomasello 1998; Whiten *et al.* 1999). Cultural transmission could permit the propagation of new ideas much more rapidly between individuals of the same or of different groups. Chimpanzees possess cultural behaviours, but such evidence is not yet conclusive for any other non-human primate. In a review of the possible differences in culture observed in chimpanzees and humans, two main differences emerge. First, humans probably make more extensive use of the cumulative evolution of cultural patterns, leading to sophisticated artefacts such as a tent or a hammer, that have been progressively improved by many different individuals over many generations (Boesch and Tomasello 1998). Second, thanks to language and, more recently, to writing, humans have succeeded in transmitting information between individuals that may never see each other or that do not live at the same time. This allows cultural changes to occur rapidly over extremely wide ranges and to expand over many different generations (Boesch and Tomasello 1998).

In conclusion, chimpanzees live in Africa in very different habitats from very dense and low visibility environments, like the tropical rainforest in Côte d'Ivoire or Gabon, to open and high visibility habitats, like the savanna in Southern Mali or the Ugalla reserve in Tanzania. Such a diversity of habitats has selected for diversity in the chimpanzees (Wrangham *et al.* 1994; McGrew *et al.* 1996; Heltne and Marquardt 1989; McGrew 1992). The environment-intelligence hypothesis predicts that we will find similar diversity in the intellectual abilities developed by members of different populations. The physical environment is quite diverse, whereas the social environment is more stable, characterized by the fission–fusion social structure, which in itself selects for flexibility (see Chapter 5). These two aspects of the environment in the chimpanzee probably had a major impact in shaping its intelligence.

Our review of the cognitive abilities of wild chimpanzees illustrates how developed some of their faculties can be in the physical domain, relating to mental representation and the notion of causality, and in the social domain, relating to social knowledge and the theory of mind. These exist in the context of three traits characteristic of the chimpanzee. They are one of the few primate species living under the fission–fusion social structure; they are, with humans, the only primates hunting regularly in groups for meat; and they are, with humans, the only animal species that regularly and flexibly uses tools. In the final chapter, we try to reconcile these facts and present a model for the evolution of chimpanzees that explains the concomitant presence of these unique chimpanzee–human characteristics.

11 *Chimpanzee and human evolution*

For centuries the quest for our identity has been a major incentive for philosophical and scientific work. Understanding the uniqueness of human beings with regard to other living beings has often been biased by the difficulty of reasoning impartially about ourselves. The literature is full of statements praising human uniqueness and claiming that we occupy the highest position in evolution. It has been claimed that humans are the only species with a reflective intelligence, the only species that cares for wounded conspecifics, or that depends on cultural products for survival. Similarly, some sets of human behaviour are claimed to be especially demanding, solely because they are performed by humans, such as tool-use, tool-making, coalition building, or language. Such circular arguments that declare a behaviour as unique to humans and therefore representing a proof of the highest evolutionary achievement, do not add much to the understanding of what distinguishes humans from other animals. Although historically such an attitude may have been justified by lack of information, it is not justified today.

An understanding of the characteristics of an animal species hinges on two rather different things. First, we need to know what most members of such a species are and do. Second, we need to know what closely related species are and do. So far, we do not have such a thorough knowledge of any species, and thus our attempt to summarize the detailed characteristics of the chimpanzee is provisional. The question of the uniqueness of a species is justifiable, for each species living nowadays is the product of its own evolutionary process: chimpanzees are unique, as are ants, guinea fowls, or humans. The more related two species are, the more they have in common, and this affects their respective uniqueness. The earth is estimated to be about 4.5 billion years old, and life appeared some 3.5 billion years ago. Multi-cellular organisms appeared about 600 million years ago, the vertebrates about 360 million years ago, and the first primates about 90 million of years ago. Since chimpanzees and humans were first distinguishable about 5–6 million years ago, they share a very long evolutionary history before each followed a distinct path.

In the first part of this chapter, we summarize the results of the ten previous chapters, with an emphasis on the major characteristics of the chimpanzee and its behavioural diversity, and include them in an evolutionary scenario. In the second part, we use these findings to reinterpret the evolution of modern man and contrast it to scenarios previously proposed.

Chimpanzee behavioural diversity

During the last three decades, knowledge of the wild chimpanzees has relied almost exclusively on observations collected on a community living in the Gombe Stream National Park by Jane Goodall and others, and on one living in the Mahale Mountains National Park by Toshisada Nishida and his colleagues. Both populations are located relatively close to each other on the eastern side of lake Tanganyika in Tanzania. They also live in relatively similar environments characterized by a mosaic of savanna, woodland, and dense shrub habitats. Whenever most people think about chimpanzees, it is these Tanzanian chimpanzees they have in mind. The Tanzanian chimpanzee became the prototype of the chimpanzee, and behavioural diversity was restricted to the few differences existing between these two populations. Anthropological and psychological literature and textbooks are still strongly biased in assuming that the Tanzanian chimpanzees' behaviour stands for all chimpanzees' behaviour.

Meanwhile we should be aware of the limitation of relying on the knowledge of an animals species drawn from only a few populations living in similar environment conditions. Modern evolutionary thinking predicts that populations of the same species living in different conditions will evolve in different ways to survive and reproduce. The response can be genetic, or behavioural based on extended learning abilities. Learning is expected to be more flexible than genetics, and species able to respond in their behaviour are expected to produce wide behavioural diversity. Chimpanzees possess extended learning abilities, as we have seen from the learning of nut-cracking and hunting behaviour (Chapters 8 and 9), and we should expect them to show many different adaptations depending on the environment in which they live. Chimpanzees have always lived in diverse environments, ranging from the tropical rainforest to some very dry savanna regions, with however at least some gallery forests (Suzuki 1969; Kano 1971; McGrew *et al.* 1979; Kortlandt 1983).

Some years after the start of the Tanzanian studies four decades ago, other chimpanzee studies were undertaken in different sites with varying types of environment throughout Africa. This resulted in a big increase of information about chimpanzee behaviour, and behavioural diversity is now acknowledged to be very wide in this species. This raises the question of how this diversity evolved.

Behavioural diversity in wild chimpanzees

Apart from the Tanzanian woodland–savanna chimpanzees, it is the ones of West Africa, Taï in Côte d'Ivoire and Bossou in Guinea, that are so far the best known. The former live in a tropical rainforest, the latter within a mixed deciduous and secondary forest. Such diverse habitats may represent a strong selective pressure to produce different adaptations. Table 11.1 summarizes the different life-history traits found in the four populations. Substantial differences are found in such traits as age of maturity, interbirth interval, and frequency of migration between communities. They appear to be related to different levels of food availability between sites, and the corresponding different levels of intra-community competition for food (Chapters 2 and 3). At this level, chimpanzees have diverse strategies in reaction to varying ecological conditions. Some data from Taï and Gombe suggest that these strategies may change over time within a population, for

Table 11.1: Life history variability in well known chimpanzee populations

	Gombe	Mahale	Taï	Bossou
Female migration	13–50%	113%	95%	–
Female age of maturity (years)	14.9	14.6	13.7	–
Interbirth interval (months)	66	72	69	61
Menopause	no	yes	no	–
Variation in IBI	⇑ for all daughters	no	⇑ for sons of dominant females	no
Maximal fertility[1]	4.56	4.23	5.39	5.29

[1] Maximal fertility estimates the number of offspring a female can produce within her maximum life span.

example rate of female migration has been seen to vary (Pusey *et al.* 1997). Variation in life-history traits has been reported within a wide variety of animal species, both between populations and over time within one population (Stearns 1992; Holekamp and Smale 1995). We also expect the behaviour of chimpanzee populations to react to different environmental conditions.

Table 11.2 lists some of the marked behavioural differences we have noted between these four chimpanzee populations. We find here extensive differences in the social domain, in hunting behaviour, and in tool-use (see also Table 9.3). This list is certainly not exhaustive, especially since we deal with only large and easily quantifiable differences. Nevertheless, it shows the amplitude of differences for only four populations and gives an impression of how wide they are likely to become when information from more populations becomes available. We have also seen from the discussion of the distribution of grooming that a behaviour may change over time within a same population, depending upon the social composition of the community. Shorter studies on other chimpanzee populations have revealed additional diversity in the use of tools (McGrew 1992), or in the social domain (McGrew *et al*, 1996).

Table 11.2 suggests that whatever behaviour patterns we look for, it is imprecise to say that the chimpanzees behave in one way. Rather each community studied at best serves as a prototype of the behaviour patterns of the chimpanzees belonging to the same region. Even so we should remain cautious. Gombe and Mahale chimpanzees are about 200 km apart and differ strikingly in many aspects. Similarly, chimpanzees west or east of the Sassandra river, in Côte d'Ivoire, only 30 to 50 km apart, differ in tool uses (Chapter 9). At present, we cannot present an exhaustive list of all the possible ways chimpanzees behave in all known populations. Many more years will be needed to obtain this knowledge, and whether we shall be able to do so in the end is not guaranteed, given the often precarious situation of wild animals and their habitats. One general conclusion is that behavioural diversity is high in wild chimpanzees, and no population can be considered as the prototype for the whole species.

Is the behavioural diversity we observe in chimpanzees wider than that found in other species? We have to be very careful here, since the knowledge we have of different primate species tends to be related to the taxonomic proximity of these species to ourselves. More is known about great apes than about any species of the Old World Monkeys, about which we nevertheless know more than we do for any prosimian species. Luckily, there are some exceptions: for macaques, baboons, and langurs several popula-

Table 11.2: Behavioural diversity in well known chimpanzee populations

	Gombe	Mahale	Taï	Bossou
Social behaviour				
Mean party size	5.6	6.1	8.3	4.0
Mixed parties	30%	52%	61%	42%
Mixed parties corrected[1]	0.37	0.79	2.4	3.5
Male/male association	24%	10%	35%	–
Female/female association	5%	5%	11%	–
Female friendship	Mother–daughter	Absent	Present	–
Dominance	One	One	Three	–
Female position in hierarchy[2]	Inferior	Inferior	Dominant (1/3)	–
Consortship	28%	8%	31%	–
Male access to fertile female	All	Alpha	All alphas	–
Grooming: male/male	37%	85%	44%	6%
Grooming: female/female	13%	15%	22%	62%
Looser support coalitions	–	17%	60%	–
Female coalition formation[3]	0	0.12	8.5	–
Infanticide[4]	w/b	w/b	w	–
Extra group paternity	yes	?	55%	25%
Hunting behaviour				
Specialization (% *C. badius*)	55%	13%	79%	–
Prey size (% adult)	11%	21%	44%	–
Yearly hunting frequency	160	–	250	–
Group hunt	36%	24%	93%	–
Collaboration	19%	0%	68%	–
Meat sharing[5]	Age, rank	Ally	Hunting tactics	–
Tool-use behaviour				
Number of tool-use types	22	14	26	15
Number of unique types	4	3	12	5
Number of tool-making types	3	3	6	–

[1] Mixed party corrected = (frequency of mixed parties/sex ratio).
[2] Some Taï females dominate most males for meat access.
[3] Frequency per month.
[4] w = within the community, b = between communities.
[5] Factor that favours the amount of meat received.

tions of each have been studied. For non-primate species, the social carnivores, such as lions and wild dogs are exceptions. But such studies are often limited to one or two topics and therefore we cannot use them to estimate behavioural diversity. One reason is that scientists have to present original, new results to secure funding and topics that do not guarantee such originality are less studied. Lions have been studied in many different Southern African countries and impressive variations in their hunting behaviour are documented (see Chapter 8), but very little is known on behavioural variations in other domains. Similarly, langurs have bee studied in many regions of the Indian continent, and they present wide variations in group composition and infanticide rate, depending on their environment (Hrdy 1977; Sommer 1994), but there is little on other behavioural differences. A preliminary conclusion is that behavioural diversity in mammals is limited to specific behaviour patterns, that they seem to have an adaptive value, and that they seem to reflect a response to ecological conditions.

Why should behavioural diversity be wider in chimpanzees than in other primates? We propose that behavioural diversity in chimpanzees is concentrated in domains that are almost entirely lacking in other primate or mammal species. The most striking behavioural divergence in chimpanzees exists in tool-use, tool-making, and in the cooperation between individuals seen during social coalitions, food-sharing, hunting, and inter-group aggressive interactions. Most of these behavioural domains are absent in other animal species, and if present, are simple. Tool-use in animals, except in chimpanzees and humans, is restricted to a few instances, and tool-making is generally absent (see Chapter 9; Beck 1980; Goodall 1970). Coalitions, food-sharing, and hunting have been observed in lions and in most primate species, but important differences exist. In these other species the collaboration between two individuals does not take account of the particularities of each individual (Harcourt 1992) and is not based on the coordination of one's actions with those of another (Chalmeau 1994; Chalmeau *et al.* 1997), and food-sharing between adult primate group members has either not been reported or is very rare. Thus, cooperation in chimpanzees is more general and more flexible. The same is true for tool-use (Chapter 9). In other words, chimpanzees' most striking behavioural diversity exists in domains that require elaborate cognitive abilities, and are based on extensive learning aptitudes. So, once learning aptitudes have achieved a certain importance, the behaviour patterns used by different populations can diverge rapidly.

Thus we should consider elaborate cognitive abilities and prolonged learning capacities as the chimpanzee key characteristics. These two abilities combined free the individual to a certain extent from environmental constraints and allow the invention of more diverse solutions. The individual is therefore not guided by the physical world alone, and this leads to both behavioural diversity and to culture. We see some continuity within the primate family in the sense that a greater array of behaviour patterns can be observed, with some species using tools and making coalitions, and that these patterns can vary between populations living in different environmental conditions. In chimpanzees, however, learning abilities have further increased and this has allowed them to develop a larger array of behaviours, sometimes independent of the habitat conditions.

Social life and the role of females in wild chimpanzees

What is the typical chimpanzee social structure? All chimpanzee populations which have been studied over a long period of time live in multi-male social groups with a fission–fusion structure. Beyond that, agreement is difficult to find. The fission–fusion social structure can vary from fluid parties changing composition every twenty minutes to parties remaining stable for a much longer time. We suggest that when community size decreases, parties become more stable and the fission–fusion structure loses its fluidity (Chapter 5). Thus, community size has an important effect on group structure. The determinants of community size are probably pressure from predation and the availability of food. In addition, long-term studies of different populations have documented important changes in community size resulting from illness (Chapter 2). In other words, the type of fission–fusion structure observed will depend on the situation prevailing in the population at the time of observation.

Female primates have a special position that is directly related to the demography observed in a species. If females remain in their natal group, they tend to form matrilines and compose the stable social network of the group. If, however, they transfer between groups, males are the stable group members and females have a less secure position

(Dunbar 1988; Pusey and Packer 1987). In agreement with this, the Tanzanian chimpanzee society has been described as male-bonded with social interactions strongly dominated by the males (Wrangham 1986; Foley 1989). This has been contrasted with the female-bonded society of the bonobos (White 1988; Kano 1992). Is this consistent with more recent evidence on chimpanzees? The main addition from Taï is the role of the females. Taï females build strong friendships and alliances with other females, are active grooming partners, take an active role in the social conflicts between the males, make coalitions with some of them, invest for many years in their adult sons, help to reinforce the meat-sharing rules and achieve some of the highest status for meat access, exercise active choice of mating partners, and are more efficient and more frequent tool-users than males (Chapters 3, 4, 5, 6, 9). All these features are in marked contrast to descriptions of the Tanzanian chimpanzees, and some of them have been proposed as being unique to the bonobos. We suggest that the proposed dichotomy between bonobos and chimpanzees in these matters is not real, and that it reflects a population difference brought about by different environmental conditions. Under certain circumstances, the social life of chimpanzees grants the females a much more substantial position than is observed in Tanzanian chimpanzees.

How have females come to gain a more important position in the Taï society than in other populations? We propose stronger intra-group competition, a highly biased sex ratio, and large party size as being at the root of both the female friendships and the more prevalent use of 'dishonest' sexual swellings observed at Taï. With longer periods of oestrus, Taï females obtain more sexual interest and protection from the males, which allows them to build longer-lasting relations with some of them. Female friendships, on the other hand, allow them to cope better with the high level of competition between group members and to achieve higher social ranks. Once that is the case, more changes follow, since for females who have strong support from other females and from males, it will also be to their advantage to reciprocate this support, and by doing so they achieve higher social status in contexts previously inaccessible to them.

In chimpanzees, the female position in society is flexible. Where there is strong intra-group competition, females achieve a social position like that found in Taï. Where there is weak competition, a more male-bounded situation like that found in Tanzanian chimpanzees prevails. Probably other solutions will be found in populations living in different conditions. It has been suggested that at Kibale, in Uganda, some females have an especially strong peripheral position (Wrangham *et al.* 1992), and that might be one of these solutions.

Comparison with the bonobos, where females have been described as having a central position in social life and as dominating the males, remains tentative due to the lack of quantitative data to support such claims. Early comparisons based on the Tanzanian chimpanzee observations have accentuated the difference between the bonobo and the chimpanzee (White 1988; Kuroda 1979; Zihlmann 1981; Tanner and Zihlmann 1976; de Waal 1997). Detailed quantification of possible differences in sexual behaviour have shown that bonobos do not copulate more promiscuously than do chimpanzees, nor do they copulate more frequently (Furuichi 1987; Takahata *et al.* 1996). No quantification of the dominance relationship between the sexes or of coalition frequency has yet been published. Nevertheless, the more central female role in bonobos agrees with our findings on

Taï females. We argue that female bonobos' greater sociality, apparent from the very stable nature of the parties in the Lomako and Wamba communities, results in strong competition between group members. Like Taï females, bonobo females may need to control this competition, and they use the same method, manipulating the length of the period of oestrus, starting oestrus well before being fertile and readily mating with males for longer periods of time (Furuichi 1987, 1989). Female bonobos seem to have gone a step further in the sense that with homosexual behaviour they directly control competition between females (de Waal 1989; Thompson-Handler *et al.* 1984). We propose that the forest environment allows or forces bonobos and chimpanzees to build larger and more cohesive parties, and that to control for the resulting higher competition between group members, females manipulate the duration of their sexual swelling, and associate and ally more with females to acquire friends.

Evolution of behavioural diversity in chimpanzee

When a population lives within a given environment it is well enough adapted to survive and reproduce there, otherwise it would be absent. It is adapted well enough to face competitors, but we should not assume that it achieves the best possible state (Williams 1966; Ridley 1996). 'The survival of the fittest' is a relative notion and not an absolute one. In other words, there is always room for improvement, and behavioural flexibility is an advantage because it allows the use of many alternatives. The fact that *Diana* monkeys in the Taï forest do not use tools does not mean they are not well adapted to their habitat. But if they did, new opportunities might open to them. Flexibility and the possibility of using all opportunities would in the long run be beneficial. Why did chimpanzees acquire such a wide behavioural diversity?

We present here a possible evolutionary scenario that makes sense of some evolutionary puzzles in primates. Why are chimpanzee and human beings the only primates in which hunting is regularly observed in all known living populations? Why is tool-use and tool-making so much more flexible and frequent in chimpanzee and human populations compared with any other wild living primate? The second question has attracted some attention, because many primate species, and especially great apes living in captive conditions, use tools much more than they do in the wild (Chapter 9; Tomasello and Call 1997). Both puzzles are important landmarks in understanding the evolution of behavioural diversity in chimpanzees.

Fission–fusion social structure

In our scenario, the first step came with the adoption of the fission–fusion social structure, which demanded great flexibility. This represents a challenge for all individuals since they rarely associate with the same group members for the whole day. It is this flexibility in the composition of parties that forces individuals to evaluate the ever-changing social forces present. Why did fission–fusion evolve in the first place? Answers to this question are not conclusive. For a long time it has been presented as the solution found by large primates that are free from predation to cope more efficiently with patchy and varying food resources (Dunbar 1988; Wrangham 1980). Six species of sub-human primates have adopted this social structure: the chimpanzee, the bonobo, the spider monkey, the red colobus, the gelada baboon, and the hamadryas baboon. However, observations of predation by large cats against chimpanzees (Boesch 1991*c*; Tsukahara 1993) have shown that fission–fusion is compatible with predation. What remains is that this mode of grouping has the great advantage of allowing a flexible response to food availability or predation pressure. It allows, especially, a simultaneous response to the conflict of finding an optimal solution for avoiding predators (larger groups profit from higher dilution and confusion effect) and for feeding competition (larger groups suffer from more competition). If there is fluctuation in predation pressure and food availability, the optimal solution is certainly to vary group size according to the fluctuations and the need for socializing: stay in larger parties when food is abundant and reduce size when food availability is low, but increase it when socializing. The costs are, however, the difficulty of socializing in ever-changing parties. According to the environment-intelligence hypothesis, we expect individuals to react to the new social structure, but this may take different forms. Three solutions have been observed:

(a) Keeping the social interactions at the usual monkey level, for example in the spider monkeys and the red colobus. Monkeys with the richest social life, baboons and macaques, all live in stable group compositions.

(b) Make party composition totally predictable by having stable sub-groups, for example the harem structure of the hamadryas and the gelada baboons that gather regularly in larger bands.

(c) Learn the required flexibility to be able to track and predict the new fluid social environment, for example the bonobo and the chimpanzee.

Fission–fusion social structures provide the individual with a social environment that continually fluctuates in both the number and identity of social partners. Thus, dominance relationships and power relations must be evaluated anew after each social change, and as potential social partners change, so do the types of possible social relationship. Many social primates regularly seek recourse in coalitions and elaborate social tactics to achieve specific goals in social life. Living in a fission–fusion group, however, implies using such tactics on a conditional basis. It requires more flexibility in applying learned rules and at the same time opens a larger window for alternative options for many social members. For example, low-ranking individuals may become dominant, depending upon the temporary composition of the parties, and they can exploit such opportunities. Such a situation favours flexible reasoning and a more elaborate understanding of causality in social situations. Here it is the social part of the environment that has the first role in improving intelligence.

Hunting in trees

The second step is the adoption of hunting as a feeding strategy. Hunting has been observed in many primate species, but a closer look at the description shows that in most cases it would better be described as gathering of live vertebrate prey, since there is often no pursuit involved. The only exceptions are some baboons and cebus populations. The baboons of Gilgil used to chase antelopes on the ground due to the exceptional personality of one adult male (Strum 1981). Group of cebus monkeys regularly chase squirrels and Caotis (Rose 1997). In non-primate animals, social carnivores hunt most by pursuing herbivores in a two-dimensional space applying the rule 'the straight line is the best and the quickest', apart for some temporary obstacles such as trees. Raptors also hunt over a two-dimensional space, where the same rule can be applied. Hunting by pursuit forces the hunter to understand the ability and the reaction of another species, the prey, to predict its behaviour, and be able to capture it. Here hunters must be able to perceive through the eyes of the prey.

A new qualitative change occurred when hunting for arboreal monkeys. Chasing after rapidly moving prey *in trees* adds a qualitatively new dimension to the hunt not faced when hunting in two-dimensions. The hunts happen on predefined and clearly limited routes (the branches of the trees) that have different properties and are differently used by the prey and the hunters. This forced the hunters to develop a new dimension in their understanding of causality, namely the understanding of the relations between external and independent objects. Such a level of causality requires an elaborate mental process to visualize the outcomes of these relations (Piaget *et al.* 1948; Visalberghi and Limongelli 1996). Such an understanding is already required for a single chimpanzee hunting for arboreal monkeys, for he has to realize that the prey possesses different physical qualities. Prey runs at different speeds, can jump larger gaps, can use paths in trees that are too dangerous for the hunter, and thus perceives distance, height, danger or escape routes differently. All hunters chasing prey could benefit from such an understanding, but in a tree this knowledge is essential to achieve a capture.

Thus, another aspect of the environment-intelligence hypothesis is stressed here. In the present evolutionary scenario, progresses achieved in the hunting domain may profit from progress achieved in the social domain, and this leads to a synergetic effect. It is difficult to state the order of events in an evolutionary scenario. Hunting has evolved independently in some primates and in carnivores, but since chimpanzees, bonobos, and humans possess a fluid social structure, we propose tentatively fission–fusion to be the first evolutionary event in our scenario within the chimpanzee–human line.

Generalization of tool-use

The third step was the development and generalization of tool-use in the common ancestors between chimpanzees and humans. Tool-use is present in many animal sprcies, as we have discussed in Chapter 9, and this ability persists in primate species thanks to social learning abilities (Byrne 1995). However, what is intriguing is the limited use of tools made by most of these species. We suggest that the impressive increase in the types and frequency of tool-use seen in all chimpanzee populations results from the elaborate notion of causality they acquired during hunting. This new ability allowed chimpanzees to perceive the benefit they could achieve if they were to use a tool in a new situation. In the Taï forest, mangabeys regularly observe chimpanzees cracking nuts with a hammer, and they may eat some kernel remains of these same nuts, but they do not 'see' that they themselves could reach the kernels of other nuts with the help of the same hammer. Similarly, captive baboons, watching conspecifics using a tool to reach a banana out of the cage, do not 'see' that they could use the same tool to reach another banana in the same situation. Once this ability to perceive the benefit of a tool is present, then tools can be applied to many new situations and their use can become very flexible as the opportunities within environments vary. At the same time, tool-use can become more complicated as the understanding of external tertiary causality acquired when pursuing prey in trees can lead to the use of combined and secondary tools or the complex modifications of objects to produce tools. This often requires bi-manual manipulation of the tools and a long apprenticeship as in nut-cracking behaviour. Such complex tool-use requires at the same time an elaborate understanding of the relation between external tertiary objects.

Then, synergetic feedback with hunting behaviour might begin, as the precise coordination of actions between different hunters and the prey becomes possible. Collaborative hunting will be effective because hunters can act as a coordinate team and predict accurately the effect of individual actions on the prey and on the position of others needed to achieve success. Thus, as expected under the environment-intelligence hypothesis, tool-use and hunting behaviour are related in the sense that both require an elaborate notion of causality and a precise ability to predict the results of interactions between external agents. Hunting would have paved the way for more complex tool-uses, which in turn would have allowed the use of more complex hunting tactics. Both are characteristics of chimpanzees and humans, and by relating them this evolutionary scenario links the two puzzles. It is difficult to test whether hunting or complex tool-use appeared first in the chimpanzees evolutionary history, but the present scenario would explain why the ability to use tools, as observed in birds and many mammal species, was not enough to guarantee the emergence of flexible and complex tool-use. Hunting is irregularly distributed in mammals, it is present in social carnivores and in primates, where it is observed only in cebus monkeys and in the chimpanzee–human clade. Here, the specificity of hunting in a three-dimensional space on predefined routes has forced chimpanzees to acquire new abilities that are not required in other hunters. Under our scenario, we should expect cebus monkeys hunting in trees to use more tools than has been described in work so far published (new observations in the field have recently been done, Boysen, personal communication; Kimberley 1998). Thus, better understanding of causality, more precise predictions and understanding of third parties, and a larger capacity of learning would result from the combined influence of hunting and tool-use. These capacities could then also be used in other social domains. We suggest that this is what we observe with the development of empathy, cooperation tactics, and flexible use of sharing rules in wild chimpanzees.

This scenario is compatible with observed differences between chimpanzees and other primates in the domain of social behaviour, hunting behaviour, and tool-use. However, bonobos have not been observed to hunt regularly, nor to use many tools in the wild. This is a problem for our scenario, because they are the closest living relative to the chimpanzees, splitting from the chimpanzees after the divergence between humans and chimpanzees. Bonobos live in fission–fusion groups, and they possess some advanced cognitive abilities like chimpanzees: for example, all great apes and only great apes recognize themselves in mirrors, and only chimpanzees, bonobos, and gorillas imitate sequences of behaviours (Byrne 1995; Povinelli 1996). Since the observations on hunting behaviour, tool-use, and tool-making in primates suggest a dichotomy between a chimpanzee–human clade and other primates, we need to explain why bonobos rarely hunt and rarely use tools.

One possibility is that the bonobos lost some traits inherited from their ancestors when diverging from the chimpanzee line: this may happen, for example, if they lived in an environment where both hunting in trees and tool-use would not be beneficial. Testing this is so far difficult, but we could speculate that the only colobus species living in their habitat, the big *Colobus angolensis*, might be too large for the slender bonobos to attack. It seems more difficult to understand why they should not use tools, as their habitat is similar to the forests of West Africa. According to this explanation, it is not surprising that captive studies have revealed in bonobos some advanced understanding of causality and abilities for cooperating and sharing food that are probably dormant in wild individuals but part of their biological heritage. The other explanation would be that chimpanzees and humans acquired such abilities independently and that the simultaneous presence of hunting, tool-use, and elaborate forms of cooperation in the two species would be fortuitous. We then would need to explain how and why the bonobos nevertheless acquired some of the same cognitive abilities.

The evolutionary scenario assumes changes in the cognitive domain to permit such evolutionary progresses. Data on intelligence in primates are still very fragmentary. To test our scenario, more precise studies are needed on the understanding of causality existing in all primates, especially those concerning the relations with external agents. Similarly more studies are needed on the diverse theories of mind existing in primate species, with an emphasis on their specificity and not on their ability to reproduce a human theory of mind. Our prediction is that chimpanzees and humans have an intelligence that is oriented towards apprehending tertiary external object-relations and a theory of mind that allows a better understanding and prediction of the physical performance of third parties.

Human evolution

Human evolution has always fascinated. The quest for clues and their interpretation, both scientific and religious has been diverse, and it still is a hotly debated topic. Anthropologists have proposed many theories of human evolution, and these have changed as findings accumulated and methods of interpreting ancient human fossils improved. Recent fossil findings have greatly changed our views on this question, and we give here a brief summary of the present state of the human evolution based on fossil findings and related theories, before turning to the contribution that the increasing knowledge on chimpanzee behaviour can add to this debate.

The systematics of the human family

Systematics relies on material from two origins. First, fossils from human ancestors are described and classified to reconstruct our evolutionary tree. Dating the fossils is done indirectly by dating the adjacent soil layers. This procedure can be difficult due to possible displacements of the fossils of such layers before they are discovered. Second, recent studies on the degree of genetic relatedness between living species have produced estimates of the dates of divergence between them, based on different assumptions about the rapidity of genetic divergence. The two methods sometimes disagree, and their confrontation has led to much progress in the understanding of the human evolution.

Early classifications of primates reflected a certain ambivalence, and humans were *a priori* classified apart from other primates (Table 11.3). *Ramapithecus*, a fossil dating some 15 million years ago, was proposed as being the ancestor of the human line, giving

Table 11.3: Two alternative classifications of the great apes and humans. The traditional view was proposed in the early 1950s, with humans quite separate from the others. The modern view is supported by the most recent morphological and molecular analysis proposing a chimpanzee–human clade.

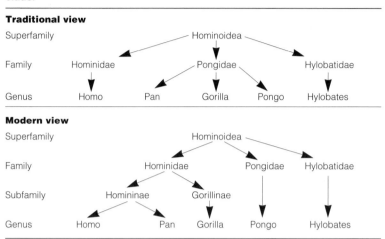

this line an early separation date. One argument presented was that the reconstruction of the upper jaw of *Ramapithecus* possessed the typical human-like V shape. Then, in 1976, molecular techniques were used to estimate the evolutionary relationship between the great apes. As a result, the anthropological world was in shock, for the data clearly proposed a much more recent date for the divergence between humans and chimpanzees (Sarich and Wilson 1967). After much discussion and revaluation, a consensus emerged on a divergence date between human and chimpanzee of around 5 million years ago, and the close relations between chimpanzees and humans became obvious. *Ramapithecus* jaws have been reanalyzed and are now considered to possess a U shape, which is typical for apes, and is now seen as an ancestor of the Asian apes (the orang-utan). Chimpanzee and human are now classified together as *Hominini*, sharing over 98.4% of their nuclear genetic material (Table 11.3). The gorilla shares 95.6% of its nuclear genetic material with the chimpanzee and human, while the orang-utan shares less than 93% with them (Jones *et al*. 1992).

Human evolution through the fossil evidence

Fossil evidence has always been the limiting factor in understanding evolution, for large periods of time are not represented by any fossils. However, a great deal of effort has been invested in the quest for human fossils, and the number discovered in the last decades is very impressive. To take one example, three new species of possible human ancestors of the *Australopithecus* genus were identified in 1994 (Brunet *et al*. 1995; Leakey *et al*. 1995; White *et al*. 1994). Table 11.4 lists some of the most important ones. It gives the impression that by now there is quite a good record of the human lineage available, certainly a better one than for many other animal lineages. The absence of any ancestor for the chimpanzee in this table is not only due to the primary interest in our own history, but to the absence of any such fossil remains in digging sites east of the Rift Valley in East Africa. West of this valley, within the tropical forests, fossilization is probably much more problematic and no systematic digging has been done. Thus, there are no fossils of this period on the chimpanzee lineage. Recently a French team digging in Chad, west of the Rift Valley, has found some proto-human remains (*A. barelghazali*) (Brunet *et al.* 1995) and it is hoped that this will stimulate more work in such regions. The results of such studies will allow us to ascertain whether human ancestors lived only east of the Rift Valley, as has been so far proposed.

Some remarks can be made about the Table 11.4. First, *Ardipithecus ramidus* appears surprisingly close to the proposed date of divergence between human and chimpanzee suggested by genetic studies. The morphological description of this fossil shows it to possess an extremely chimpanzee-like morphology; some anthropologists have even considered the possibility that it might be an ape (Wood 1994). Divergence of opinion about classification is to be expected for a common ancestor between chimpanzee and human and thus, *Ardipithecus ramidus* might be a viable candidate.

Second, the morphological analysis of all *Australopithecus* species reveals that our human ancestors looked strikingly like chimpanzees in many of their features for 3 million years. Descriptions of *Australopithecus afarensis* show variations that overlap regularly with chimpanzee morphology (Johansen and Edey 1981). Similarly, recent measures of cranial capacity show that *Australopithecus africanus* had a range of brain sizes very close to those of the chimpanzees, and that the expansion of brain size may have been quite slow in early hominid evolution (Conroy *et al.* 1998). This has led anthropologists to suggest that our ancestors look more and more like chimpanzees the further back one goes in the fossil record, and that the common ancestor between chimpanzee and human must have looked very much like a chimpanzee (Zihlmann 1996). In

Table 11.4: Chronological list of the most important ancestors of the modern human which have been found. Arrows indicate the length of time they are suspected to have existed. Time is given in million years before our time.

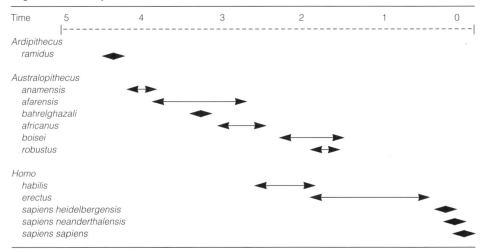

other words, the chimpanzee lineage is the conservative one, and there is not much evidence that it changed much morphologically after its divergence from the human lineage. This may reflect the fact that chimpanzees spent most of their time in the tropical rainforest, a stable environment. Thus, for more than two million years neither chimpanzees nor human ancestors evolved very much, nor diverged substantially.

Third, the morphological evidence suggests that our ancestors did evolve but not in the way we would have predicted. The most obvious changes occurred in the post-cranial features: arms and especially legs and the pelvis. Bipedalism was clearly the first qualitative trait differentiating our ancestors from apes, and this appeared some four million years ago with *Australopithecus afarensis* and possibly *A. anamensis*. However, precise analysis of their locomotory apparatus indicates that their hands, fingers, and torso retained much of the ape's adaptation for climbing (Stern and Susman 1983; Susman *et al*. 1984, Susman 1988; McHenry and Berger 1996, 1998). At the same time, the brain capacities of the *Australopithecus* sp. remained stable for as long as they lived, at around some 350 to 500 cm^3, which is not much of an increase over the chimpanzee brain size of about 300 to 400 cm^3. Brain structure for all *Australopithecus* sp. also remained very chimpanzee-like, a clear shift occurring only with the emergence of *Homo*, when average brain size increased to about 600 cm^3 (Falk 1991).

Fourth, while the genus *Homo* has a long history, having appeared about 2.6 million years ago, modern humans (*Homo sapiens sapiens*) appeared very recently, some 0.1 million years ago (Bräuer *et al*. 1997). However, before its appearance, our direct ancestors had produced some tremendous changes that are visible in the much more developed lithic technology to be seen in the Mousterian culture. *Homo erectus*, for more than one million years, had been making the same tools, of the Acheulean type, which were a clear improvement over tools produced by *Homo habilis*, but remained similar for a startlingly long period of time. Thus, looking at the history of our lineage on an evolutionary scale, the cultural products of our ancestors were very stable and simple for most of our history.

We can conclude that after the divergence from the common ancestor we shared with the chimpanzee, our ancestors remained chimpanzee-like for about half of the hominization process. Once *Homo* appeared, brain size increased, and a new and relatively crude stone-tool culture was observed. *Homo habilis* still had less than half our brain size. *Homo erectus* conquered the world from Africa, but its cultural products were limited and changed very little. Hominization was for the most part a slow and stable process. A clear acceleration occurred with the appearance of Neanderthals and modern humans some 100 000 years ago. If the morphological evolution is disappointing, we may still hope that on the behavioural level our ancestors rapidly became distinctly human. However, as behaviour is difficult to reconstruct from fossils, there is a choice of many theories to account for modern humans.

Scenarios of human evolution

The basic savanna model, proposed first by Dart (1925), suggests that our ancestors were forced to live in a drier and more open habitat east of the Rift Valley, and that this accelerated the evolution of so-called human characteristics. They had to adapt to these new conditions, and for this reason started to move bipedally, use tools, hunt for meat, share food, and adopt home bases. Elements of this model have remained in many more recent theories (Leakey 1980; Isaac 1978), for it seems supported by the fact that all fossils of early hominids were found east or south of the Rift Valley and most living African apes live west of the Rift Valley. The theory has only recently been challenged by the discovery of a new hominid fossil in Chad, far west of the Rift Valley (Brunet *et al.* 1995). New evidence about the environment that prevailed east of the Rift Valley some four to three million years ago indicates that it was in fact less dry than previously thought, but was nevertheless much more open and dry than the tropical rainforest that prevailed west of the Rift Valley (Rayner *et al.* 1993).

The details of how the savanna model could work is at the origin of a wide ongoing debate. In the 1960s, the hunting hypothesis prevailed under the 'Man the hunter' heading (Leakey 1961). In the 1970s, the food-sharing hypothesis (Isaac 1978) dominated the debate. In the 1980s, the hunting abilities of our ancestors were challenged and the scavenging hypothesis (Shipmann 1986) and the female-gatherer hypothesis (Tanner and Zihlmann 1976) dominated. It remains extremely difficult to test such hypotheses directly, as all there is to rely on to reconstruct behaviour are traces on fossil remains at early hominid sites. To use these to answer whether our ancestors hunted or scavenged is a hazardous game with little reliable evidence. Other hypotheses have been proposed, like the stone cache hypothesis (Potts 1984), suggesting that our ancestors kept their stone tools in regular caches. To account for the early evolution of bipedalism, many hypotheses have been proposed speculating on the various possible advantages of this way of locomotion: freeing the hands, increasing predator detection, and transporting tools, food, and offspring. A consensus has not been reached, although everyone seems to agree that our ancestors lived in an open environment, in contrast to the more forested regions where apes still remain.

By 2.5 millions years ago, the appearance of the first clearly identifiable stone tools coincide closely with the earliest evidence of the genus *Homo*. By then, stone tools occurred both as isolated scatters and in association with concentrations of animal bones. Thanks to cut-marks on bone fossils it could be shown that some of the tools were used in relation to the meat. To progress with questions about the behaviour of our ancestors, we need to supplement the fossil evidences with information from other areas. Two have been proposed: first, use information from modern hunter–gatherers, as this way of life is supposed to be the one closest to that of our ancestors. This is certainly an interesting avenue, but has the major problem of providing information on modern humans that we know to be morphologically very different from our ancestors. The second is to supplement fossil information with behavioural observations from our closest living relatives, the chimpanzees and the bonobos. Like the first alternative, it suffers from a weakness, the fact that the observations come from another species. However, the discussion of the early stages of human evolution has made it clear that for these periods of hominization, chimpanzees were more similar to our ancestors than is any living human being. Therefore, we think that relying on our increasing knowledge of the chimpanzee and bonobo can shed some light on the questions of how our ancestors behaved. The central question is then to discover at which stage they became too different to be a relevant model. This depends upon the priority one gives to the various factors: brain size, stone tools, culture type, brain-growth pattern, and so forth.

The chimpanzee model in human evolution

The earliest human ancestors

Theoretical models of hominization have specified which behaviour patterns are human characteristics. Behaviour patterns that were considered classically to be a unique part of the human stock, like tool-use, tool-making, cooperation, and food sharing have been suggested to be also characteristic of the chimpanzee (Parker and Gibson 1977; McGrew 1992; Boesch and Boesch 1989, 1990; Boesch-Achermann and Boesch 1994). New evidence coming from different fields further strengthens this proposition. Morphological similarities, genetic relatedness, and similarities in behaviour all suggest the reality of a chimpanzee–human clade, and thus a long common evolutionary history for these two species.

We expect *Australopithecus* populations to have reached at least the same level of sophistication as chimpanzees. In other words, *Ardiphithecus ramidus* and *Australopithecus anamensis*, *afarensis* and *africanus* regularly hunted and used tools. Only if one assumes that the environmental changes that occurred east of the Rift Valley made these behaviour patterns useless or difficult to perform would this proposition be unrealistic. Nobody would suggest that, with a brain size equal to the chimpanzee and a similar brain organization, *Australopithecus* was less intelligent than a chimpanzee. Could the environment prevailing in East Africa four to three million years ago prevent our ancestors from using tools and hunting small mammals? Chimpanzees living in open environments like the Gombe National Park, in Tanzania, or an even drier environment like the Niokolo-Koba National Park in Senegal (McGrew *et al.* 1978, 1979) use tools and hunt small and medium-size mammals. Because the new evaluation of environmental conditions some three million years ago suggests a similar structure and a similar amount of water, these observations are directly relevant to the question. Lastly, evaluations of the *Australopithecus'* locomotory activity suggest that they still retained some climbing abilities and were therefore able to hunt for monkeys. They can be presumed to have hunted and used tools to at least the same extent as chimpanzees do. The main reason why anthropologists have not considered this possibility is that they have based their arguments on findings of stone tools. As long as tools did not appear in relation with meat, such an idea was not regarded as a valid hypothesis. The chimpanzee model shows convincingly that tools are used in many contexts other than in relation to meat (Chapter 9), and this is also the case in contemporary hunter–gatherers. Because we now know that chimpanzees mainly use wooden tools, the priority placed traditionally on stone tools to conclude the presence of an activity seems too restrictive. *Australopithecus* can be supposed to have used tools to enrich their diet with insects and possibly also in social contexts. Nut-cracking might have become an important part of their foraging activity, with for example nuts like the mongongo nuts cracked by !Kung Bushmen (Lee 1973). Stones used to crack nuts do not necessarily produce traces obvious enough to be recognized by anthropologists millions of years later, and such an activity might have remained unnoticed. Similarly, *Australopithecus* was in a position to hunt for monkeys in the woodlands of its home ranges, where colobus monkeys were shown to have lived (Leakey *et al.* 1995). New results of an analysis of the enamel layers of the teeth of *Australopithecus afarensis* suggest that they might have eaten meat regularly (Sponheimer and Lee-Thorp 1999). No tools were necessary to hunt such prey, and again archaeological traces of such an activity would simply be missing. Thus, the application of the chimpanzee model suggests an important rehabilitation of our *Australopithecus* ancestors and a recognition of their hunting and tool-using abilities.

Why should we prefer a chimpanzee model to the bonobo model proposed by Zihlmann (1981), which placed the emphasis on gathering activities and a low sexual dimorphism? One reason is that chimpanzee and human are the only primates regularly using complex toolkits and hunting small mammals for meat. Favouring a bonobo-model would require that tool-use and hunting evolved twice independently in the chimpanzee and the human lineage, whereas the chimpanzee model would require only one such event. The second reason is that with recent observations of more chimpanzee populations, the new version of the chimpanzee model as described above differs from earlier ones. Previously, the chimpanzee model was presented as a rigid male-dominated society where male activities such as hunting, inter-group aggression, and coalitions prevailed (Wrangham 1986; Foley 1989). Now, however, the chimpanzee model emphasizes flexibility in social interactions, hunting, and tool-use, within a fluid and bi-sexual society where females have a flexible position depending upon prevailing conditions, but allowing them to be active in social interactions, in food-sharing, in choosing male partners, and having a leading role in tool-using. In some situations, chimpanzee society is then very similar to the bonobo society. We believe that this flexible model is one that has advantages when we think about human evolution. At present, seven species of *Australopithecus* have been identified and only a flexible model can account for such a variety of species, which were able to successfully exploit different environments around 4 to 2.5 million years ago.

Australopithecine individuals were not full-blown chimpanzees, since we suspect that ancestral chimpanzees still differed in some respects from modern chimpanzees. At a minimum, the fission–fusion social structure was present with some simple tool-using and hunting behaviours. Therefore, in the course of millions of years with australopithecines as the only representatives of our lineage, they followed a behavioural evolution that paralleled the morphological one. That is they made only limited, small changes not differentiating notably from those made by the chimpanzee ancestors. Bi-manual tool use as well as secondary tools were discovered and hunting became progressively more sophisticated in the sense that cooperation between individuals became progressively more important.

Evolution of the *Homo* genus

Around 2.6 million years ago something happened that explained the emergence of *Homo habilis* and the synchronous appearance of the first recognized stone-tool culture (the Oldowan industry or the pebble-tool culture) (Semaw *et al.* 1997). In our perspective, stone tools that had been used for ages by australopithecines in foraging activities were suddenly associated with meat. It is probable that here an evolutionary event occurred that chimpanzee ancestors were never confronted with. In the tropical rainforest of Africa, the humidity is always close to 100% and any dead mammal decomposes extremely quickly. We observed the meat of poached elephants to be completely eaten by fly maggots within one week. In the drier environment of the savanna, dead bodies last much longer: leopards routinely keep their quarries for days before eating them. Thus, a window became available in dry environments that did not exist in wet forests: the exploitation of the meat of large dead mammals (Blumenshine and Cavallo 1992).

However, exploiting this new food source required a kind of tool that had not been required before: sharp-edged tools to cut hide and meat. Such an evolutionary innovation requires much time to achieve; it requires new thinking about the shaping of material, new actions to be performed on objects, and long practice. While tools can be used even by birds, the innovation of regularly used flexible tools was achieved only by chimpanzees. Sharp-edged stone tools were invented by *Homo habilis* some three million years after our common ancestor with the chimpanzee used tools systematically. Scavenging complemented hunting and tool-use in the *Homo* ancestors for the first time and gave them a decisive evolutionary advantage over *Australopithecus*. Such a stone culture persisted for about one million years.

This evolutionary scenario is a dynamic one, for it proposes that different aspects have different roles at different periods in this process. The food-sharing hypothesis (Isaac 1978), with the important role attributed to exchange, reciprocity, subsistence, and technology would apply neatly to our common ancestor, to the ancestral chimpanzee and to the australopithecine period. For some time, the evolution of chimpanzees and hominids was very similar, and environmental differences explain the small difference revealed by the morphology. Hominization and chimpanzee evolution ran in parallel. Hunting was, however, always part of it, and complex use of hunting tactics requires complex rules of meat sharing. Thus, hunting implies food sharing. The 'food-sharing' and 'Man the hunter' hypotheses (Leakey 1961) would act in a complementary way. Similarly, the 'woman the gatherer' hypothesis (Zihlmann 1981) would apply as well, because in our scenario hunting and tool-use interacted. Females played a predominant role in using and developing tools, because their constant need to think about the interests of their offspring and themselves predisposed them to think in terms of third parties, which is needed for complex tool-use. Once the understanding of causality became sophisticated enough, and once the environment became dry enough for large dead mammals to be exploitable, the 'scavenging' (Shipman 1986) and the 'stone cache' hypothesis (Potts 1984) would apply, and tool-using techniques became further elaborated.

Homo was then still similar to Australopithecus, and the brain-size increase is mainly explained by an increase in learning and memory abilities that we expect to develop along with the use of more sophisticated tools. An analysis of the stone artefacts of the Oldowan industry produced by *Homo habilis* shows that the basic cognitive abilities required to make them, symmetry, transitivity, and rotation abilities (Wynn 1981), are within the range of abilities of modern chimpanzees. However, the learning of such new techniques was certainly quite demanding and time consuming, and this produced strong pressure to increase the storage capacities of the brain. Learning abilities and memory increased further with the appearance of the Acheulian industry and *Homo erectus*. This new industry represents a sophistication similar to the Oldowan industry (Wynn 1979), and not a radical change. With the emergence of modern humans some 100 000 years ago, some qualitatively important changes are seen that can no longer be attributed to learning and memory. Most probably the emergence of language permitted such striking qualitative changes.

Our recent discovery of the diversity of wild chimpanzee populations thus allows to propose a new evolutionary scenario for the early stages of hominization. This scenario takes into account the evidence coming from primatology, that is the unique importance of fission–fusion grouping patterns, hunting, and tool-use in the chimpanzee–human clade, and from anthropology, that is the very chimpanzee-like morphology of most australopithecines. It suggests a much more important and earlier role for both hunting and tool-use in our evolutionary history. Scavenging would at a later stage favour the evolution of more elaborate forms of tool-use. It was only once modern humans had appeared that qualitative, radical changes were observed in our lineage, and this is possibly due to the appearance of language.

Recent information available on a larger sample of chimpanzee populations provides us with a much more complicated image of what chimpanzees are. Behavioural diversity seems to be the general rule, making generalization about the chimpanzee prototype a complex enterprise, since most behaviour patterns differ in important ways within populations. We should remain very cautious when making comparisons between species, for our knowledge of population variability in chimpanzees is heavily restricted by the small number of populations under study.

Nevertheless, we suggest that three characteristics are typical of all chimpanzee populations and distinguish this species from all other primates, except humans:

- A fission–fusion social system,
- Hunting behaviour in trees,
- Flexible and numerous types of tool uses.

In the evolutionary scenario we propose, each of these characteristic selects for specific higher cognitive abilities that are all related to better understanding of causality and of third-party intention and competence. At a second stage, each characteristic profits from the increase in intelligence needed to perform the other, and they are further elaborated in a synergetic way. This evolution, that could occur in the absence of any environmental changes, was a slow process and might well have started before the divergence between the chimpanzee and human lineage.

After the divergence from the common ancestors, australopithecines followed an evolutionary route parallel to chimpanzees but in a different environment. This led to the limited documented morphological and inferred behavioural changes. Australopithecine species were tool-users and active hunters of small mammals, behaviours that anthropologists have proposed to appear in the human lineage some two million years later. The strong association of stone artefacts with skeleton remains of large mammals with the emergence of the *Homo* genus suggests that scavenging had started to complement the diet of our ancestors and this required new types of tools.

We suggest a chimpanzee–human clade based on behavioural similarities found mainly in the social, technical, and cognitive domains. This corroborates similar conclusions coming from many morphological comparisons as well as from most recent molecular analysis. In the last decade, our perception of the abilities of this species, our closest living relative on earth, has changed in an impressive way thanks to the new observations of a few more chimpanzee populations. We are, however, convinced that our knowledge is still fragmentary and similar changes are to come once observations of new populations are reported. Thus, the limiting factor in our quest for human and chimpanzee identity is their survival. The fight for the survival of the chimpanzees in nature should more than ever become an absolute priority.

Appendix

Table A.1: List of all chimpanzees that we identified in the Taï community. As a general convention, female names are in capital letters whereas male names have only the first letter in capital. Birth dates followed by an asterisk are estimates. Infants that could not be sexed were not named and are marked by an x, whereas those sexed but not named are marked by the initial of the mother followed by x.

Name of mother (Birth/death date)	Name of infant	Birth date	Death date
AGATHE (1977*/11–92)	APHRO	29–11–91	11–92
AURORE (1969*/6–87)	AZUR	1982	6–87
AWA (1966*/10–84	Ali	1979*	11–92
BELLE (1976)	BAGHERRA	18–9–94	3–03–97
BICHE (1963*/12–83)	BELLE	1976*	
	BX	1983	6–84
BIJOU (1975*/31–5–94)	Bambou	6–2–89	25–3–91
	Baloo	7–4–92	3–6–94
CANNELLE (1970*/12–83)	x	?	1983
CHANEL (1942*/10–88)	COCO	1975*	emigrate 7–87
	CHOUCHOU	3–81	
	CX	7–1988*	23–10–88
CASTOR (1976*/)	Pollux	4–90	11–11–90
	Congo	28–1–92	4–3–93
	CACAO	14–3–94	
DILLY (1978*/)	DORRY	24–11–91	
ELLA (1956*/5–90)	Kendo	1969*	11–94
	Fitz	1975*	11–94
	Gérald	1983	11–92
	Louis	8–88	7–89

FANNY (1969*/11–94)	Marius	4–82	
	MANON	10–9–87	15–11–92
	FOUTOU	25–9–92	11–94
FOSSEY (1979*/)	DIANE	13–2–92	23–2–92
	FEDORA	12–11–93	
GALA (1964*/10–94)	GX	1977*	?
	x	6–83	10–4–84
	GX	2–85	1–87
GAULOISE (1971*/8–88)	Gallus	7–84	8–88
GITANE (1949*/10–94)	Gipsy	10–80	10–94
	GOYAVE	7–85	
	GUINI	23–06–93	07–07–93
GOMA 1973*	GOSHU	10–3–86	19–8–92
	Gargantua	21–9–91	
	GISELE	18–5–96	
HERA 1965*	Haschich	4–78	11–87
	Eros	1982	8–88
	Homère	1987	20–2–90
	Hector	10–12–90	
	HELENE	4–8–95	6–96
JOJO (1964*/2–84)	x	1977*	1984
	x	1983	1984
KIRI (1969*/7–2–92)	Kummer	8–82	6–87
	KANA	5–6–87	
LOLA (1953*/9–88)	Rousseau	1966*	8–93
	Lorenz	5–74*	7–87
	LOLITA	5–79	7–87
	LEILA	8–86	9–88
LOUKOUM (1972*)	LYCHEE	1–3–86	22–11–92
	Lefkas	7–10–91	
MALIBU (1971*/7–90)	x	15–11–84	19–1–85
	FIDJI	1–1–88	7–90
MARLENE (1971*/2–89)	Brando	7–84	1–91
MOMO (1969*/8–88)	Molière	2–82	2–89
MYSTERE (1975*)	Magic	14–12–88	10–6–89
	MOGNIE	31–7–90	
	Mozart	11–95	
NANA (1970*/7–87)	x		
	NABU	5–3–85	22–8–87
	x	5–4–86	16–6–86
	x	25–4–87	7–87
NOVA (1966*/3-90)	Négus	1982	7–88
	NX	3–6–89	7–89

ONDINE 1954*/19–11–92)	CLEO	1975*	emigrated 12–86
	OX	1979	7–83
	OREE	10–83	20–10–88
	SIRENE	11–87	
	Ovide	11–92	19–11–92
PERLA (1976*)	Papot	19–11–89	1–08–97
	PANDORA	2–9–95	
POKOU (1956*/11–89)	Darwin	1969*	12–93
	Px	1977*	12–83
	PITCHOU	7–82	7–87
	PX	7–2–88	11–89
POUPEE (1973*/11–94)	x	5–6–85	27–6–85
	PX	5–86	6–1–87
	Popeye	26–11–87	4–12–89
	PIMENT	12–2–91	11–94
RICCI 1963*	NINETTE	1976*	emigrated 6–85
	NINA	1982	6–87
	Nino	2–3–88	
	ROXANE	8–94	
SALOME (1958*/2–9–90)	Snoopy	1970*	8–88
	Sartre	1980*	11–94
	SIMONE	4–2–88	26–8–90
SAPHIR (1970*/2–88)	RUBIS	7–83	8–88
TOSCA (1956*/1–89)	SANDRA	1975*	emigrated 86
	TINA	1979*	6–2–89
	Tarzan	2–84	1–91
VENUS 1978*	VANILLE	30–5–91	
VOLGA (1968*–1989)	VX	1982	?
X	Ulysse	1965*	12–90
	Clyde	1971*	1–84
	BONNIE	1977*	emigrated 7–90
XERES (1970*/7–93)	x	2–1983	7–5–83
	XINDRA	8–84	emigrated 93
	Don Quichotte	20–11–90	7–93
ZOE (1965*/12–89)	ZIRCE	1978*	emigrated 11–87
	Zorba	8–83	8–88
	x	1989	12–89

Individuals with no known relatives:

Name	Birth date	Death date
Males		
Balzac	1960*	2–84
Brutus	1951*	
Falstaff	1948*	11–87

Individuals with no known relatives (*cont.*):

Name	Birth date	Death date
Le Chinois	1966*	2–84
Macho	1964*	
Pistache	1958*	2–84
Schubert	1962*	12–86
Wotan	1950*	2–84
Robi	1970*	2–84
Ténor	1975*	2–84
Females		
ARIANE	1982*	11–94
IMI	1976*	26–11–89
NARCISSE	1984*	
VERA	1971*	1985
ZERLINA	1982*	10–94

Table A.2: List of the fauna living in the territory of the chimpanzees in the Taï forest with some indications about their frequency based on our own observations. (1 = seen once so far since 1979; 2 = seen 1–2 times per year; 3 = seen many times per year; 4 = seen almost every day spent in the forest). For flying squirrels, the frequency is given as number of days for a period of 53 days [Gartshore *et al.* 1995].

Latin name	English name	Frequency
Primates		
Galago demidovii	Dwarf galago	3
Perodicticus potto	Bosman's Potto	2
Cercopithecus petaurista	Lesser spot-nosed monkey	4
Cercopithecus diana	Diana monkey	4
Cercopithecus mona	Mona monkey	4
Cercopithecus nictitans	White-spotted nose monkey	3
Cercocebus atys	Sooty mangabey	4
Colobus badius badius	Red colobus	4
Colobus polykomos	Black and white colobus	4
Procolobus verus	Olive colobus	3
Pan troglodytes verus	Chimpanzee	4[1]
Carnivora		
Mellivora capensis	Honey badger	3
Lutra maculicollis	White-throated otter	2
Aonyx capensis	Cape clawless otter	2
Nandinia binotata	African palm civet	1
Poiana richardsoni	African Linsang	1
Genetta pardina	Pardine genet	2
Viverra civetta	Civet cat	2
Crossarchus obscurus	Long-nose mongoose (Cusimance)	3
Atilax paludinosus	River mongoose	*
Liberiictis kuhni	Liberian mongoose	1
Felis aurata	Golden cat	2
Panthera pardus	Leopard	2
Artiodactyla		
Choeropsis liberiensis	Pygmy hippopotamus	2
Potamochoerus porcus	Bushpig	3
Hylochoerus meinertzhageni	Giant forest hog	2
Hyemoschus aquaticus	Water chevrotain	*
Boocerus euryceros	Bongo	2
Neotragus pygmeaus	Royal antelope	2

Cephalophus silvicultor	Yellow-backed duiker	1
Cephalophus jentinki	Jentink duiker	2
Cephalophus niger	Black duiker	*
Cephalophus zebra	Zebra duiker	3
Cephalophus dorsalis	Bay duiker	4
Cephalophus monticola	Blue duiker	4
Cephalophus ogilbyi	Ogilby duiker	3
Syncerus caffer nanus	African buffalo	3[2]

Pholidota

| *Manis gigantea* | Giant pangolin | 1 |
| *Manis tetradactyla* | Long-tailed pangolin | 1 |

Proboscidea

| *Loxodonta africana* | Forest elephant | 3[2] |

Hyracoidae

| *Dendrohyrax dorsalis* | Tree hyrax | 2 |

Orycteropodidae

| *Orycteropus afer* | Ant bear (Aardvark) | 1 |

Rodentia

Funisciurus pyrrhopus	Red-footed squirrel	3
Funisciurus lemniscatus	Four-banded squirrel	3
Protoxerus stangeri	Giant forest squirrel	4
Epixerus ebii	Temminck's giant squirrel	3
Anomalurus derbianus	Lord Deby's flying squirrel	2
Anomalurus peli	Pel's flying squirrel	3
Anomalurus beecrofti	Beecroft's flying squirrel	3
Anomalurus pusillus	Little flying squirrel	2
Cricetomuys emini	Giant Gambian rat	3
Atherurus africanus	African brush-tailed porcupine	3

Chiroptera

Hipposideros sp.	Leaf-nosed bat	2/53
Hipposideros commersoni	Commerson's leaf-nosed bat	5/53
Rhinolophus sp.	Horseshoe-faced bat	1/53
Epomops buettikoferi	Buettikofer's fruit bat	4/53
Hypsignathus monstrosus	Hammer-headed fruit bat	2/53
Myonycteris torquata	Little collared fruit bat	1/53
Eptesicus sp.	Serotine	1/53
Nycteris sp.	Slit-faced bat	1/53

[1] Chimpanzees, being the centre of our study, were seen much more frequently than normal.

[2] Species that were regularly observed at the start of our study in 1979 but that had in the meantime been so heavily poached that they are not, or most rarely, seen any more.

* Species known in region close to the territory of the chimpanzees but never seen within it.

References

Aké Assi, L. and Pfeffer, P. (1975). *Parc National de Taï, Inventaire de la flore et de la faune*. BDPA, Paris.

Alberts, S. and Altmann, J. (1995). Balancing costs and opportunities: dispersal in male baboons. *American Naturalist*, **145**, 279–306.

Alcock, J. (1972). The evolution of the use of tools by feeding animals. *Evolution*, **26**, 464–73.

Alexander, R. (1974). The evolution of social behavior. *Annual Review of Ecology and Systematics*, **5**, 325–83.

Alexander, R. (1979). *Darwinism and human affairs*. Washington Press, Seattle.

Alexander, R. and Noonan, K. (1979). Concealment of ovulation, parental care, and human social evolution. In *Evolutionary biology and human social behavior: an anthropological approach* (ed. N. Chagnon, and W. Irons), pp. 436–53. Duxbury Press, North Scituate.

Alp, R. (1997). 'Stepping-sticks' and 'seat-sticks': New types of tools used by wild chimpanzees (*Pan troglodytes*) in Sierra Leone. *American Journal of Primatology*, **41**, 45–52.

Altmann, J. (1974). Observational study of behaviour: sampling methods. *Behaviour*, **49**, 227–65.

Altmann, J. (1980). *Baboon mothers and infants*. Harvard University Press, Cambridge, Massachussetts.

Altmann, S.A. and Altmann, J. (1970). *Baboon ecology: African field research*. University of Chicago Press, Chicago.

Andersson, M. (1994). *Sexual selection*. Princeton University Press, Princeton.

Andrews, P. (1989). Palaeoecology of Laetoli. *Journal of Human Evolution*, **18**, 173–81.

Andrews, P. (1992). Evolution and environment in the Hominidae. *Nature*, **360**, 641–6.

Aubreville, A. (1959). *La flore forestière de Côte d'Ivoire* (2nd edition). C.T.F.T., Nogent-sur-Marne, France.

Axelrod, R. and Dion, D. (1988). The further evolution of cooperation. *Science*, **242**, 1385–90.

Axelrod, R. and Hamilton, W.D. (1981). The evolution of cooperation. *Science*, **211**, 1390–6.

Badrian, A. and Badrian, N. (1984). Social organisation of Pan paniscus in the Lomako Forest, Zaire. In *The pygmy chimpanzee* (ed. R. Sussman), pp. 325–46. Plenum Press, New York.

Bailey, K.G. (1987). *Human paleopsychology: applications to aggression and pathological processes*. L. Erlbaum, London.

Bailey, R., Head, G., Jenike, M., Owen, B., Rechtmann, R., and Zechenter, E. (1989). Hunting and gathering in tropical rain forest: is it possible? *American Anthropologist*, **91**(1), 59–82.

Baker, R. and Bellis, M. (1993). Human sperm competition: ejaculate manipulation by females and a function for the female orgasm. *Animal Behaviour*, **46**, 887–909.

Bates, E., Benigni, L., Bretherton, I., Camaioni, L., and Volterra, V. (1979). *The emergence of symbols*. Academic Press, New York.

Bauer, H.R. (1979). Agonistic and grooming behavior in the reunion context of Gombe Stream chimpanzees. In *The great apes* (ed. D.A. Hamburg and E. McCown), pp. 395–404. Benjamin/Cummings, Menlo Park.

Beatty, H. (1951). A note on the behavior of the chimpanzee. *Journal of Mammology*, **32**, 118.

Beck, B. (1973). Observational learning of tool use by captive Guinea baboons (Papio papio). *American Journal of Physical Anthropology*, **38**, 579–82.

Beck, B. (1980). *Animal tool behavior*. Garland Press, New York.

Bednarz, J. (1988). Cooperative hunting, in Harris' hawks (*Parabuteo unicinctus*). *Science*, **239**, 1525–7.

Begon, M., Harper, J.L., and Townsend C.R. (1990). *Ecology: individuals, populations and communities* (2nd edn). Blackwell Scientific Publications, Oxford.

Begon, M., Mortimer, M., and Thompson, D. (1996). Population ecology: an *unified study of Animals and plants* (2nd Edn). Blackwell Science, Oxford.

Bercovitch, F.B. (1988). Coalitions, cooperation and reproductive tactics among adult male baboons. *Animal Behaviour*, **36**, 1198–1209.

Bercovitch, F.B. (1995). Female cooperation, consortship maintenance, and male mating success in savanna baboons. *Animal Bahviour*, **50**, 137–49.

Bernstein, I.S. (1988). Metaphor, cognitive belief and science. *Behavioral and Brain Sciences*, **11**, 247–8.

Birkhead, T. (1995). Sperm competition: evolutionary causes and consequences. *Reproduction and fertility development*, **7**, 755–75.

Birkhead, T. and Møller, P. (1992). *Sperm competition in birds: evolutionary causes and consequences*. Academic Press, London.

Blumenschine, R.J. and Cavallo, J.A. (1992). Scavenging and human evolution. *Scientific American*, **267**, 70–6.

Blurton-Jones, N., Smith, L., O'Connell, J., Hawkes, K. and Kamuzora, C. (1992). Demography of the Hadza, an increasing and high density population of savanna foragers. *American Journal of Physical Anthropology*, **89**, 159–81.

Boehm, C. (1992). Segmentary warfare and the management of conflict: comparison of East African chimpanzees and patrilineal–patrilocal humans. *In Coalitions and alliances in humans and other animals* (eds. A., Harcourt, and F., De Waal), pp. 137–73. Oxford University Press, Oxford.

Boesch, C. (1978). Nouvelles observations sur les chimpanzés de la forêt de Taï (Côte d'Ivoire). *Terre et Vie*, **32**, 195–201.

Boesch, C. (1991*a*). Teaching in wild chimpanzees *Animal Behaviour*, **41**(3): 530–2.

Boesch, C. (1991*b*). Symbolic communication in wild chimpanzees? *Human Evolution* **6**(1): 81–90.

Boesch, C. (1991*c*). The effect of leopard predation on grouping patterns in forest chimpanzees. *Behaviour*, **117**(3–4): 220–42.

Boesch, C. (1992). New elements of a theory of mind in wild chimpanzees. *Behavioral and Brain Sciences*, **15**(1), 149–50.

Boesch, C. (1993). Toward a new image of culture in chimpanzees. *Behavioral and Brain Sciences*, **16**(3), 514–15.

Boesch, C. (1994*a*). Chimpanzees — red colobus: a predator-prey system. *Animal Behaviour*, **47**, 1135–48.

Boesch, C. (1994*b*). Cooperative hunting in wild chimpanzees. *Animal Behaviour*, **48**, 653–67.

Boesch, C. (1995). Innovation in wild chimpanzees. *International Journal of Primatology*, **16**(1), 1–16.

Boesch, C. (1996*a*). Three approaches for assessing chimpanzee culture. In *Reaching into thought* (ed. A. Russon, K. Bard, and S. Parker), pp. 404–29. Cambridge University Press, Cambridge.

Boesch, C. (1996*b*). The emergence of cultures among wild chimpanzees. In *Evolution of social behaviour patterns in primates and man* (ed. W. Runciman, J. Maynard-Smith and R. Dunbar), pp. 251–68. British Academy, London.

Boesch, C. (1996*c*). Social grouping in Taï chimpanzees. In *Great apes societies* (ed. W. McGrew, T. Nishida, and L. Marchant), pp. 101–13. Cambridge University Press, Cambridge.

Boesch, C. (1997). Evidence for dominant mothers investing more in sons among wild chimpanzees. *Animal Behaviour,* **54**, 811–15.

Boesch, C. and Boesch, H. (1981). Sex differences in the use of natural hammers by wild chimpanzees: a preliminary report. *Journal of Human Evolution*, **10**, 585–93.

Boesch, C. and Boesch, H. (1983). Optimization of nut-cracking with natural hammers by wild chimpanzees. *Behaviour*, **83**, 265–86.

Boesch, C. and Boesch, H. (1984*a*). Mental map in wild chimpanzees: an analysis of hammer transports for nut cracking. *Primates*, **25**, 160–70.

Boesch, C. and Boesch, H. (1984*b*). Possible causes of sex differences in the use of natural hammers by wild chimpanzees. *Journal of Human Evolution*, **13**, 415–40.

Boesch, C. and Boesch, H.(1989). Hunting behavior of wild chimpanzees in the Taï National Park. *American Journal of Physical Anthropology*, **78**, 547–73.

Boesch, C. and Boesch, H. (1990). Tool use and tool making in wild chimpanzees. *Folia primatologica*, **54**, 86–99.

Boesch, C., Marchesi, P., Marchesi, N., Fruth, B., and Joulian, F. (1994). Is nut cracking in wild chimpanzees a cultural behaviour? *Journal of Human Evolution*, **26**, 325–38.

Boesch, C. and Tomasello, M. (1998). Chimpanzee and human cultures. *Current Anthropology*, **39**(5), 591–614.

Boesch-Achermann, H. and Boesch, C. (1994). Hominization in the rainforest: the chimpanzee's piece of the puzzle. *Evolutionary Anthropology*, **3**(1), 9–16.

Bräuer, G., Yokoyama, Y., Falguères, C. and Mbua, E. (1997). Modern human origins backdated. *Nature*, **386**, 337–8.

Brownell, C. and Carriger, M. (1990). Changes in cooperation and self-other differentiation during the second year. *Child Development*, **61**, 1164–74.

Brunet, M., Beauvilain, A., Coppens, Y., Heintz, E., Moutaye, H. and Pilbeam, D. (1995). The first australopithecine 2500 kilometres west of the Rift Valley (Chad). *Nature*, **378**, 273–4.

Bshary, R. (1995). 'Rote Stummelaffen, Colobus badius, und Dianameerkatzen, Cercopithecus diana, im Taï Nationalpark, Elfenbeinküste: Wozu assoziieren sie?' PhD thesis, Ludwig-Maximilien Universität, München.

Buffon, de G. (1792). *Histoire naturelle des quadrupèdes*, Vol. 9. Nouvelle Société Typographique, Berne.

Busse, C. (1977). Chimpanzee predation as a possible factor in the evolution of red colobus monkey social organization. *Evolution*, **31**, 907–11.

Busse, C. (1978). Do chimpanzees hunt cooperatively? *American Naturalist*, **112**, 767–70.

Bygott, J. (1972). Cannibalism among wild chimpanzees. *Nature*, **238**, 410–11.

Bygott, J. (1974). Agonistic behaviour in Gombe chimpanzees. PhD. Thesis. University of Cambridge.

Bygott, J. (1979). Agonistic behavior, dominance, and social structure in wild chimpanzees of the Gombe national Park. In *The great apes* (ed. D.A. Hamburg and E. McCown), pp. 405–28. Benjamin/Cummings, Menlo Park.

Byrne, R. (1995) *The thinking ape*. Oxford University Press, Oxford.

Byrne, R. (1997). The technical intelligence hypothesis: an additional evolutionary stimulus to intelligence? In *Machiavellian intelligence II: Extensions and evaluations* (ed. A. Whiten and W. Byrne), pp. 289–311. Cambridge University Press, Cambridge.

Call, J. and Tomasello, M. (1996). The effect of humans on the cognitive development of apes. In: *Reaching into thought: the minds of the great apes* (ed. A. Russon, K. Bard and S. Parker), pp. 371–403. Cambridge University Press, Cambridge.

Caraco, T. and Wolf, L.L. (1975). Ecological determinants of group sizes of foraging lions. *American Naturalist*, **109**, 343–52.

Caro, T.M. (1994). Cheetahs of the Serengeti plains: group living in an asocial species. The University of Chicago Press, Chicago.

Caro, T.M. and Hauser, M.D. (1992). Is there teaching in nonhuman animals? *Quarterly Review of Biology*, **67**(2), 151–74.

Carruthers, P. and Smith, P. (1996). *Theories of theories of mind*. Cambridge University Press, Cambridge.

Caughley, G. (1977). *Analysis of vertebrate populations*. Wiley, New York.

Chagnon, N.A. (1988). Life histories, blood revenge and warfare in a tribal population. *Science*, **239**, 985–92.

Chalmeau, R. (1994). Do chimpanzees cooperate in a learning task? *Primates*, **35**, 385–92.

Chalmeau, R., Visalberghi, E., and Gallo, A. (1997). Capuchin monkeys, *Cebus apella*, fail to understand a cooperative task. *Animal Behaviour*, **54**, 1215–25.

Chapman, C., Whites, F., and Wrangham, R. (1994). Party size in chimpanzees and bonobos: a reevaluation of theory based on two similarly forested sites. In *Chimpanzee culture* (ed. R. Wrangham, W. McGrew, F. de Waal and P. Heltne), pp. 41–57. Harvard University Press, Cambridge.

Charnov, E. and Berrigan, D. (1993). Why do female primates have such long lifespans and so few babies? Or life in the slow lane. *Evolutionary Anthropology*, **1**(6), 191–4.

Cheney, D. and Seyfarth, R. (1990). *How monkeys see the world: inside the mind of another species*. Chicago University Press, Chicago.

Cheney, D. and Wrangham, R. (1897). Predation. In *Primates societies* (ed. B. Smuts, D. Cheney, R. Seyfarth, R. Wrangham and T. Struhsaker), pp. 227–39. Chicago University Press, Chicago.

Chevalier-Skolnikoff, S. (1977). A Piagetian model for describing and comparing socialization in monkey, ape, and human infants. In *Primate biosocial development: biological, social and ecological determinants* (ed. S. Chevalier-Skolnikoff and F. Poirier), pp. 158–87. Garland Publishing, New York.

Chevalier-Skolnikoff, S. (1988). Spontaneous tool use and sensorimotor intelligence in Cebus compared with other monkeys and apes. *Behavioral and Brain Sciences*, **12**, 561–27.

Clark, A. (1978). Sex ratio and local resource competition in a prosimian primate. *Science*, **201**, 163–5.

Clements, K. and Stephens, D. (1995). Testing models of non-kin cooperation: mutualism and the Prisoner's Dilemma. *Animal Behaviour*, **50**, 527–35.

Clutton-Brock, T. (1988). *Reproductive success*. Chicago University Press, Chicago.

Clutton-Brock, T. (1991). *The evolution of parental care*. Princeton University Press, Princeton.

Clutton-Brock, T. and Harvey P.H. (1976). Evolutionary rules and primate societies. In *Growing points in ethology* (eds. P.P.G. Bateson and R. Hinde), pp. 195–237. Cambridge University Press, Cambridge.

Clutton-Brock, T., Albon, S.D., and Guiness, F.E. (1984). Maternal dominance, breeding success, and birth sex ratios in red deer. *Nature*, **308**, 358–60.

Collins, A. (1981). Social behaviour and patterns of mating among adult yellow baboons (*Papio c. cynocephalus*). Ph.D. Thesis, University of Edinburgh.

Collins A. and McGrew W. (1987). Termite fauna related to differences in tool-use between groups of chimpanzees (*Pan troglodytes*). *Primates*, **28**(4), 457–71.

Collins A. and McGrew W. (1988). Habitats of three groups of chimpanzees (*Pan trolodytes*) in western Tanzania compared. *Journal of Human Evolution*, **17**, 553–74.

Conroy, G., Weber, G., Seidler, H., Tobias, P., Kane, A., and Brunsden, B. (1998). Endocranial capacity in an early hominid cranium from Sterkfontein, South Africa. *Science*, **280**, 1730–1.

Cooper, S.M. (1990). The hunting behaviour of spotted hyenas (*Crocuta crocuta*) in a region containing both sedentary and migratory populations of herbivores. *African Journal of Ecology*, **28**, 131–41.

Cooper, S.M. (1991). Optimal hunting group size: the need for lions to defend their kills against loss to spotted hyenas. *African Journal of Ecology*, **29**, 130–6.

Cords, M. (1987). Forest guenons and patas monkeys: male-male competition in one-male groups. In *Primate societies* (ed. B. Smuts, D. Cheney, R. Seyfarth, R. Wrangham and T. Struhsaker), pp. 98–111. Chicago University Press, Chicago.

Creel, S. (1997). Cooperative hunting and group size: assumptions and currencies. *Animal Behaviour*, **54**, 1319–24.

Creel, S. and Creel, N.M. (1995). Communal hunting and pack size in african wild dogs, *Lycaon pictus*. *Animal Behaviour*, **50**, 1325–39.

Dart, R. (1925). *Australopithecus africanus*, the man-ape of South Africa. *Nature*, **115**, 195–9.

Darwin, C. (1871). The descent of man and selection in relation to sex. Murray, London.

Dasen, P.R. (1975). Concrete operational development in three cultures. *Journal of Cross-cultural Psychology*, **6**(2), 156–72.

Dasen, P.R. (1982). Cross-cultural data on operational development: Asymptotic development curves. In *Regressions in mental development: basic phenomena and theories* (ed. T. Bever), pp. 221–32. L. Erlbaum Associates, Hillsdale.

Dasen, P.R. and Heron, A. (1981). Cross-cultural tests of Piaget's theory. In *Handbook of cross-cultural psychology: Developmental psychology* (ed. H. Trandis, and A. Heron), pp: 295–341. Allyn and Bacon, inc., Boston.

Dasen, P.R., Inhelder, B., Lavallée, M., and Reitschitzki, J. (1978). *La naissance de l'intelligence chez l'enfant Baoulé de Côte d'Ivoire*. Hans Huber, Berne.

Dennen, van der J. (1995) *The origin of war*. Origin Press, Groningen.

Diamond, J. (1991). *The rise and fall of the third chimpanzee*. Hutchinson Radius, London.

Dind, F. (1995).'Etude d'une population cible de léopards (*Panthera pardus*) en forêt tropicale humide (Parc National de Taï, Côte d'Ivoire).' Travail de diplôme, Université de Lausanne.

Dixson, A. (1998). *Primate sexuality*. Oxford University Press, Oxford.

Doran, D. (1997). Influence of seasonality on activity patterns, feeding behaviour, ranging and grouping patterns in Taï chimpanzees. *International Journal of Primatology*, **18**(2), 183–206.

Doran, D. and McNeilage, A. (1998). Gorilla ecology and behavior. *Evolutionary Anthropology*, **6**, 120–131.

Dugatkin, L. (1997). *Cooperation among animals*. Oxford University Press, Oxford.

Dunbar, R. (1988). *Primate social systems*. Cornell University Press, New York.

Dunbar, R. (1992). Neocortex size as a constraint on group size in primates. *Journal of Human Evolution*, **20**, 469–93.

Dunbar, R. and Colishaw, G. (1992). Dominance and reproductive success? *Animal Behaviour*, **44**, 1171–3.

Dunn, P. and Roberston, P. (1993). Extra-pair paternity in polygynous tree swallows. *Animal Behaviour*, **45**, 231–9.

Durham, W. (1976). Resource competition and human aggression, part 1: a review of primitive war. *Quarterly Review of Biology*, **51**, 385–415.

Durham, W. (1991). *Coevolution: genes, culture and human diversity*. Stanford University Press, Stanford.

Eberhart, W. (1996). *Cryptic female choice*. Princeton University Press, Princeton.

Eddy, T., Gallup, G., and Povinelli, D. (1996). Age differences in the ability of chimpanzees to distinguish mirror-images of self from video images of others. *Journal of Comparative Psychology*, **110**(1), 38–44.

Egger, L. and Boesch, C. in preparation. Ranging pattern and mating opportunities in wild chimpanzees.

Endler, J.A. (1992). Interactions between predators and prey. In: *Behavioural ecology: an evolutionary approach* (ed. J.R. Krebs and N.B. Davies), pp. 169–96. Blackwell Scientific Publications, Oxford.

Estes, R.D. and Goddard, J. (1967). Prey selection and hunting behaviour of the African wild dog. *Journal of Wildlife Management*, **31**, 52–70.

Fanshawe, J.H. and Fitzgibbon, C.D. (1993). Factors influencing the hunting success of an African wild dog pack. *Animal Behaviour*, **45**, 479–90.

Falk, D. (1991). 3.5 million years of hominid brain evolution. *Seminar of Neuroscience*, **3**, 409–16.

Fentress, J., Ryon, J., McLeod, P., and Havkin, G. (1986). A multidimensional approach to agonistic behavior in wolves. In *Man and wolf: advances, issues, and problems in captive wolf research* (ed. H. Frank), pp. 253–73. Junk Publishers, Dordrecht.

Fernandez, M. (1991). Tool use and predation of oysters (*Crassostrea rhizophorae*) by the tufted capuchin, *Cebus apella apella*, in brackish water mangrove swamp. *Primates*, **32**, 529–31.

Field, D., Chemnick, L., Robbins, M., Garner, K., and Ryder, O. (1998). Paternity determination in captive lowland gorillas and orangutans and wild mountain gorillas by microsatellite analysis. *Primates*, **39**(2), 199–209.

Fisher, R. (1930). *The genetical theory of natural selection*. Oxford University Press, Oxford.

Foley, R.A. (1989). The evolution of hominid social behaviour. In *Comparative socioecology: the behavioural ecology of humans and other mammals* (ed. V. Standen and R.A. Foley), pp. 473–94. Blackwell Scientific Publications, Oxford

Formenty, P., Boesch, C., Wyers, M., Steiner, C., Donati, F., Dind, F., Walker, F., and Le Guenno, B. (1999). Ebola outbreak in wild chimpanzees living in a rainforest of Côte d'Ivoire. *Journal of Infectious Diseases*, **179** (Suppl.1), 120–126.

Fossey, D. (1983). *The gorillas in the mist*. Houghton Mifflin, Boston.

Fouts, R.S. (1983). Chimpanzee language and elephant tails: a theoretical synthesis. In *Language in primates; perspectives and implications* (ed. J. de Luce and H.T. Wilder), pp. 63–75. Springer Verlag, New York.

Frause, J. and Godin, J-G. (1995). Predator preferences for attacking particular group sizes: consequences for predator hunting success and prey predation risk. *Animal Behaviour*, **50**, 465–73.

Fruth, B. and Hohmann, G. (1994). Comparative analyses of nest building behavior in bonobos (*Pan paniscus*) and chimpanzees (*Pan troglodytes*). In *Chimpanzees cultures* (ed. R. Wrangham, W. McGrew, F. de Waal and P. Heltne), pp. 109–28. Harvard University Press, Cambridge.

Funk, M. (1985). 'Werkzeuggebrauch beim öffnen von Nüssen: Unterschiedliche Bewältigungen des problems bei Schimpansen und Orang-Utans.' Masters thesis, University of Zürich

Furuichi, T. (1987). Sexual swelling, receptivity and grouping of wild pygmy chimpanzee females at Wamba, Zaire. *Primates*, **28**(3), 309–18.

Furuichi, T. (1989). Social interactions and the life history of female *Pan paniscus* in Wamba, Zaire. *International Journal of Primatology*, **10**(3), 173–97.

Gagneux, P., Woodruff, D., and Boesch, C. (1997). Furtive mating in female chimpanzees. *Nature*, **387**, 358–9.

Gagneux, P., Boesch, C., and Woodruff, D. (1999). Female reproductive strategies, paternity, and community structure in wild West African chimpanzees. *Animal Behaviour*, **57**, 19–32.

Gagneux, P., Wills, C., Gerloff, U., Tautz, D., Morin, P., Boesch, C., Fruth, B., Hohmann, G., Ryder, O., and Woodruff, D. (1999). Mitochondrial sequences show diverse evolutionary histories of African hominoids. *Proceedings of the National Academy of Science*, **96**, 5077–5082.

Galat, G. and Galat-Luong, A. (1978). *Comparaison de l'abondance relative et des associations plurispécifiques des primates diurnes de deux zones du parc national de Taï, Côte d'Ivoire*. ORSTOM: Adiopodoumé.

Galat, G. and Galat-Luong, A. (1985). La communauté de primates diurnes de la forêt de Taï, Côte d'Ivoire. *Revue d'Ecologie (Terre Vie)*, **40**, 30–32.

Galef, B. (1990). Tradition in animals: field observations and laboratory analyses. In *Interpretation and explanation in the study of animal behavior* (ed. M. Bekoff, and D. Jamieson), pp. 74–95. Westview Press, Boulder.

Galef, B. and Heyes C. (1996). *Social learning in animals: the roots of culture*. Academic Press, New York.

Gallup, G. (1970). Chimpanzee: self-recognition. *Science*, **167**, 86–7.

Gardner, B.T., and Gardner, R.A. (1971). Two-way communication with an infant chimpanzee. In *Behavior of nonhuman primates* (ed. A.M. Schrier and F. Stollnitz), pp. 117–83. Academic Press, New York.

Garrett, J. (1995). *The coming plague*. Penguin Books, New York.

Gartshore, M., Taylor, P., and Francis, I. (1995). *Forest birds in Côte d'Ivoire*. Study report 58. Birdlife International, Cambridge.

Gauthier-Hion, A., Quris, R., and Gauthier, J.-P. (1983). Monospecific versus polyspecific life: a comparative study of foraging and antipredator tactics in a community of *Cercopithecus* monkeys. *Behavioral Ecology and Sociobiology*, **12**, 325–335.

Gavan, J.A. (1971). Longitudinal, postnatal growth in chimpanzee. In *The chimpanzee* (Ed. G.H. Bourne), Vol. 4, pp. 46–102. S. Karger, Basel.

Gerald, N. (1995). 'Demography of the Virunga mountain gorilla (*Gorilla gorilla beringei*).' Masters thesis, Princeton University.

Ghiglieri, M.P. (1984). *The chimpanzees of the Kibale forest*. Columbia University Press, New York.

Gigerenzer, G. (1997). The modularity of social intelligence. In *Machiavellian intelligence II: extensions and evaluations* (ed. A. Whiten and R. Byrne), pp. 264–88. Cambridge University Press, Cambridge.

Godfray, H. (1995). Evolutionary theory of parent–offspring conflict. *TREE*, **376**, 133–8.

Gonder, K., Oates, J., Disotell, T., Forstner, M., Morales, J., and Melnick, D. (1997). A new west African chimpanzee subspecies? *Nature*, **388**, 337.

Goodall, J. (1963). Feeding behaviour of wild chimpanzees: a preliminary report. *Symposium of the Zoological Society, London*, **10**, 39–48.

Goodall, J. (1968). Behaviour of free-living chimpanzees of the Gombe Stream area. *Animal Behaviour Monograph*, **1**, 163–311.

Goodall, J. (1970a). Tool-using in primates and other vertebrates. In *Advances in the study of behavior* (ed. D.S. Lehrmannn, R.A. Hinde, and E. Shaw), Vol. 3, pp. 195–249. Academic Press, New York.

Goodall, J. (1970b). *In the Shadow of Man*. Collins, London.

Goodall, J. (1973). Cultural elements in a chimpanzee community. In *Precultural primate behaviour* (ed. E. Menzel), Fourth IPC Symposia Proceedings, Vol. 1, pp. 195–249. Karger, Basel.

Goodall, J. (1983). Population dynamics during a fifteen-year period in one community of free-living chimpanzees in the Gombe National Park, Tanzania. *Zeitschrift für Tierpsychologie*, **61**, 1–60.

Goodall, J. (1986). *The chimpanzees of Gombe: Patterns of behavior*. The Belknap Press of Havard University Press, Cambridge.

Goodall, J. (1995). *Through a window*. Houghton Press, New York.

Goodall, J., Bandura, A., Bergmann, E., Busse, C., Matam, H., Mpongo, E., Pierce, A., and Riss, D. (1979). Inter-community interactions in the chimpanzee populations of the Gombe National Park. In *The great apes* (ed. D. Hamburg and E. McCown), pp. 13–53. Benjamin/Cummings, Menlo Park, Calif.

Gopnik, A. and Graf, P. (1988). Knowing how you know: young children's ability to identify and remember the sources of their beliefs. *Child Development*, **59**, 1366–71.

Gopnik, A. and Meltzoff, A. (1994). Minds, bodies and persons: Young children's understanding of the self and others as reflected in imitation and theory of mind research. In *Self-awareness in animals and humans* (ed. S. Parker, R. Mitchell, and M. Boccia), pp. 166–86. Cambridge University Press, Cambridge.

Gowaty, P.A. (1993). Differential dispersal, local resource competition and sex ratio variation in birds. *American Naturalist*, **141**, 263–280.

Gowaty, P.A. and Lennartz, M. (1985). Sex ratios of nestling and fledgling red-cockaded woodpeckers (*Picoides borealis*) favor males. *American Naturalist*, **126**(3), 347–353.

Greenfield, P. (1984). A theory of the teacher in the learning activities of everyday life. In *Everyday cognition: its development in social context* (ed. B. Rogoff and J. Lave), pp. 117–38. Harvard University Press, Cambridge.

Greenfield, P., Brazelton, B., and Childs, C. (1989). From birth to maturity in Zinacantan: ontogenesis in cultural context. In *Ethnographic encounters in Southern Mesoamerica:*

celebratory essays in honor of Evan Vogt (ed. V. Bricker and G. Gossen), pp. 177–216. Institute of Mesoamerican Studies (SUNY), Albany.

Grinnell, J., Packer, C., and Pusey, A.E. (1995). Cooperation in male lions, kinship, reciprocity or mutualism? *Animal Behaviour*, **49**, 95–105.

Guillaumet, J.-L. (1967). Recherches sur la végétation et la flore de la région du Bas-Cavally (Côte d'Ivoire). Mémoire ORSTOM, Vol. 20, Paris.

Günther, M. and Boesch, C. (1993). Energetic cost of nut-cracking behaviour in wild chimpanzees. In *Evolution of hands* (ed. D. Chivers and H. Preuschoft), pp. 109–29. Gustav Fisher Verlag, Stuttgart.

Hala, S. and Chandler, M. 1996. The role of strategic planning in accessing false-belief understanding. *Child Development*, **67**, 2948–2966.

Hall, K. (1963). Tool-using performances as indicators of behavioural adaptability. *Current Anthopology*, **4**(5), 479–87.

Halperin, S. (1979). Temporary association patterns in free ranging chimpanzees: an assessment of individual grouping preferences. In *The Great Apes* (Ed. D.A. Hamburg and E. McCown), pp. 491–9. Benjamin/Cummings, Menlo Park.

Haltenorth, T. and Diller, H. (1977). *Säugetiere Afrikas und Madagaskars*. BVL Verlagsgesellschafts Gmbh, München.

Hamai, M., Nishida, T., Takasaki, H, and Turner, L. (1992). New records of within-group infanticide and cannibalism in wild chimpanzees. *Primates*, **33**(2), 151–62.

Hamilton, A. (1982). *Environmental history of East Africa: a study of the quaternary*. Academic Press, London.

Hamilton, W.D. (1964). The genetical theory of social behaviour (I and II). *Journal of Theoretical Biology*, **7**, 1–32.

Hamilton, W.D. (1967). Extraordinary sex ratios. *Science*, **156**, 477–88.

Hamilton, W.D. (1971). Geometry for the selfish herd. *Journal of Theoretical Biology*, **7**, 295–311.

Hamilton, W. (1984). Significance of paternal investment by primates to the evolution of male-female associations. In *Primate paternalism* (ed. D. Taub), pp. 309–335. Van Nostrand Reinhold, New York.

Hannah, A. and McGrew, W.C. (1987). Chimpanzees using stones to crack open oil palm nuts in Liberia. *Primates*, **28**(1), 31–46.

Harcourt, A. (1989). Social influences on competitive ability: alliances and their consequences. In *Comparative socioecology: the behavioural ecology of humans and other mammals* (ed. V. Standen and R.A. Foley), pp. 223–42. Blackwell Scientific Publications, Oxford.

Harcourt, A. (1992). Coalitions and alliances: are primates more complex than non-primates? In *Cooperation in competition in animals and humans* (ed. A. Harcourt and F. de Waal), pp. 445–71. Oxford University Press, Oxford.

Harcourt, A.H. and de Waal, F. (1992). *Cooperation in competition in animals and humans*. Oxford University Press, Oxford.

Harcourt, A., Fossey, D., Stewart, K., and Watts, D. (1980). Reproduction in wild gorillas and some comparisons with chimpanzees. *Journal of Reproduction and Fertility Supplement*, **28**, 59–70.

Hart, D. and Karmel, M. (1996). Self-awareness and self-knowledge in humans, apes, and monkeys. In *Reaching into thought: the minds of the great apes* (ed. A. Russon, K. Bard, and S. Parker), pp. 325–47. Cambridge University Press, Cambridge.

Hasegawa, T. (1989). Sexual behavior of immigrant and resident female chimpanzees at Mahale. In *Understanding chimpanzees* (ed. P. Heltne and L. Marquardt), pp. 90–103. Harvard University Press, Cambridge.

Hasegawa, T. and Hiraiwa-Hasegawa, M. (1990). Sperm competition and mating behavior. In *The chimpanzees of Mahale Mountains: sexual and life history strategies* (ed. T. Nishida), pp. 115–32. University of Tokyo Press, Tokyo.

Hayaki, H., Huffman, M., and Nishida, T. (1989). Dominance among male chimpanzees in the Mahale Mountains National Park, Tanzania: a preliminary study. *Primates*, **30**, 187–97.

Heinsohn, R., Packer, C. and Pusey, A. (1996). Development of cooperative territoriality in juvenile lions. *Proceedings of the Royal Society. London*, B, **263**, 475–79.

Heltne, P. and Marquardt, L. (1989). *Understanding chimpanzees*. Harvard University Press, Cambridge.

Hemelrijk, C. (1990). Models of, and tests for, undirectionality and other social interaction patterns at a group level. *Animal Behaviour*, **39**, 1013–29.

Herbinger, I., Boesch, C. and Rothe, H. in press. Territory characteristics among three neighbouring chimpanzee communities in the Taï National Park. *International Journal of Primatology*.

Heyes, C.M. (1993). Anecdotes, training, trapping and triangulating: do animals attribute mental states? *Animal Behaviour*, **46**, 177–88.

Heyes, C.M. (1994). Imitation, culture and cognition. *Animal Behaviour*, **46**, 999–1010.

Heyes, C.M. (1998). Theory of mind in nonhuman primates. *Behavioral and Brain Sciences*, **21**(1), 101–34.

Hill, K. and Hurtado, M. (1996). *Ache life history: the ecology and demography of a foraging people*. Walter de Gruyter, New York.

Hill, K., Boesch, C., Pusey, A., Williams, J., and Wrangham, R. Demography of wild chimpanzees. [In preparation.]

Hiraiwa-Hasegawa, M. (1998). Adaptive significance of infanticide in primates. *TREE*, **3**(5), 102–5.

Holekamp, K.E. and Smale, L. (1995). Rapid change in offspring sex ratios after clan fission in the spotted hyena. *American Naturalist*, **145**(2), 261–78.

Holenweg, A., Noë, R., and Schabel, M. (1996). Waser's gas model applied to associations between red colobus and diana monkeys in the Taï National Park, Ivory Coast. *Folia Primatologica*, **67**, 125–36.

Hoppe-Dominik, B. (1984). Etude du spectre des proies de la panthère, *Panthera pardus*, dans le Parc National de Taï en Côte d'Ivoire. *Mammalia* **48**(4), 477–87.

Howell, N. (1979). *Demography of the Dobe !Kung*. Academic Press, New York.

Hrdy, S. (1977). *The langurs of Abu: female and male strategies of reproduction*. Harvard University Press, Cambridge, Ma.

Hrdy, S. and Whitten, P. (1987). Patterning of sexual activity. In *Primates Societies* (ed. B. Smuts, D. Cheney, R. Seyfarth, R. Wrangham and T. Struhsaker), pp. 370–84. Chicago University Press, Chicago.

Humphrey, N.K. (1976). The social function of intellect. In *Growing points in ethology* (ed. P.P. Bateson and R. Hinde), pp. 303–17. Cambridge University Press, Cambridge.

Hunt, G. (1996). Manufacture and use of hook-tools by New Caledonian crows. *Nature*, **379**, 249–51.

Ihobe, H. (1992). Male–male relationships among wild bonobos (*Pan paniscus*) at Wamba, Republic of Zaire. *Primates*, **33**(2), 163–79.

Ingmanson, E. (1996). Tool-using behavior in wild *Pan paniscus*: social and ecological considerations. In *Reaching into thought* (ed. A. Russon, K. Bard, and S. Parker), pp. 190–210. Cambridge University Press, Cambridge.

Isaac, G. (1978). The food sharing behavior of protohuman hominids. *Scientific American*, **238**, 90–108.

Isabirye-Basuta, G. (1989). Feeding ecology of chimpanzees in the Kibale forest, Uganda. In *Understanding chimpanzees* (ed. P. Heltne, and L. Marquardt), pp. 116–27. Harvard University Press, Cambridge.

Isbell, L. and Young, T. (1993). Human presence reduces predation in a free-ranging vervet monkey population in Kenya. *Animal Behaviour*, **45**, 1233–1235.

Jenny, D. (1996). Spatial organization of leopards, *Panthera pardus*, in Taï National Park, Ivory Coast. *Journal of Zoology*, **240**, 427–440.

Johanson, D. and Edey, M. (1981). *Lucy: the beginnings of humankind*. Simon and Schuster, New York.

Jolly, A. (1966). Lemur social intelligence and primate intelligence. *Science*, **155**, 501–6.

Jones, S., Martin, R., and Pilbeam, D. (1992). *Human evolution*. Cambridge University Press, Cambridge.

Kano, T. (1971). The chimpanzees of Filabanga, western Tanzania. *Primates*, **12**, 229–46.

Kano, T. (1992). *The last ape: pygmy chimpanzee behavior and ecology*. Stanford University Press, Stanford.

Kaplan, H. and Hill, K. (1985). Hunting ability and reproductive success among male Ache foragers: preliminary results. *Nature*, **26**(1), 131–3.

Kaplan, H. and Hill, K., Lancaster, J., and Hurtado, A. Learning skills in two hunter-gatherers societies. [In press.]

Kawanaka, K. (1990). Alpha males' interactions and social skills. In *The Chimpanzees of the Mahale Mountains: sexual and life history strategies* (ed. T. Nishida), pp. 149–70. University of Tokyo Press, Tokyo.

Kawanaka, K. and Nishida, T. (1974). Recent advances in the study of inter-unit-group relationships and social structure of wild chimpanzees of the Mahale Mountains. In *Proceedings of the 5th Congress of the International Primatological Society*, (ed. S. Kondo, M. Kawai, A. Ehara and S. Kawamura), pp. 173–85. Japan Science Press, Tokyo.

Keller, L., Acese, P., Smith, J., Hochachka, W., and Stearns, S. (1995). Selection against inbred song sparrows during a natural population bottleneck. *Nature*, **372**, 356–357.

Kempenaers, B., Verheyen, G., Broeck, M., Burke, T., Broeckhoven, C., and Dhondt, A. (1992). Extra-pair paternity from female preference for high-quality males in blue tit. *Nature*, **357**, 494–66.

Kimberley, P. (1998). Tool use in wild capuchin monkeys. *American Journal of Primatology*, **46**, 259–61.

Kortlandt, A. (1962). Chimpanzees in the wild. *Scientific American*, **206**, 128–38.

Kortlandt, A. (1983). Marginal habitats of chimpanzees. *Journal of Human Evolution*, **12**, 231–78.

Kortlandt, A. and Holzhaus, E. (1987). New data on the use of stone tools by chimpanzees in Guinea and Liberia. Primates, **28**(4), 473–96.

Krebs, C. (1994). *Ecology: The experimental analysis of distribution and abundance*. (4th ed.) Harper Collins College Publishers, New York.

Krebs, J.R. and Davies, N.B. (1993). *An introduction to behavioural ecology*. (3rd ed.) Blackwell Scientific Publications, Oxford.

Kruuk, H. (1972). *The spotted hyena*. University Press of Chicago, Chicago.

Kummer, H. (1971). *Primate societies: group techniques of ecological adaptation*. Aldine, Chicago.

Kummer, H., Götz, W., and Angst, W. (1974). Triadic differentiation: an inhibitory process protecting pair bonds in baboons. *Behaviour*, **49**, 62–87.

Kuroda, S. (1979). Grouping of the pygmy chimpanzee. *Primates*, **20**, 161–83.

Kuroda, S., Suzuki, S., and Nishihara, T. (1996). Preliminary report on predatory behavior and meat sharing in Tschego chimpanzees (*Pan troglodytes troglodytes*) in the Ndoki Forest, Northern Congo. *Primates*, **37**(3), 253–9.

Langer, J. (1996). Heterochrony and the evolution of primate cognitive development. In *Reaching into thought: the mind of the great apes* (ed. A. Russon, K. Bard, and S. Parker), pp. 257–77. Cambridge University Press, Cambridge.

Leakey, L.S.B. (1961). *The progress and evolution of man in Africa*. Oxford University Press, London.

Leakey, M., Feibel, C., McDougali, I., and Walker, A. (1995). New four-million-year-old hominid species from Kanapoi and Allia Bay, Kenya. *Nature*, **376**, 565–71.

Leakey, R. (1980). *The making of mankind*. Book Club Associates, London.

Leakey, R. and Lewin, R. (1977). *Origins*. McDonald and Jones, London.

Lee, R. (1973). Mongongo: The ethnography of a major wild food resource. *Ecology of Food and Nutrition*, **2**, 307–21.

Lee, R.B. and DeVore, I. (1968). *Man the hunter*. Aldine-Atherton, Chicago.

Le Guenno, B., Formenty, P., Wyers, M., Gounon, P., Walker, F., and Boesch, C. (1995). Isolation and partial characterisation of a new strain of Ebola virus. *Lancet*, **345**, 1271–4.

Lethmate, J. (1982). Tool using skills in orang-utans. *Journal of Human Evolution*, **11**, 49–64.

Lifjeld, J., Dunn, P., Robertson, R., and Boag, P. (1993). Extra-pair paternity in monogamous tree swallows. *Animal Behaviour*, **45**, 213–29.

Ligon, D. and Zwartjes, P. (1995). Female red junglefowl choose to mate with multiple males. *Animal Behaviour*, **49**, 127–35.

Lovejoy, C.O. (1981). The origin of man. *Science*, **211**, 341–350.

Lovejoy, C.O., Meindl, R., Pryzbeck, T., Barton, T., Heiple, K. and Kotting, D. (1977). Paleodemography of the Libben site, Ottawa county, Ohio. *Science*, **198**, 291–3.

Luce, J. de, and Wilder, H.T. (1983). Introduction. In *Language in primates: perspectives and implications* (ed. J. de Luce and H.T. Wilder), pp: 1–17. Springer Verlag, New York.

McGrew, W. (1974). Tool use by wild chimpanzees in feeding upon driver ants. *Journal of Human Evolution*, **3** 501–8.

McGrew, W. (1977). Socialization and object manipulation of wild chimpanzees. *In Primate bio-social development: biological, social and ecological determinants* (ed. S. Chevalier-Skolnikoff and E. Poirier), pp. 159–87. Garland Publishing, New York.

McGrew, W. (1979). Evolutionary implications of sex differences in chimpanzee predation and tool use. In *The great apes* (ed. D.A. Hamburg and E. McCown), pp. 440–63. Benjamin/Cummings, Menlo Park, Ca.

McGrew, W. (1992). *Chimpanzee material culture: implications for human evolution*. Cambridge University Press, Cambridge.

McGrew, W.C., P.J. Baldwin, and Tutin, C. (1979). Chimpanzees, tools and termites: Cross cultural comparisons of Senegal, Tanzania and Rio Muni. *Man*, **14**, 185–214.

McGrew, W.C., Tutin, C.E.G., Baldwin P.J., Sharman M.J. and Whiten, A. (1978). Primates preying upon vertebrates: new records from West Africa. *Carnivores*, **1**, 41–5.

McGrew, W., Nishida, T., and Marchandt, L. (1996). *Great apes societies*. Cambridge University Press, Cambridge.

McHenry, H. and Berger, L. (1996). Ape-like body proportions in *Australopithecus africanus* and their implication for the origin of the genus *Homo*. *American Journal of Physical Anthropology*, **22**, 163–4.

McHenry, H. and Berger, L. (1998). Body proportions in *Australopithecus afarensis* and *A. africanus* and the origin of the genus *Homo*. *Journal of Human Evolution*, **35**(1), 1–22.

Malenky, R. and Stiles, E. (1991). Distribution of terrestrial herbaceous vegetation and its consumption by *Pan paniscus* in the Lomako Forest, Zaïre. *American Journal of Primatology*, **23**, 153–69.

Malinowski, B. (1941). An anthropological analysis of war. *American Journal of Sociology*, **46**, 521–50.

Manson, J. and Wrangham, R. (1991). Intergroup aggression in chimpanzees and humans. *Current Anthropology*, **32**(4), 369–90.

Marchesi, P., Marchesi, N., Fruth, B., and Boesch, C. (1995). Census and distribution of chimpanzees in Côte d'Ivoire. *Primates*, **36**(4), 591–607.

Marchesi, P., Marchesi, N., and Boesch, C. Protected forest in Côte d'Ivoire and satellite pictures. [In press.]

Martin, C. (1989). *Die Regenwälder Westafrikas: Oekologie, Bedrohung und Schutz*. Birkhäuser Verlag, Basel.

Martin, R.D. (1990). *Primate origins and evolution: a phylogenetic reconstruction*. Chapman and Hall, New York.

Martin, R.D. (1992). Female cycles in relation to paternity in primate societies. In *Paternity in primates: genetic tests and theories* (ed. R.D. Martin, A.F. Dixson and E.J. Wickings), pp. 238–74. Karger, Basel.

Mason, W. (1978). Social experience and primate development. In *The Development of behavior: comparative and evolutionary aspects* (ed. G.M. Burghardt and M. Bekoff), pp. 233–51. Garland Press, New York.

Matsuzawa, T., Sakura, O., Kimura, T., Hamada, Y. and Sugiyama, Y. (1990). Case report on the death of a wild chimpanzee (*Pan troglodytes verus*). *Primates*, **31**(4), 635–641.

Matsuzawa, T. and Yamakoshi, G. (1996). Comparison of chimpanzee material culture between Bossou and Nimba, West Africa. In *Reaching into thought* (ed. A. Rousson, K. Bard, and S. Parker), pp. 211–32. Cambridge University Press, Cambridge.

Maynard-Smith, J. (1982). *Evolution and the theory of games*. Cambridge University Press, Cambridge.

Mech, D.L. (1970). *The Wolf*. Natural History Press, New York.

Meindl, R.S. (1992). Human population before agriculture. In *Human Evolution* (ed. J. Stones, R. Martin, and D. Pilbeam), pp. 406–10. Cambridge University Press, Cambridge.

Melnick, D. and Pearl, M. (1987). Cercopithecines in multimale groups: genetic diversity and population structure. In *Primates Societies* (ed. B. Smuts, D. Cheney, R. Seyfarth, R. Wrangham and T. Struhsaker), pp. 121–34. Chicago University Press, Chicago.

Meltzoff, A. (1995). Understanding the intentions of others: re-enactment of intended acts by 18-month-old children. *Development Psychology*, **31**(5), 838–50.

Meltzoff, A. (1996). The human infant as imitative generalist: a 20-year progress report on infant imitation with implications for comparative psychology. In *Social learning in animals: the roots of culture* (ed. B. Galef and C. Heyes), pp. 347–70. Academic Press, New York.

Menzel, C. (1974). A group of young chimpanzees in a one-acre field: leadership and communication. In *Behavior of nonhuman primates* (ed. D. Hamburg and E. McCown), pp. 359–71, Benjamin Cummings, Menlo Park, Ca.

Mills, M.G.L. (1990). *Kalahari hyenas: the comparative behavioural ecology of two species*. Chapman and Hall, London.

Milton, K. 1981. Distribution patterns of tropical plant foods as an evolutionary stimulus to primate mental development. *American Anthropologist*, **83**, 534–48.

Morbeck, M. and Zihlmann, A. (1989). Body size and proportions in chimpanzees, with special reference to *Pan troglodytes schweinfurthii* from Gombe National Park, Tanzania. *Primates*, **30**(3), 369–82.

Morin, P.A., Moore, J.J. Chakraborty, R., Jin, L. Goodall, J., and Woodruff, D.S. (1994a). Kin selection, social structure, gene flow and the evolution of chimpanzees. *Science*, **265**, 1193–1201.

Morin, P.A., Wallis, J.J. Moore, J.J. and Woodruff, D.S. (1994b). Paternity exclusion in a community of wild chimpanzees using hypervariable simple sequence repeats. *Molecular Ecology*, **3**, 469–78.

Morris, K. and Goodall, J. (1977). Competition for meat between chimpanzees and baboons of the Gombe National Park. *Folia Primatologica*, **28**, 109–21.

Myers, N. (1976). *The leopard Panthera pardus in Africa*. Monograph 5, IUCN, Morges.

Neel, J. and Chagnon, N. (1968). The demography of two tribes of primitive relatively unacculturated American Indians. *Proceeding of the National Academy of Science Symposium*, **59**, 680–9.

Nishida, T. (1968). The social group of wild chimpanzees in the Mahali Mountains. *Primates*, **9**, 167–224.

Nishida, T. (1979). The social structure of chimpanzees of the Mahale Mountains. In *The great apes* (ed. D.A. Hamburg and E. McCown), pp. 73–122, Benjamin Cummings, Menlo Park, Ca.

Nishida, T. (1983). Alpha status and agonistic alliance in wild chimpanzees. *Primates*, **24**, 318–36.

Nishida, T. (1987). Local traditions and cultural transmission. In *Primate Societies* (ed. S.S. Smuts, D.L. Cheney, R.M. Seyfarth, R.W. Wrangham and T.T. Strusaker), pp. 462–74. Chicago University Press, Chicago.

Nishida, T. (1989). Social interactions between resident and immigrant female chimpanzees. In *Understanding chimpanzees* (ed. P. Heltne, and L. Marquardt), pp. 68–89. Harvard University Press, Cambridge, Ma.

Nishida, T. (1990). *The Chimpanzees of the Mahale Mountains: Sexual and Life History Strategies*. University of Tokyo Press, Tokyo.

Nishida, T. (1997). Sexual behavior of adult male chimpanzees of the Mahale Mountains National Park, Tanzania. *Primates* **38**(4), 379–98.

Nishida, T. and Hiraiwa, M. (1982). Natural history of a tool-using behaviour by wild chimpanzees in feeding upon wood-boring ants. *Journal of Human Evolution*, **11**, 73–99.

Nishida, T. and Hosaka, K. (1996). Coalition strategies among adult male chimpanzees of the Mahale Mountains, Tanzania. In *Great Apes Societies*. (ed. W. McGrew, T. Nishida, and L. Marchandt), pp. 114–34. Cambridge University Press, Cambridge.

Nishida, T. and Kawanaka, K. (1985). Within-group cannibalism by adult male chimpanzees. *Primates* **26**(3), 274–84.

Nishida, T. and Turner, L. (1996). Food transfer between mother and infant chimpanzees of the Mahale Mountains National Park, Tanzania. *International Journal of Primatology*, **17**(6), 947–68.

Nishida, T. and Uehara, U. (1980). Chimpanzees, tools and termites: another example from Tanzania. *Current Anthoropology*, **21**, 671–2.

Nishida, T., Uehara, S. and Nyondo R. (1983). Predatory behavior among wild chimpanzees of the Mahale Mountains. *Primates* **20**, 1–20.

Nishida, T. and Hiraiwa-Hasegawa, M., Hasegawa, T., and Takahata, Y. (1985). Group extinction and female transfer in wild chimpanzees in the Mahale National Park, Tanzania. *Zeitschrift für Tierpsychologie*, **67**, 284–301.

Nishida, T., Takasaki, H. and Y. Takahata. (1990). Demography and reproductive profiles. In *The chimpanzees of the Mahale Mountains: sexual and life history strategies* (ed. T. Nishida), pp. 63–97. University of Tokyo Press, Tokyo.

Nishida, T., Hasegawa, T., Hayaki, H., Takahata, Y., and Uehara, S. (1992). Meat-sharing as a coalition strategy by an alpha male chimpanzee? In *Topics in Primatology:* vol. 1. *Human Origins* (ed. T. Nishida, W.C. McGrew, P. Marler, M. Pickford, and F. de Waal), pp. 159–74. Karger AG, Basel.

Noé, R. (1990). A veto game played by baboons: a challenge to the use of the Prisoner's Dilemma as a paradigm for reciprocity and cooperation. *Animal Behaviour*, **39**(1), 78–90.

Noé, R. (1992). Alliance formation among male baboons: shopping for profitable partners. *In cooperation in competition in animals and humans* (ed. A.H. Harcourt, and F. de Waal), pp. 285–321. Oxford University Press, Oxford.

Noé, R. and Bshary, R. (1997). The formation of red colobus–diana monkey associations under predator pressure from chimpanzees. *Proceeding of the Royal Society, London*, Series B, **264**, 253–9.

Norikoshi, K. (1982). One observed case of cannibalism among wild chimpanzees of the Mahale Mountains. *Primates* **23**(1), 66–74.

Oates, J. (1985). Action plan for African Primates Conservation: 1986–1990. IUCN, Gland.

Olsson, M., Shine, R., Madsen, T., Gullberg, A., and Tegelstöm, H. (1996). Sperm selection by females. *Nature*, **383**, 585.

Orsdol, K. van (1994). Foraging behaviour and hunting success of lions in Queen Elizabeth National Park, Uganda. *African Journal of Ecology*, **22**, 79–99.

Packer, C. (1977). Reciprocal altruism in Papio anubis. *Nature*, **265**, 441–3.

Packer, C. and Ruttan, L. (1988). The evolution of cooperative hunting. *American Naturalist*, **132**(2), 159–98.

Packer, C., Scheel, D., and Pusey, A.E. (1990). Why lions form groups: food is not enough. *American Naturalist*, **136**, 1–19.

Packer, C., Gilbert, D.A., Pusey, A.E., and O'Brien, S.J. (1991). A molecular genetic analysis of kinship and cooperation in African lions. *Nature*, **351**, 562–5.

Pagel, M. (1994). The evolution of conspicuous advertisement in Old World monkeys. *Animal Behaviour*, **47**, 721–723.

Parker, G.A. (1985). Models of parent–offspring conflict. V. Effects of the behaviour of the two parents. *Animal Behaviour*, **33**, 519–33.

Parker, G.A. and MacNair, M.R. (1979). Models of parent–offspring conflict. IV. Suppression: Evolutionary retaliation by the parent. *Animal Behaviour*, **27**, 1210–35.

Parker, S.T. (1977). Piaget's sensorimotor period series in an infant macaque: a model for comparing unstereotyped behavior and intelligence in human and nonhuman primates. In *Primate biosocial development: biological, social and ecological determinants* (ed. S. Chevalier-Skolnikoff and F. Poirier), pp. 43–112. Garland Publishing, New York.

Parker, S.T. and Gibson, K.R. (1977). Object manipulation, tool-use and sensorimotor intelligence as feeding adaptations in cebus monkeys and great apes. *Journal of Human Evolution*, **6**, 623–41.

Parker, S.T. and Gibson, K.R. (1979). A developmental model of the evolution of language and intelligence in early hominids. *Behavioral and Brain Sciences*, **2**, 367–408.

Passingham, R. (1982). *The human primate*. Freeman, London.

Paturel, J., Servat, E., Kouame, B., Boyer, J., Lubes, H., and Masson, J. (1995). Manifestations de la sécheresse en Afrique de l'Ouest non sahélienne: Cas de la Côte d'Ivoire, du Togo et du Bénin. *La Sécheresse*, **6**(1), 95–102.

Peters, C. (1987). Nut-like oil seeds: food for monkeys, chimpanzees, humans and probably ape-man. *American Journal of Physical Anthropology*, **73**, 333–63.

Pfluger, T., Nohon, G. and Boesch, C. Maternal investment and weaning in wild chimpanzees: Influence of rank and infant sex. [In press.]

Piaget, J. (1935). *La Naissance de l'Intelligence chez l'Enfant*. Delachaux et Niestlé, Neuchâtel.

Piaget, J. (1945). *La Formation du Symbole chez l'Enfant*. Delachaux et Niestlé, Neuchâtel.

Piaget, J. and Inhelder, B. (1947). *La Représentation de l'Espace chez l'Enfant*. Presses Universitaires de France, Paris.

Piaget, J., Inhelder, B. and Szeminska, A. (1948). *La Géométrie Spontanée de l'Enfant*. Presses Universitaires de France, Paris.

Pilbeam, D.R. (1980). Major trends in human evolution. In *Current argument on early Man* (ed. L.K. Konigsson), pp. 261–85. Pergamon Press, Oxford.

Potts, R. (1984). Home bases and early hominids. *American Scientific*, **72**, 338–47.

Povinelli, D. (1994). What chimpanzees (might) know about the mind. In *Chimpanzees cultures* (ed. R. Wrangham, W. McGrew, F. de Waal and P. Heltne), pp. 285–300. Harvard University Press, Cambridge, Ma.

Povinelli, D. (1996). Chimpanzee theory of mind? The long road to strong inference. In *Theories of Theories of Mind* (ed. P. Carruthers and P. Smith), pp. 293–329. Cambridge University Press, Cambridge.

Povinelli, D. and Cant, J. (1995). Arboreal clambering and the evolution of self-conception. *The Quarterly Review of Biology*, **70**(4), 393–421.

Povinelli, D., Nelson, K. and Boysen, S. (1990). Inferences about guessing and knowing by chimpanzees (*Pan troglodytes*). *Journal of Comparative Psychology*, **104**, 203–10.

Povinelli, D., Nelson, K. and Boysen, S. (1992). Comprehension of role reversal in chimpanzees: evidence of empathy? *Animal Behaviour*, **43**, 633–40.

Povinelli, D., Rulf, A., and Bierschwale, D. (1994). Absence of knowledge attribution and self-recognition in young chimpanzees (*Pan troglodytes*). *Journal of Comparative Psychology*, **108**(1), 74–80.

Povinelli, D., Gallup, G., Eddy, T., Bierschwale, D., Engstrom, M., Perilloux, H. and Toxopeus, I. (1997). Chimpanzees recognize themselves in mirrors. *Animal Behaviour*, **53**, 1083–8.

Power, M. (1991). *The Egalitarians: Human and Chimpanzee*. Cambridge University Press, Cambridge.

Premack, D. and Premack, A.J. (1983). *The Mind of an ape*. Norton and Company, New York.

Premack, D. and Woodruff, G. (1978). Does the chimpanzee have a theory of mind? *Behavioral and Brain Sciences*, **4**, 515–26.

Prosterman, R. (1972). *Surviving to 3000: An introduction to the study of lethal conflict*. Duxbury Press, North Scituate.

Pulliam, H.R. and Caraco, T. (1984). Living in groups: Is there an optimal group size? In *Behavioural ecology: an evolutionary approach*. (ed. J.R. Krebs and N.B. Davis), pp. 122–47. Blackwell Science, Oxford.

Pusey, A. and Packer, C. (1987). Dispersal and philopatry. In *Primate Societies* (ed. B. Smuts, D. Cheney, R. Seyfarth, R. Wrangham and T. Struhsaker), pp. 250–66. Chicago University Press, Chicago.

Pusey, A., Williams, J., and Goodall, J. (1997). The influence of dominance rank on reproductive success of female chimpanzees. *Science*, **277**, 828–31.

Rahm, U. (1971). L'emploi d'outils par les chimpanzés de l'ouest de la Côte d'Ivoire. *La Terre et la Vie*, **25**, 506–9.

Rayner, R.J., Moon, B.P., and Masters, J.C. (1993). The Makapansgat australopithecine environment. *Journal of Human Evolution*, **24**, 219–31.

Reynolds, V. (1965). *Budongo: a forest and its chimpanzees*. Methuen, London.

Reynolds, V. The structure of temporary subgroupings in Budongo Forest chimpanzees. [In press.]

Reynolds, V. and Reynolds, F. (1965). Chimpanzees of the Budongo Forest. In *Primate Behavior* (ed. I. DeVore), pp. 368–424. Holt, Rinehart, and Winston, New York.

Ridley, M. (1996). *Evolution*. (2nd ed.) Blackwell Science, Oxford.

Ristau, C.A. and Robbins, D. (1982). Language in great apes: a critical review. In *Advances in the study of behavior* (ed. J. Rosenblatt, R. Hinde, C. Beer, and M. Busnel), pp. 141–255. Academic Press, New York.

Robbins, M. (1995). A demographic analysis of male history and social structure of mountain gorillas. *Behaviour*, **132**, (1–2), 21–47.

Roduit, G. (1999). Le parasitisme du chimpanzé de la forêt de Taï en Côte d'Ivoire: une étude coprologique qualitative et quantitative. Ph.D. thesis. University of Bern, Switzerland.

Rogoff, B., Chavajay, P. and Matusov, E. (1993). Questioning assumptions about culture and individuals. *Behavioral and Brain Sciences*, **16**(3), 533–4.

Van Rompaey, R. (1993). Forest gradients in West Africa: a spatial gradient analysis. Ph.D. thesis, Wageningen.

Rose, L. (1997). Vertebrate predation and food-sharing in *Cebus* and *Pan*. *International Journal of Primatology*, **18**(5), 727–765.

Roth, D. and Leslie, A. (1998). Solving belief problems: Toward a task analysis. *Cognition*, **66**, 1–31.

Russon, A.E. and Galdikas, B. (1995). Constraints on great apes' imitation: model and action selectivity in rehabilitant orangutan imitation. *Journal of Comparative Psychology*, **109** (1), 5–17.

Sabater Pi, J. (1974). An elementary industry of the chimpanzees in the Okorobikó Mountains, Rio Muni (Republic of Equatorial Guinea), West Africa. *Primates* **15**(4), 351–64.

Sabater Pi, J., Bermejo, M., Illera, G., and Vea, J. (1993). Behavior of bonobos (*Pan paniscus*) following their capture of monkeys in Zaire. *International Journal of Primatology*, **14**(5), 797–804.

Sade, D. (1967). Determinants of dominance in a group of free-ranging rhesus monkeys. In *Social communication among primates* (ed. S. Altmann), pp. 99–114. University of Chicago Press, Chicago.

Sakura, O. (1994). Factors affecting party size and composition of chimpanzees (*Pan troglodytes verus*) at Bossou, Guinea. *International Journal of Primatology*, **15**, 167–83.

SAS. (1985). *SAS user's guide: basics*, Version 5 Edition. SAS Institute Inc., Cary, NO.

Sarich, V. and Wilson, A. (1967). Immunological time scale for hominid evolution. *Science*, **179**, 1144–7.

Savage, T.S. and Wyman, J. (1843/44). Observations on the external characters and habits of *Troglodytes niger*, Geoff. and on its organization. *Boston Journal Of Natural History*, **4**, 362–86.

Savage-Rumbaugh, E.S., Rumbaugh, D.M., and Boysen S. (1978). Linguistically mediated tool use and exchange by chimpanzees (*Pan troglodytes*). *Behavioral and brain sciences*, **201**, 641–4.

Schaik, C. van, (1983). Why are diurnal primates living in groups? *Behaviour* **87**, 120–44.

Schaik, C. van, (1996). Social evolution in primates: the role of ecological factors and male behaviour. In *Evolution of social behaviour patterns in primates and Man* (ed. W. Runciman, J. Maynard-Smith, and R. Dunbar), pp. 9–31. British Academy, London.

Schaik, C. van, and Hrdy, S. (1991). Intensity of local resource competition shapes the relationship between maternal rank and sex ratios at birth in cercopithecine primates. *American Naturalist*, **138**, 1555–62.

Schaik, C. van, Fox, E., and Sitompul, A. (1996). Manufacture and use of tool in wild Sumatran orangutans: Implications for human evolution. *Naturwissenchaften*, **83**, 186–8.

Schaik, C. van, Deaner, R., and Merill, M. The conditions for tool use in primates: implications for the evolution of material culture. [In press.]

Schaller, G.B. (1972). *The Serengeti Lion*. University of Chicago Press, Chicago.

Schaller, G. and Lowther, G. (1969). The relevance of carnivore behaviour to the study of early hominids. *Southwestern Journal. Anthropology*, **25**, 307–41.

Scheel, D. and Packer, C. (1991). Group hunting behaviour of lions: a search for cooperation. *Animal Behaviour*, **41**, 697–709.

Schwartz, A. (1993). *Sous-Peuplement et Développement dans le Sud-Ouest de la Côte d'Ivoire; Cinq Siècles d'Histoire Economique et Sociale*. Editions de l'ORSTOM, Paris.

Segall, M., Dasen, P., Berry, J., and Poortinga, Y. (1990). *Human Behavior in Global Perspective: an Introduction to Cross-Cultural Psychology*. Pergamon Press, New York.

Semaw, S., Renne, P., Harris, J., Feibel, C., Bernor, R., Fesseha, N., and Mowbray, K. (1997). 2.5-million-year-old stone tools from Gona, Ethiopia. *Nature*, **385**, 333–6.

Servat, E., Paturel, J., Lubès, H., Kouamé, B., Ouedraogo, M., and Masson, J. (1997). Climatic variability in humid Africa along the Gulf of Guinea, Part 1: detailed analysis of the phenomenon in Côte d'Ivoire. *Journal of Hydrology*, **191**, 1–15.

Seyfarth, R. (1977). A model of social grooming among adult female monkeys. *Journal of Theoretical Biology*, **65**, 671–98.

Shipman, P. (1986). Scavenging or hunting in early hominids: theoretical framework and tests. *American Anthropologist*, **88**, 27–43.

Silk, J.B. (1978). Patterns of food sharing among mother and infant chimpanzees at Gombe National Park, Tanzania. *Folia Primatologica*, **29**, 129–41.

Silk, J. (1983). Local resource competition and facultative adjustment of sex ratios in relation to competitive abilities. *American Naturalist*, **121**(1), 56–66.

Silk, J. (1988). Maternal investment in captive bonnet macaques (*Macaca radiata*). *American Naturalist*, **132**(1), 1–19.

Silk, J. (1992). Patterns of intervention in agonistic contest among male bonnet macaques. In *Coalitions and alliances in humans and other animals* (ed. A. Harcourt and F. De Waal), pp. 215–32. Oxford University Press, Oxford.

Silk, J., Cherny, D., and Seyfarth, R. (1996). The form and function of post-conflict interactions between female baboons. *Animal Behaviour*, **52**, 259–268.

Simpson, M. (1973). The social grooming of male chimpanzees. In *Comparative ecology and behaviour in primates* (ed. R. Michael and J. Crook), pp. 411–505. Academic Press, New York.

Smith, P. (1996). Language and the evolution of mind-reading. In *Theories of theories of mind* (ed. P. Carruthers and P. Smith), pp. 344–54. Cambridge University Press, Cambridge.

Sommer, V. (1994). Infanticide among the langurs of Jodhpur: testing the sexual selection hypothesis with a long-term record. In *Infanticide and Parental Care* (ed. S. Parmigiani and F. vom Saal), pp. 155–98. Harwood Academic Publishers, London.

Sponheimer, M. and Lee-Thorp, J. (1999). Isotopic evidence for the diet of an early hominid, *Australopithecus africanus*. *Science*, **283**, 368–70.

Stamps, J., Clark, A., Arrowood, P., and B. Kus. (1985). Parent–offspring conflict in budgerigars. *Behaviour*, **94**, 1–39.

Stander, P.E. (1992). Cooperative hunting in lions: the role of the individual. *Behavioral Ecology and Sociobiology*, **29**, 445–54.

Stander, P.E. and Albon, S.D. (1993). Hunting success of lions in a semi-arid environment. *Symposium of the Zoological Society, London*, **65**, 127–43.

Stanford, C. (1995). The influence of chimpanzee predation on group size and anti-predator behaviour in red colobus monkeys. *Animal Behaviour*, **49**, 577–87.

Stanford, C., Wallis, J., Matama, H., and Goodall, J. (1994a). Patterns of predation by chimpanzees on red colobus monkeys in Gombe National Park, Tanzania, 1982–1991. *American Journal of Physical Anthropology*, **94**, 213–29.

Stanford, C., Wallis, J., Mpongo, E., and Goodall, J. (1994b). Hunting decisions in wild chimpanzees. *Behaviour*, **131**, 1–20.

Stearns, S. (1992). *Evolution of life histories*. Oxford University Press, Oxford.

Steiner, C., Kpazahi, H., and Boesch, C. Female social interactions in Taï chimpanzees. [In preparation]

Stern, J.T. and Susman, R.L. (1983). The locomotor anatomy of *Australopithecus afarensis*. *American Journal of Physical Anthropology*, **60**, 279–317.

Struhsaker, T.T. (1975). *The red colobus monkey*. University of Chicago Press, Chicago.

Struhsaker, T.T. and Hunketer, P. (1971). Evidence of tool-using by chimpanzees in the Ivory Coast. *Folia Primatologica*, **15**, 212–19.

Struhsaker, T.T. and Leakey, M. (1990). Prey selectivity by crowned hawk-eagles on monkeys in the Kibale Forest, Uganda. *Behavioral Ecology and Sociobiology*, **26**, 435–43.

Strum, S.C. (1981). Processes and products of change: baboon predatory behavior at Gilgil, Kenya. In *Omnivorous primates: gathering and hunting in human evolution* (ed. R.S.O. Harding and G. Teleki), pp. 255–302. Columbia University Press, New York.

Sugargjito, J. and Nuhuda, N. (1981). Meat eating behavior in wild orang-utans. *Primates*, **22**(3), 414–16.

Sugiyama, Y. (1968). Social organization of chimpanzees in the Budongo Forest, Uganda. *Primates*, **9**, 225–58.

Sugiyama, Y. (1981). Observations on the population dynamics and behavior of wild chimpanzees of Bossou, Guinea, 1979–1980. *Primates*, **22**, 435–44.

Sugiyama, Y. (1984). Population dynamics of wild chimpanzees at Bossou, Guinea, between 1976 and 1983. *Primates*, **25**, 391–400.

Sugiyama, Y. (1985). The brush-stick of chimpanzees found in south-west Cameroon and their cultural characteristics. *Primates*, **26**(4), 361–74.

Sugiyama, Y. (1988). Grooming interactions among adult chimpanzees at Bossou, Guinea, with special reference to social structure. *International Journal of Primatology*, **9**(5), 393–407.

Sugiyama, Y. (1993). Local variation of tools and tool use among wild chimpanzee. In *The use of tools by human and non-human primates* (ed. A. Berthelet and J. Chavaillon), pp. 175–87. Oxford University Press, Oxford.

Sugiyama, Y. (1994*a*). Tool use by wild chimpanzees. *Science*, **367**, 327.

Sugiyama, Y. (1994*b*). Age-specific birth rate and lifetime reproductive success of chimpanzees at Bossou, Guinea. *American Journal of Primatology*, **32**, 311–18.

Sugiyama, Y. and Koman J. (1979*a*). Social structure and dynamics of wild chimpanzees at Bossou, Guinea. *Primates*, **20**, 323–39.

Sugiyama, Y. and Koman J. (1979*b*). Tool-using and making behavior in wild chimpanzees at Bossou, Guinea. *Primates*, **20**, 513–24.

Sugiyama, Y. and Koman, J. (1987). A preliminary list of chimpanzees' alimentation at Bossou, Guinea. *Primates*, **28**(1), 133–47.

Sugiyama, Y., and Koman, J., and Bhoye Sow, M. (1988). Ant-catching wands of wild chimpanzees at Bossou, Guinea. *Folia Primatologica*, **51**, 56–60.

Sugiyama, Y., Kawamoto, S., Takenaka, O., Kumazaki, K., and Miwa, N. (1993). Paternity discrimination and inter-group relationship of chimpanzees at Bossou. *Primates*, **34**(4), 545–52.

Sumita, K., Kitahara-Frisch J., and Norikoshi, K. (1985). The acquisition of stone-tool use in captive chimpanzees. *Primates*, **26**, 168–81.

Susman, R. (1984). *The Pygmy chimpanzee*. Plenum Press, New York.

Susman, R. (1986). Pygmy chimpanzees and common chimpanzees: models for the behavioral ecology of the earliest hominids. In *The ecology of human behavior: primate models* (ed. W.G. Kinzey), pp. 72–86. Suny Press, New York.

Susman, R. (1988). Hand of Paranthropus robustus from Member 1, Swartkrans: Fossil evidence for tool behavior. *Science*, **240**, 781–4.

Susman, R., Stern, J.T., and Jungers, W.L. (1984). Arboreality and bipedality in the Hadar Hominids. *Folia Primatologica*, **43**, 113–56.

Suzuki, A. (1969). An ecological study of chimpanzees in savanna woodland. *Primates*, **10**, 103–48.

Suzuki, A. (1971). Carnivority and cannibalism observed among forest-living chimpanzees. *Journal of the Anthropological Society Nippon*, **79**, 30–48.

Takahata, Y. (1985). Adult male chimpanzees kill and eat a male newborn infant: newly observed intragroup infanticide and cannibalism in Mahale National Park, Tanzania. *Folia Primatologica*, **44**, 161–70.

Takahata, Y. (1990*a*). Adult males' social relations with adult females. In *The chimpanzees of the Mahale Mountains: sexual and life history strategies* (ed. T. Nishida), pp. 133–48. University of Tokyo Press, Tokyo.

Takahata, Y. (1990*b*). Social relationships among adult males. In *The chimpanzees of the Mahale Mountains: sexual and life history strategies* (ed. T. Nishida), pp. 149–70. University of Tokyo Press, Tokyo.

Takahata, Y., Hasegawa, T., and Nishida, T. (1984). Chimpanzee predation in the Mahale Mountains from August 1979 to May 1982. *International Journal of Primatology*, **5**, 213–33.

Takahata, Y., Ihobe, H., and Idani, G. (1996). Comparing copulations of chimpanzees and bonobos: do females exhibit proceptivity or receptivity? In *Great apes societies* (ed. W. McGrew, T. Nishida, and
L. Marchandt). pp. 146–55. Cambridge University Press, Cambridge.

Tanner, N.M. and Zihlmann, A. (1976). Women in evolution. I. Innovation and selection in human origins. *Ligns*, **1**, 585–608.

Taylor, C.R. and Rowntree, V.J. (1973). Running on two or on four legs: which consumes more energy? *Science*, **179**, 186–7.

Teleki, G. (1973). *The predatory behavior of wild chimpanzees*. Bucknell University Press, Brunswick.

Teleki, G. (1974). Chimpanzee subsistence technology: materials and skills. *Journal of Human Evolution,* **3**, 575–94.

Teleki, G. (1975). Primate subsistence patterns: collector–predators and gatherer–hunters. *Journal of Human Evolution,* **4**, 125–84.

Teleki, G. (1989). Population status of wild chimpanzees *(Pan troglodytes)* and threats to survival. In *Understanding chimpanzees* (ed. P. Heltne and L. Marquardt), pp. 312–53. Harvard University Press, Cambridge, Ma.

Teleki, G., Hunt, E.E., and Pfifferling, J.H. (1976). Demographic observations (1963–1973) on the chimpanzees of the Gombe National Park, Tanzania. *Journal of Human Evolution*, **5**, 559–98.

Terborgh, J. (1983). *Five new world primates: a study in comparative ecology*. Princeton University Press, Princeton, N.J.

Thompson-Handler, N., Malenky, R., and Badrian, N. (1984). Sexual behavior of *Pan paniscus* under natural conditions in the Lomako forest, Equateur, Zaire. In *The pygmy chimpanzee* (ed. R. Sussman), pp. 347–68. Plenum Press, New York.

Tomasello, M. (1990). Cultural transmission in tool use and communicatory signalling of chimpanzees? In *Comparative developmental psychology of language and intelligence in Primates* (ed. S. Parker and K. Gibson), pp. 274–311. Cambridge University Press, Cambridge.

Tomasello, M. (1998). Uniquely primate, uniquely human. *Developmental Science,* **1**(1), 1–30.

Tomasello, M. and Call, J. (1997). *Primate cognition*. Oxford University Press, Oxford.

Tomasello, M., Kruger, A., and Ratner, H. (1993). Cultural learning. *Behavioral and Brain Sciences*, **16**(3), 450–88.

Tooby, J. and DeVore, I. (1987). The reconstruction of hominid behavioral evolution through strategic modelling. In *The evolution of human behavior: primate models* (ed. W. Kinzey), pp. 183–237. State University of New York Press, Albany.

Toth, N. and Schick, K. (1993). Early stone industries and inferences regarding language and cognition. In *Tools, language and intelligence: evolutionary implications* (ed. K. Gibson, and T. Ingold), pp. 346–62. Cambridge University Press, Cambridge.

Trivers, R.L. (1971). The evolution of reciprocal altruism. *Quarterly Review of Biology*, **46**, 35–57.

Trivers, R.L. (1974). Parent–offspring conflict. *American Zoologist*, **14**, 249–64.

Trivers, R.L. (1985). *Social evolution*. Benjamin Cummings, Menlo Park, California.

Trivers, R.L. and Willard, D.E. (1973). Natural selection of parent ability to vary the sex ratio of offspring. *Science*, **179**, 742–6.

Tsukahara, T. (1993). Lions eat chimpanzees: the first evidence of predation by lions on wild chimpanzees. *American Journal of Primatology*, **29**, 1–11.

Tutin, C. (1979). Mating patterns and reproductive strategies in a community of wild chimpanzees (*Pan troglodytes schweinfurthii*). *Behavioral Ecology and Sociobiology*, **44**, 225–82.

Tutin, C. (1996). Reproductive success story: variability among chimpanzees and comparisons with gorillas. In: *Chimpanzees cultures* (ed. R. Wrangham, W. McGrew, F. de Waal and P. Heltne), pp. 181–93. Harvard University Press, Cambridge, Ma.

Tutin, C. and Fernandez, M. (1984). Nationwide census of gorilla (*Gorilla g. gorilla*) and chimpanzee (*Pan t. troglodytes*) populations in Gabon. *American Journal of Primatology*, **6**, 313–36.

Uehara, S. and Nishida, T. (1987). Body weights of wild chimpanzees (*Pan troglodytes schweinfurthii*) of the Mahale Mountains National Park, Tanzania. *American Journal of Physical Anthropology* **72**, 315–21.

Uehara, S., Nishida, T., Hamai, M., Hasegawa, T., Hayaki, H., Huffman, M., Kawanaka, K., Kobayashi, S., Mitani, J., Takahata, Y., Takasaki, H., and Tsukahara, T. (1992). Characteristics of predation by the chimpanzees in the Mahale Mountains National Park, Tanzania. In: *Topics in Primatology*, Vol: 1: *Human Origins* (ed. T. Nishida, W.C. McGrew, P. Marler, M. Pickford, and F. de Waal), pp. 143–58. University of Tokyo Press, Tokyo.

Van Orsdol, K.G. (1984). Foraging behaviour and hunting success of lions in Queen Elizabeth National Park, Uganda. *African Journal of Ecology*, **22**, 79–99.

Visalberghi, E. and Fragaszy, D. (1990). Do monkeys ape? In *Language and intelligence in monkeys and apes: comparative developmental perspectives*. (ed. S. Parker and K. Gibson), pp. 247–73. Cambridge University Press, Cambridge.

Visalberghi, E. and Limongelli, L. (1996). Acting and understanding: tool use revisited through the minds of capuchin monkeys. In: *Reaching into thought: the minds of the great apes* (ed. A. Russon, K., Bard and S. Parker), pp. 57–79. Cambridge University Press, Cambridge.

Visalberghi, E. and Trinca, L. (1989). Tool use in capuchin monkeys: distinguishing between performing and understanding. *Primates*, **30**(4), 511–21.

Waal, F. de (1989). Dominance 'style' and primate social organization. In *Comparative socioecology: the behavioural ecology of humans and other mammals* (ed. V. Standen and R.A. Foley), pp. 243–63. Blackwell Scientific Publications, Oxford.

Waal, F. de (1992). Coalitions as part of reciprocal relations in the Arnheim chimpanzee colony. In *Coalitions and alliances in humans and other animals* (ed. A. Harcourt and F. de Waal), pp. 233–57. Oxford University Press, Oxford.

Waal, F. de (1996). Conflict as negotiation. In *Great apes societies* (ed. W. McGrew, L. Marchand and T. Nishida), pp. 159–72. Cambridge University Press, Cambridge.

Waal, F. de (1997). *Bonobo: the forgotten ape*. University of California Press, Berkeley.

Waal, F. de and Roosmalen von, A. (1979). Reconciliation and consolation among chimpanzees. *Behavioral Ecology and Sociobiology*, **5**, 55–66.

Waal, F. de and Luttrell, L. (1988). Mechanisms of social reciprocity in three primates species: symmetrical relationship characteristics or cognition? *Ethology and Sociobiology*, **9**, 101–18.

Wallis, J. (1995). Seasonal influence on reproduction in chimpanzees of Gombe National Park. *International Journal of Primatology*, **16**(3), 435–51.

Wallis, J. (1997). A survey of reproductive parameters in the free-ranging chimpanzees of Gombe National Park. *Journal of Reproductive Physiology*, **109**, 297–307.

Washburn, S.L. (1978). The evolution of man. *Scientific American*, **239**, 146–54.

Watson, P. (1998). Multi-male mating and female choice increase offspring growth in spider *Neriene litigiosa (Linyphiidae)*. *Animal Behaviour*, **55**, 387–403.

Watts, D. (1996). Comparative socio-ecology of gorillas. In *Great apes society* (ed. W. McGrew, L. Marchant, and T. Nishida), pp. 16–28. Cambridge University Press, Cambridge.

White, F. (1988). Party composition and dynamics in *Pan paniscus*. *International Journal of Primatology*, **9**(3), 179–93.

White, F.J. and Wrangham R. (1988). Feeding competition and patch size in the chimpanzee species *Pan paniscus* and *Pan troglodytes*. *Behaviour*, **105**(1–2): 148–64.

White, T., Suwa, G., and Asfaw, B. (1994). *Australopithecus ramidus*, a new species of early hominid from Aramis, Ethiopia. *Nature*, **371**, 306–12.

Whiten, A. (1996). When does smart behaviour-reading become mind-reading? In *Theories of Theories of mind* (ed. P. Carruthers and P. Smith), pp. 277–92. Cambridge University Press, Cambridge.

Whiten, A. and Byrne R.W. (1988). Tactical deception in primates. *Behavioral and Brain Sciences*, **11**, 233–73.

Whiten, A. and Ham, R. (1992). On the nature and evolution of imitation in the animal kingdom: reappraisal of a century of research. In *Advances in the Study of Behavior* (ed. P. Slater, J. Rosenblatt and C. Beer), pp. 239–83. Academic Press, New York.

Whiten, A. and Milner, P. (1984). The educational experiences of Nigerian infants. In *Nigerian children: development perspectives* (ed. Valerie H. Curran), pp. 34–73. Routledge and Kegan Paul, London.

Whiten, A., Custance, D., Gomez, J.-C., Teixidor, P., and Bard, K. (1996). Imitative learning of artificial fruit processing in children (*Homo sapiens*) and chimpanzees (*Pan troglodytes*). *Journal of Comparative Psychology*, **110**, 3–14.

Whiten, A., Goodall, J., McGrew, W., Nishida, T., Reynolds, V., Sugiyama, Y., Tutin, C., Wrangham, R., and Boesch, C. (1999). Culture in chimpanzee. *Nature*, **399**. 682–685.

Whitesides, G.H. (1985). Nut cracking by wild chimpanzees in Sierra Leone, West Africa. *Primates*, **26**(1), 91–4.

Williams, G. (1966). *Adaptation and natural selection: a critique of some evolutionary thought* Princeton University Press, Princeton.

Wimmer, H., Hogrefe, G., and Perner, J. (1988). Children's understanding of informational access as source of knowledge. *Child Development*, **59**, 386–96.

Wood, B. (1994). The oldest hominid yet. *Nature*, **371**, 280–281.

Wood, D., Bruner, J.S., and Ross, G. (1976). The role of tutoring in problem-solving. *Journal of Child Psychology and Psychiatry*, **17**(2), 89–100.

Wood, J., Johnson, P. and Campbell, K. (1985). Demographic and endocrinological aspects of low natural fertility in highland New Guinea. *Journal of Biosocial Sciences*, **17**, 57–79.

Woodruff, G., Premack, D., and Kennel, K. (1978). Conservation of liquid and solid quantity by the chimpanzee. *Science*, **202**, 991–4.

Wrangham, R. (1974). Artificial feeding of chimpanzees and baboons in their natural habitat. *Animal Behaviour*, **22**, 83–93.

Wrangham, R. (1975). The behavioural ecology of chimpanzees in the Gombe Stream National Park. Ph.D. thesis, University of Cambridge, Cambridge.

Wrangham, R. (1980). An ecological model of female-bonded primates. *Behaviour*, **75**, 262–300.

Wrangham, R. (1986). Ecology and social relationships in two species of chimpanzee. In *Ecological aspects of social evolution: birds and mammals* (ed. D.I. Rubenstein, and R. Wrangham), pp. 352–378. Princeton University Press, Princeton, N.J.

Wrangham, R. Lethal raiding among adult chimpanzees and humans: the imbalance-of-power hypothesis. *Yearbook of Physical Anthropology*. [In press.]

Wrangham, R. and van Bergmann-Riss, E. (1990). Rates of predation on mammals by Gombe chimpanzees. *Folia Primatologica*, **31**, 157–70.

Wrangham, R., Clark, A., and Isabirye-Basuta, G. (1992). Female social relationships and social organization of Kibale forest chimpanzees. In *Topics of primatology*: Vol. 1, *Human origins* (ed. T. Nishida, W. McGrew, P. Marler, M. Pickford, and F. de Waal), pp. 81–98. University of Tokyo Press, Tokyo.

Wrangham, R., McGrew, W., Waal de F., and Heltne P. (1994). *Chimpanzees cultures*. Harvard University Press, Cambridge, Ma.

Wrangham, R., Chapman, C., Clark-Arcadi, A., and Isabirye-Basuta, I. (1996). Social ecology of Kanyawara chimpanzees; implications for understanding the costs of great ape groups. In *Great apes societies* (ed. W. McGrew, L. Marchant and T. Nishida), pp. 45–57. Cambridge University Press, Cambridge.

Wright, Q. (1942). *A study of war*. Vol. 1. University of Chicago Press, Chicago.

Wu Leong, W. (1968). *Food compostion table for use in Africa*. Public Health Service, U.S. Department of Health, Education, and Welfare and Nutrition Division, Food and Agriculture Organization of the United States.

Wynn, T. (1979). The intelligence of later Acheulean hominids. *Man*, **14**, 371–91.

Wynn, T. (1981). The intelligence of Oldowan hominids. *Journal of Human Evolution*, **10**, 529–41.

Zabel, C., Glickman, S., Frank, L., Woodmansee, K., and Keppel, G. (1992). Coalitions formation in a colony of prepubertal spotted hyenas. In *Cooperation and Competition in Animals and Humans* (ed. A.H. Harcourt, and F. de Waal), pp. 114–35. Oxford University Press, Oxford.

Zemel, A. and Lubin, Y. (1995). Inter-group competition and stable group sizes. *Animal Behaviour*, **50**, 485–8.

Zihlmann A. (1981). Women as shapers of the human adaptation. In: *Woman the gatherer* (ed. F. Dahlberg), pp. 75–120. Yale University Press, New Haven.

Zihlmann A. (1996). Reconstructions reconsidered: chimpanzee models and human evolution. In *Great apes societies* (ed. W. McGrew, T. Nishida and L. Marchant), pp. 293–304. Cambridge University Press, Cambridge.

Index